Decomposer basidiomycetes

Mycena galopus, a common decomposer basidiomycete of many temperate woodlands, growing on leaf litter of *Quercus* and causing a characteristic white rot (Frankland: Chapter 14). Photo: J.K. Adamson.

Decomposer basidiomycetes: their biology and ecology

SYMPOSIUM OF
THE BRITISH MYCOLOGICAL SOCIETY
HELD AT QUEEN MARY COLLEGE
LONDON, MARCH 1979

EDITED BY
JULIET C. FRANKLAND, J. N. HEDGER
& M. J. SWIFT

CAMBRIDGE UNIVERSITY PRESS
CAMBRIDGE
LONDON NEW YORK NEW ROCHELLE
MELBOURNE SYDNEY

CAMBRIDGE UNIVERSITY PRESS
Cambridge, New York, Melbourne, Madrid, Cape Town, Singapore, São Paulo, Delhi

Cambridge University Press
The Edinburgh Building, Cambridge CB2 8RU, UK

Published in the United States of America by Cambridge University Press, New York

www.cambridge.org
Information on this title: www.cambridge.org/9780521106801

First published 1982
This digitally printed version 2009

A catalogue record for this publication is available from the British Library

Library of Congress Catalogue Card Number: 81–18145

ISBN 978-0-521-24634-7 hardback
ISBN 978-0-521-10680-1 paperback

Contents

Contributors

T. Basuki, *Department of Botany and Microbiology, University College of Wales, Aberystwyth SY23 3DA, UK.*
D. J. Dickinson, *Department of Botany, Imperial College, Prince Consort Road, London SW7 2BB, UK.*
P. Fenn, *USDA Forest Service, Forest Products Laboratory, PO Box 5130, Madison, Wisconsin 53705, USA.*
J. C. Frankland, *Institute of Terrestrial Ecology, Merlewood Research Station, Grange-over-Sands, Cumbria LA11 6JU, UK.*
E. B. Gareth Jones, *Department of Biological Sciences, Portsmouth Polytechnic, King Henry I Street, Portsmouth PO1 2DY, UK.*
J. N. Hedger, *Department of Botany and Microbiology, University College of Wales, Aberystwyth SY23 30A, UK.*
T. F. Hering, *Department of Physiology and Environmental Studies, School of Agriculture, University of Nottingham, Sutton Bonington, Loughborough LE12 5RD, UK.*
V. Hintikka, *Institute of General Botany, University of Helsinki, Viikki SF-00710, Helsinki 71, Finland.*
D. H. Jennings, *Hartley Botanical Laboratories, University of Liverpool, PO Box 147, Liverpool L69 3BX, UK.*
T. K. Kirk, *USDA Forest Service, Forest Products Laboratory, P.O. Box 5130, Madison, Wisconsin 53705, USA.*
C. L. Kramer, *Division of Biology, Kansas State University, Manhattan, Kansas 66506, USA.*
J. F. Levy, *Department of Botany, Imperial College, Prince Consort Road, London SW7 2BB, UK.*
P. C. Mercer, *Forestry Commission, Alice Holt Lodge, Wrecclesham, Farnham GU10 4LH, UK.*
R. A. P. Montgomery, *c/o* J. F. Levy.
A. D. M. Rayner, *School of Biological Sciences, University of Bath, Claverton Down, Bath BA2 7AY, UK.*

M. J. Swift, *Department of Plant Biology and Microbiology, Queen Mary College, University of London, Mile End Road, London E1 4NS, UK.*

N. K. Todd, *Department of Biological Sciences, University of Exeter, Washington Singer Laboratories, Perry Road, Exeter, UK.*

R. Watling, *Royal Botanic Garden, Edinburgh EH3 5LR, UK.*

Foreword

The toadstools of woodland basidiomycetes have always been as noticeable to the countryman as the flowering shoots of green plants, and many of them have been distinguished by vernacular names. Yet for many years, the basidiomycetes remained the least understood group of fungi, in terms of all that remained to be discovered. This was because of their complexity, both in their remarkable degree of mycelial organisation and in the wide span of their biochemical abilities; discovery of the intricacy of their outbreeding systems has equally delighted the fungal geneticist. Such complex fungal mechanisms must obviously take longer to elucidate than the simpler mechanisms of many of the common saprophytic sugar fungi and related facultative pathogens. For autecological studies, which are an essential complement to the synecological investigation of fungal communities in soil, woodland litter, composts and standing timber, it is essential to get a fungal species into axenic culture and to be able to maintain and manipulate it there. Because many of the basidiomycetes grow only rather slowly on nutrient agar, their competitive saprophytic ability on such a substrate is poor, and they are difficult to isolate, so that selective methods have had to be developed. Secondly, fungal breakdown of insoluble substrates such as cellulose and lignin is more difficult to assess than is the uptake and metabolisation of simple sugars.

As a reading of this book will show, many of the varied resources of modern science have had to be deployed to reach our present level of understanding as to how basidiomycetes decompose the complex structure of the lignified plant cell-walls that make up their natural substrates. Considerable progress has also been reported in elucidation of the ecological niche of basidiomycetes in microbial successions in woodland litter and composts, in parallel with the simpler successions of fungi in decaying timber. This bringing together in one symposium of work on litter decomposition and on timber decay is one of this book's unusual features. Just so, no microbiologist concerned with the breakdown of either cellulose or lignin can afford to ignore the contents of this book. As one of them, I can strongly recommend it as an aid to further research, and equally so for teaching.

S. D. Garrett

Preface

The basidiomycetes are the most visible of fungi – partly because of their macroscopic structures, partly because of the considerable impact they have had on man's economy. The basidiomycetes may be divided on nutritional grounds into the *biotrophs, saprotrophs* and *necrotrophs* (Lewis, 1973). The first of these categories comprises those fungi dependent on other living organisms (hosts) for their nutrition; they include major groups of parasite such as the rusts and smuts, and mutualistic symbionts of higher plants, such as the mycorrhizal basidiomycetes. In contrast, the saprotrophs obtain their carbon, energy and mineral nutrients from dead organic matter of plants, animals and other micro-organisms. The necrotrophs also utilise dead organic matter, but in addition possess the capacity to parasitise and kill living tissue prior to its exploitation. Whilst the biotrophic activities of the basidiomycetes have been reviewed extensively, the saprotrophic and necrotrophic members of this group with a decomposer function have received relatively scant attention. This volume seeks to redress this imbalance.

The importance of decomposer organisms as agents of biodegradation, impinging, as they do, on man's economy, has been appreciated for many years – including the rôle of basidiomycetes in the decay of timber. Recent years however have seen a great increase of research into the ecological rôle of decomposition processes in both natural and managed ecosystems. This has arisen largely because of the fresh interest in ecology, management and conservation at the ecosystem level, which has developed because of the pressing need for better management of our environment and natural resources. Investigation at the ecosystem level quickly focused attention on the key importance of the fluxes of nutrient and energy through the detrital compartments of the ecosystem, but the significance of the decomposer basidiomycetes in these processes has received less attention than that of other major groups of fungi and prokaryotic micro-organisms.

Within the wide taxonomic range of micro-organisms involved in decomposition in terrestrial ecosystems, the basidiomycetes have, in our opinion, a central and uniquely important rôle. It is the aim of this book to describe and illuminate that rôle. The contents are the product of a Symposium initiated by the third editor and held by the British Mycological Society at Queen Mary College, University of London, on 28–30 March, 1979. The participants were given the challenge that 'basidiomycetes are the major agents of decomposition of plant litter in a variety of terrestrial habitats'. The following chapters represent the response of various authors to that challenge.

The first five chapters of the book set the scene by discussing various aspects of the general biology of the decomposer basidiomycetes. The first chapter, by Roy Watling, is concerned with their taxonomy; it is something of a maxim in fungal ecology that you can know either the identity of a basidiomycete or something of its activity but not both. Watling, therefore, examines possible correlations between taxonomic groupings and ecological function. The ecological success of a fungus depends in part on its ability to colonise new substrates competitively; one feature of competitive ability rests on the efficiency of spore production and dispersal, the topic that is reviewed in Chapter 2 by Charles Kramer. This subject links naturally with those of Chapters 6 and 7 in which Alan Rayner and Norman Todd review various aspects of the structure of basidiomycete populations within decaying wood. Colonisation by basidiospores carries the potential for the establishment of a genetically diverse population, and these authors explore how this potential is realised.

The three intervening chapters discuss various aspects of the biochemistry and physiology of the decomposer basidiomycetes and attempt to evaluate their success in terms of substrate exploitation. In Chapters 3 and 4, Richard Montgomery and Kent Kirk with Patrick Fenn review our current understanding of polysaccharidase and ligninolytic enzymes, respectively. If basidiomycetes have a claim to being highly adapted to the decomposer function, it must rest largely on their possession of the most diverse and efficient battery of depolymerising enzymes of any group of living organisms. A further selective feature imposed on many decomposer organisms, particularly those associated with wood decay, is the relatively nutrient-poor nature of their food source. In Chapter 5, David Jennings shows how the formation of hyphal strands by the wood decay fungus *Serpula lacrimans* may help to overcome problems of nutrient limitation and the high C:N ratio of timber. Again, the factors considered here provide some insight into the competitive ability of the decomposer basidiomycetes. A link may also be seen between this chapter and the later one by Juliet Frankland (Chapter 14) in which she considers the ability of basidiomycetes to capture nutrients in the competitive situation of the natural microhabitats in litter and soil.

The remaining chapters are more directly ecological in content. Four consider the biology of basidiomycetes inhabiting wood in a variety of habitats. These range from the living tree in the forest with necrotrophic basidiomycetes discussed by Peter Mercer (Chapter 8) to that of cut timber in outdoor, indoor and aquatic situations. John Levy and David Dickinson (Chapters 9 and 10) alert us to the immense economic as well as ecological importance of basidiomycetes in wood rot. Decay of cut and seasoned timber is restricted to a relatively few species of basidiomycetes, and we need to understand how these fungi differ from the diverse range of decomposers which exist in forest habitats but do not succeed in cut timber. The patterns of colonisation of timber described by Levy provide further information for our picture of the competitive adaptations of the decomposer basidiomycetes. Gareth Jones (Chapter 11) follows by summarising our current knowledge, including taxonomy, of decomposer basidiomycetes on submerged materials particularly wood, and he suggests reasons for their apparent lack of success as a group in the aquatic environment.

The next three chapters take us back into more natural ecosystems. They are all concerned with various aspects of basidiomycete activity in the litter layer of forests. Tom Hering in Chapter 12 brings together several features of the general biology of the decomposer basidiomycetes and discusses the rôle of these fungi as agents of leaf-litter decomposition. Veikko Hintikka (Chapter 13) again takes up the question of colonisation and also compares the physiology of lignicolous and litter-decomposing basidiomycetes in relation to various chemical features of their environment. Whereas most of the preceding authors are concerned with the action of the basidiomycete decomposers on wood and leaf litter, Juliet Frankland in Chapter 14 reverses the view and considers the amount of basidiomycete mycelium produced during decomposition of leaf litter, particularly in temperate woodlands. She then assesses what this biomass represents in terms of mineral nutrients 'locked up' or released by these fungi. It is only from estimates of this kind that we can deduce the importance of the basidiomycete contribution to energy flow and nutrient cycling in an ecosystem.

The final two chapters are synthetic in nature, attempting to assess the rôle of the decomposer basidiomycetes as components of total systems. In Chapter 15, John Hedger and Triadi Basuki discuss the basidiomycetes that colonise a simplified decomposition system: compost. They show that their importance in composts can be linked to their biochemical capabilities, and to their interactions with other micro-organisms – attributes which could partially determine their success in natural ecosystems. Finally, Michael Swift considers the more complex system of the forest and also tries to weave together some of the threads of earlier chapters which help to characterise this remarkable group of fungi.

Terminology

The editors have not aimed at complete uniformity between chapters, and we hope that sufficient definitions have been given to avoid confusion when an author has departed from a common usage or viewpoint. Variety of concept and term is surely a sign that mycology is an active and evolving science, and should give the reader food for thought. However, we did reduce nine alternatives for the same phenomenon (carpophore, basidiocarp, basidiome, fruiting body, fruit(-)body, sporocarp, sporophore and toadstool) to one – fruit body! Our choice is not necessarily the best; Roy Watling preferred 'basidiome', but this term has not yet appeared in *Ainsworth & Bisby's Dictionary of the Fungi*. Other potentially conflicting terms to note are: 'saprotroph' and 'saprophyte'; 'substrate' and 'resource'.

Acknowledgements

The editors thank all those who helped to organise the Symposium at Queen Mary College on which this volume is based, especially Lynne Boddy, Lesley Falconer, Fred Lawes and Rebecca Thomas.

To Roy Watling we are indebted for much taxonomic advice and for checking the Systematic Index. We thank also Andrew Bailey and Jan Poskitt for redrawing many of the Figures, and the typists at Merlewood Research Station, especially Christine Benson and Cathie Kay. The first editor is particularly grateful too for all the support she has received from the Institute of Terrestrial Ecology and from her husband, Raven Frankland, throughout the production of this book.

Juliet Frankland, John Hedger, Michael Swift.

References

Ainsworth, G. C., James, P. W. & Hawksworth, D. L. (1971). *Ainsworth & Bisby's Dictionary of the Fungi.* Kew: Commonwealth Mycological Institute.

Lewis, D. H. (1973). Concepts in fungal nutrition and the origin of biotrophy. *Biological Reviews*, **48**, 261–78.

1 Taxonomic status and ecological identity in the basidiomycetes

R. WATLING
Royal Botanic Garden, Edinburgh EH3 5LR, Scotland

Introduction

The taxonomy and ecology of the larger fungi are both so much in a state of flux that at first thought it might be considered neither possible to make a meaningful contribution nor indeed to wed these two studies together. No real gauge exists for measuring how the ecological life-style of a particular species influences our decision on how we place it in our man-made hierarchy. Nevertheless, this is an attempt at some kind of analysis to determine whether convincing correlations do really exist between the classification of fungi and their possible rôles as wood rotters or litter decomposers. To my satisfaction good correlations have in fact been found throughout the basidiomycetes (Basidiomycotina), and several families of higher fungi exhibiting distinct ecological patterns are discussed.

Until recently little challenge has been made either to Donk's outline of the Aphyllophorales (1964) or to Singer's presentation (1951, 1962, 1975) of the Agaricales, and although these classifications were revolutionary in their time they should now give way to still more natural groupings. Unfortunately such improvements are far removed from the general purpose classification proposed by Ainsworth (1966) and also, therefore, differ considerably from those of familiar textbook mycology. One must remember, however, that the taxonomy and nomenclature of the larger fungi as we know them today started with Elias Fries some seventy-five years after the publication of Linnaeus's *Species plantarum*, so it is only right to compare current mycological advancements with those of the phanerogamic botanists at the beginning of the century. The present paper is therefore a first effort, albeit shaky in its evidence, at analysing the problem as a whole, basing my thoughts on the proposals made by Oberwinkler (1977, 1978) for the Aphyllophorales (= Polyporales p.p.) and by Arpin & Kühner (1977) and Kühner (1978–80) for the Agaricales.

The rusts and smuts

Figure 1 shows the orders of basidiomycetes that are currently considered to be distinct. The Uredinales and Ustilaginales (including the Tilletiales) are often united into the Teliomycetes because predominantly they are plant parasites and possess a reduced fruit body (better termed the basidiome) and a superficially similar life cycle: however, some fungi currently placed in the Ustilaginales have been known for generations as yeast fungi and only recently have been shown by cultural techniques to have perfect states similar to some smut fungi. The presence of a yeast-phase is not unique, being paralleled in some of the jelly fungi and in many ascomycetes. The Uredinales, the familiar rust fungi, apart from those members on pteridophytes which may have evolved quite recently, exploit gymnosperm and angiosperm hosts and their life cycles are tailored to the availability in time and space of suitable host tissue. The Uredinales and Ustilaginales are not considered further here, especially as the latter may contain a mixture of rather different, even non-basidiomycete, elements (Kreisel, 1969; Moore, 1972). Certainly mitosis in those Ustilaginales so far studied, in both filamentous and yeast forms, appears to be a much more complicated process (Heath, 1978) than that found in the groups to be discussed. The lack of a dolipore–parenthesome complex in the septa of the hyphae of these fungi led Moore (1972) to erect a new division of higher fungi, the Ustomycota.

Jelly fungi – a heterogeneous group

Under the term 'jelly fungi' are brought together those basidiomycetes which have either the correlated characters of a multicellular basidium (phragmobasidium or heterobasidium) arranged in a distinct hymenium and a gelatinised fruiting structure, or superficial similarities to organisms with these correlated characters. This corporate heading covered in the classical literature four orders: Auriculariales, Septobasidiales, Tremellales, and Dacrymycetales; the last lacks the septate basidium but possesses the gelatinised thallus. The term Phragmobasidiomycetes has been coined for this group. There is, however, evidence that this is not a natural group and that the constituent orders are less related to one another than they are to other major orders; thus the

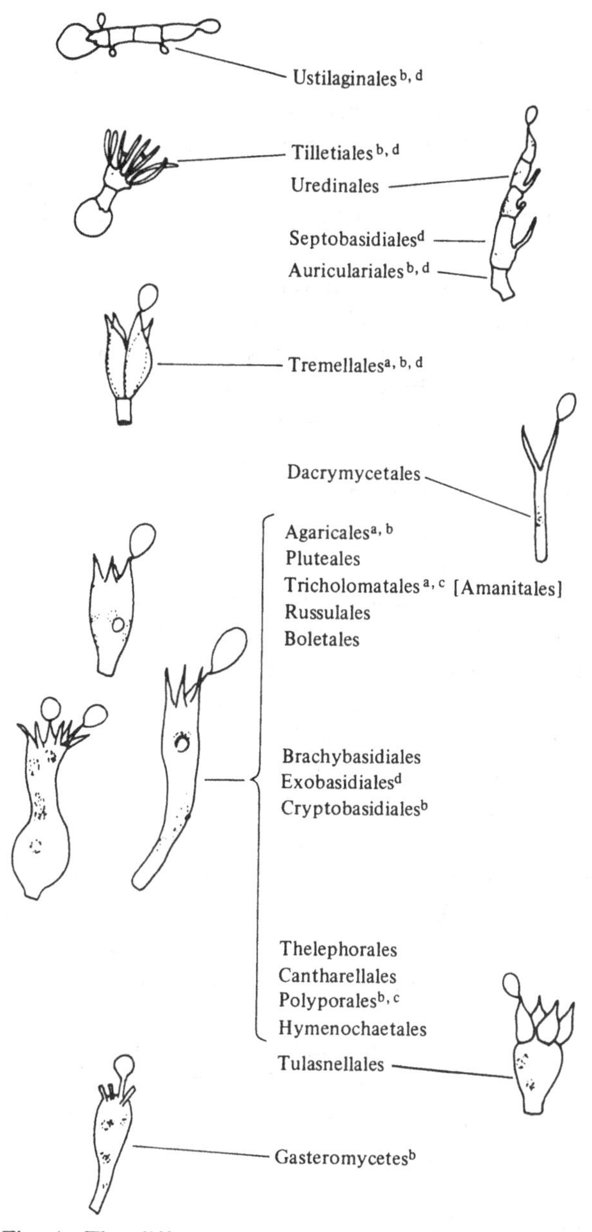

Fig. 1. The different basidial types exhibited by the orders of the Basidiomycotina, the Amanitales being marginally separate as an order. The range of shape is indicated in the orders possessing a holobasidium. [a]Orders possessing cyphellaceous members (disk or cup-shaped fruit bodies); [b]members with passive spore-dispersal; [c]basidiolichens present; [d]yeast stages present. Heterobasidiomycetous yeasts are excluded. (Based partly on Oberwinkler, 1978.)

Dacrymycetales is characterised by bisporic, undivided, basidia and is more related to the polypores and their allies. However, there are several other characters which characterise the group and give some support for its uniformity, e.g. presence of yeast-like stages, often quite elaborate conidial stages, a tendency for the basidiospore to be septate, and germination of the basidiospore by repetitive, often ballistosporic, meiospores. The part played in the ecology of the members by this additional array of characters has never been considered critically.

The first two orders, Auriculariales and Septobasidiales, possess transversely septate basidia similar to those produced by the germinating teleutospores of members of the Uredinales; they are therefore morphologically undoubtedly parallel in all ways with the Uredinales, and it would be exciting to confirm a close relationship by finding similar, or the same, mitotic patterns and processes as those which apparently characterise the Uredinales (Heath, I. B. & Heath, M. C., 1976; Kubai, 1978).

The third order, the Tremellales, in contrast possesses longitudinally divided basidia. Donk (1966) prefers to join the traditionally separate Tremellales and Auriculariales together under the grouping hymenomycetous Heterobasidiae in contrast to the Teliomycetes, but his concept is not followed in this account; indeed I believe that the two groups are phylogenetically a considerable distance apart.

The Septobasidiales is a particularly intriguing group because all the members apparently achieve intimate relationships with scale insects. Although uncommon in European temperate areas, the number of species increases in the American tropics and subtropics. They are very specialised fungi and probably more highly evolved than the Auriculariales, the order which perhaps most closely resembles the common ancestor of both the Septobasidiales and Uredinales. The most familiar European members of the Auriculariales grow equally well as saprophytes and parasites, and a few, e.g. *Helicobasidium brebissonii*, are economically important. Others are basically saprophytes, commonly occurring as members of the floras of standing-dead or dying trees and fallen wood, e.g. *Hirneola auricula-judae*; their activity is minimal in temperate Europe, but they apparently play an important rôle as wood rotters in subtropical areas of the world. A few less familiar species are rather specialised parasites of mosses, ferns, rushes and even other fungi.

The Tremellales, as presently defined, is surprisingly heterogeneous. The members are traditionally united solely on the fact that they possess longitudinally septate basidia, quite irrespective as to whether they possess hydnoid, poroid, clavarioid, pustular, cup-shaped or resupinate fruit bodies. Perhaps the recent studies by Moore (1978) go some way to confirm this heterogeneity and, furthermore, indicate which members, e.g. *Tremellodendron* and *Aphelaria*, should be examined in the future using the electron microscope. The Tremellales although possessing a divided basidium are very close to the Polyporales, and several intermediates between the two orders, e.g. *Tulasnella, Waitea, Thanatephorus* and *Uthatobasidium*, are to be found. Many of the larger, more familiar, members of the Tremellales are saprophytes, but several with highly reduced and often inconspicuous thalli are parasites of other fungi, and I am sure many more will be found. Once considered to be only wood rotters, many of these micro-forms have now been shown to colonise fruit bodies of resupinate fungi or 'discomycetes', and recently a lichenicolous member of the order has been discovered. Thus, one must not jump to conclusions concerning the ecological preferences of a particular species from superficial field observations alone.

All members of the Dacrymycetales break down woody debris, particularly rather old, fallen, trunks and rotting stumps of coniferous or frondose trees. They appear to fruit late in the decay succession, often on decorticated wood, but, although the genera and species involved are comparatively few in number, their widespread nature and common occurrence indicate that they play a fairly substantial rôle in the woodland community.

Mushrooms and toadstools – agarics

The Tremellales leads the account not evolutionarily but morphologically on to the resupinate fungi and their allies (mushrooms, toadstools and shelf-fungi), i.e. those fungi producing a well-formed or diffuse palisade (hymenium) of single-celled basidia (holobasidia) in a variety of hymenial configurations and formally termed the Hymenomycetes. They are now arranged in ten orders (Tables 1 and 2) although until a few years ago they were to be found in only two: Agaricales (mushrooms and toadstools) and Aphyllophorales (polypores and their allies). The true agarics are now distributed by Arpin & Kühner (1977) and Kühner (1978–80) in three orders: the Agaricales, Pluteales, and Tricholomatales, with the fungi related to *Russula* in the Russulales (= Asterosporales) and the boleti in the Boletales (Table 1). The Agaricales is restricted except for one genus (*Lepiota* and its close allies) to those species possessing both strongly pigmented basidiospores, with a complex wall structure, and a well-developed velar construction; the

Table 1. *Agaricoid fungi*

Order	Family
AGARICALES	Agaricaceae Bolbitiaceae Coprinaceae Cortinariaceae[l,m] Lepiotaceae Strophariaceae
TRICHOLOMATALES	Hygrophoraceae[part m] Pleurotaceae[l] Tricholomataceae[l,m] (including Cyphellaceae) ? Physalacriaceae
(AMANITALES	Amanitaceae[m])
PLUTEALES	Pluteaceae[l] Entolomataceae
RUSSULALES	Russulaceae[m]
BOLETALES	Boletaceae[m] (including Strobilomycetaceae) Gomphidiaceae[m] Paxillaceae[l,m] (Coniophoraceae)[l]

The Cyphellaceae is believed to be composed mainly of reduced members of the Tricholomataceae although a few can be assigned to the Cortinariaceae. The Coniophoraceae normally placed in the Polyporales shows relationships with the boleti and their relatives.
[l] Families which are dominantly wood-rotters and litter decomposers; [m] primarily mycorrhiza formers; ?, taxonomic position uncertain. (Based partly on Kühner, 1978.)

Pluteales has less strongly developed basidiospore pigmentation but a complex spore-wall, and the Tricholomatales is characterised in general by pale-coloured spores, a less highly specialised fruit-body development, and usually by rather primitive patterns of primordial organisation. There is some evidence to suggest that the last order should not be restricted exclusively to species with non-pigmented basidiospores.

The Amanitales has been introduced to cover those agarics (see Moore, 1978) related to the fly agaric (*Amanita muscaria*) and death cap (*A. phalloides*) with schizohymenial development, i.e. where the hymenium does not develop in a preformed gill space but is cut out of the background tissue, leaving attachments to pileus and stipe tissues until a late stage in maturation. The latter character is correlated with the bilateral hymenophoral trama. The Amanitales is a superfluous order as both the constituent genera are very close to several members of the Tricholomatales. However, if accepted, the order is a mixture of strict mycorrhizal formers (e.g. the subgenus *Amanita*) and undoubted non-mycorrhizal taxa (e.g. the subgenus *Lepidella*).

The boletes (Boletaceae) and their allies (Paxillaceae and Gomphidiaceae) have hymenophoral tramas which appear to be superficially similar to those of members of the Amanitaceae. However, this hymenophoral trama is formed in a rather different way and is termed 'divergent'. The chemistry of the boletes is very different from that of other putrescent fungi, involving many complex phenolic and pulvinic acid compounds not found in other agarics (Edwards & Elsworthy, 1967; Beaumont & Edwards, 1971; Edwards, 1977); the members all have modifications of a basic gymnocarpic development in which the hymenium is exposed from its inception. They are placed in their own order, the Boletales. In a parallel way, members of the genera *Russula* and *Lactarius* along with certain Gasteromycetes possess anatomical and chemical attributes, e.g. a heteromerous trama and reaction with sulpho-aldehydes, not found elsewhere amongst the agarics, and both genera are now placed in the Russulales. Both this order and the Boletales contain important ectotrophic mycorrhizal organisms, and in both tropical and subtropical regions some species form tissues over wounds (cicatrizers) or even possess a sheath which is capable of litter decomposition; in temperate areas a very restricted number are undoubtedly litter decomposers (Frankland: Chapter 14). In Britain, examples of wood rotters are *Paxillus atrotomentosus* and *Boletus sphaerocephalus*. Parasitism is very rare in the Boletales and Russulales, but examples are *Xerocomus radicicola* (Singer, 1978) and *Russula* species, section Pelliculariae (Heim, 1973).

Mycorrhizal formers are found in all of the three orders of agarics and include members of such genera as *Cortinarius, Inocybe* and *Hebeloma* (Cortinaricaeae), *Amanita* (Amanitaceae) and *Tricholoma sensu stricto* (Tricholomataceae). However, in the same orders many genera are predominantly humicoles, being active and efficient litter decomposers, e.g. *Pluteus, Volvariella, Mycena* and *Collybia* (Hering: Chapter 12); others break down wood and similar substrates, e.g. *Pleurotus*, and yet others are characteristic of rather more specialised habitats, e.g.

Panaeolus and *Coprinus* on dung, or *Psilocybe* on highly nitrogenous soils. Even the familiar dung fungi can be considered to be litter decomposers, utilising herbaceous debris in a highly nitrogenous flux, and are probably of recent origin, their evolution tightly bound to that of the herbivorous mammals.

As in other pigeon-hole techniques, wide sweeping generalisations swamp many minor groupings and distinct ecological units are lost. Thus within *Collybia* there exists a very distinct group of species characteristically found on dead and decaying toadstools, i.e. the *Collybia tuberosa* group; these parallel in their specialisation those fungi growing on soils containing high concentrations of urea and similar nitrogenous compounds (ammonogenous) and placed in the genus *Tephrocybe*, the carbonicoles characteristic of the sites of old woodland fires, and members of the genus *Termitomyces* and related genera intimately associated with termites. It is becoming clear that the more complex the developmental construction, the greater the number of species suspected to have mycorrhizal relationships and the more groups of species found to be restricted to rather specialised angiosperm habitats, e.g. a group of *Naucoria* species with alder (*Alnus glutinosa*). Those species with simpler developmental organisation appear to have become adapted to a saprophytic life-pattern, often growing under very specialised environmental conditions.

The majority of the basidiomycetes with a cup-shaped fruit body, formerly placed close to the polypores and their allies, are derived from agarics. Thus some species can be related to *Marasmius* and others to *Mycena*, both tricholomataceous genera. Although the majority of this small group possess hyaline basidiospores, a few have distinctly coloured spores and anatomically are more akin to members of the genus *Crepidotus*. All these species, formerly placed in the Cyphellaceae, have a reduced hymenium which is undoubtedly an adaptation for colonising herbaceous and leafy litter, twiggy detritus, etc. Such reduced forms are found in the Tremellales as well as in the agarics, as recently a few cyphellas were shown to possess a longitudinally septate basidium.

Other unmistakable agarics, especially some members of the Tricholomataceae, are very important litter decomposers; thus *Mycena*, particularly the *M. galopus* group, and *Collybia*, particularly *C. peronata*, have been estimated to decompose a considerable proportion of the annual leaf fall in British *Quercus* woodlands (Hering, 1972). Certainly species of *Marasmius* are important twig and leaf rotters, e.g. *M. rotula* (Fig. 2*a*), but many have become very highly specialised as to the

Fig. 2. (*a*) *Marasmius rotula* (Tricholomatales, Tricholomataceae) on twiggy debris. (*b*) *Pleurotus ostreatus* (Tricholomatales, Pleurotaceae) on a hardwood stump. Photographs by A. W. Brand (Royal Botanic Garden, Edinburgh collection).

substrate on which they fruit, e.g. *M. buxi* on *Buxus* and *M. hudsonii* on *Ilex*. This degree of specialisation parallels that found in *Strobilurus* where most members are found on fallen 'cones' either of conifers or primitive angiosperms.

The larger and more obvious tricholomatoid agarics of the autumn, such as *Clitocybe nebularis* and *Lepista nuda*, are important leaf litter decomposers judging from the degree of bleaching they produce in the piles of leaves on which they fruit, but the other facets of the part played by this group of fungi in the economy of a woodland are only now unfolding. Some species solubilise inorganic materials, and introduce phosphates into the system by so doing; others are known to play a secondary rôle as fungal parasites. The mobilisation of minerals may be achieved through the production of calcium oxalate (Frankland: Chapter 14; D. Jones, personal communication), a common metabolite of innumerable agaricoid fungi.

In contrast to the mycorrhizal formers, wood-rotting fungi are few amongst the true agarics although those that are lignicolous are very familiar and often prominent in our flora, e.g. *Flammulina velutipes* (Tricholomataceae) and *Hypholoma fasciculare* and *Pholiota squarrosa* (Strophariaceae). One group which is characteristically lignicolous includes the members of the so-called Pleurotaceae, but perhaps here we have allowed ecology and the single character of a reduced or absent stipe to influence our classification too much. The Pleurotaceae, as outlined in the British Check List (Dennis, Orton & Hora, 1960), is a mixture of elements, many tricholomatoid and some, e.g. *Pleurotus ostreatus* (Fig. 2*b*), apparently just as closely related to the polypores, a whole group which one always associates with wood rot, both of seasoned timber and standing trees. It is only as a result of studies of the development and anatomy of the fruit bodies of members of the Pleurotaceae that reallocation to existing genera with parallel structures could be achieved. If parallelisms can be drawn from other organisms, however, anatomy might be expected to be correlated with ecological preferences. Thus, the irregularly arranged and compacted trama of the marasmioid and pleurotoid agarics allows a retention of sufficient water to maintain essential life-processes until the next supply of moisture is available for spore production, whereas the convergent or inverse hymenophoral trama, found in the Pluteaceae, corresponds to a large-scale expansion of the fruit body with economy of structural material. This economy is found also in the Bolbitiaceae and Coprinaceae where in many of the constituent species an explosive development is found with most nutrients channelled into the production of basidia-producing tissue, and therefore into basidiospores.

Hedgehog fungi, fairy clubs and their allies

Although more closely related to the Polyporales, the two orders Brachybasidiales and Exobasidiales, which cover between them ten or so genera, are similar to the rusts in that they are obligate leaf parasites. Members of both orders lack a distinct fruit body, fruiting through the stomata in the former and involving most of the cortical layers in the latter. The genus *Exobasidium* is particularly interesting in that the majority of species in Europe are found on members of the Ericaceae (e.g. *E. rhododendri*), with a few species on other isolated phanerogamic genera, e.g. *Citrus, Laurus* and *Saxifraga*; in the region formerly considered to constitute Gondwanaland, however, members are widespread on additional host families. The Cryptobasidiaceae differs from the Exobasidiaceae in the sessile basidiospores; Oberwinkler (1978) has raised this family to the rank of order. The members are also plant parasites. Although possessing a single-celled basidium, they have been united with the hymenomycetous Heterobasidiae (Donk, 1966), probably because of many similarities in lifestyle, septation of the basidiospore and presence of a yeast-like stage. There is little doubt that the yeast stage eases infection and colonisation of leaf surfaces and is therefore correlated with the ecology of these plant parasites.

The polypores, hedgehog fungi and fairy-club fungi, originally placed in the single order Polyporales, have been divided to accentuate their macro- and microscopic features, the first group characterised by a poroid fruit body, the second by teeth on which the basidiospores are formed replacing the gills or pores in the more familiar fungi, and the third by a club- or coral-shaped fruit body (Table 2). Some of the orders in which they are now placed cut completely across all the traditional views based on the final morphological shape of the fruit body and follow the proposals Donk had already made when defining some of his families (1964).

Table 3 indicates the families now recognised within the Polyporales and their allies, and the 'grave' of those traditional families formerly assigned to them; even in this restricted sense, the Polyporales may still be heterogeneous. The orders which now comprise this group are: the Cantharellales for *Cantharellus cibarius* and its allies; the Hymenochaetales for the resupinate *Hymenochaete*, the poroid *Phellinus* and their allies; the Thelephorales for members possessing gills, pores, teeth

Table 2. *Polypores and their allies – the orders and families of the fairy clubs, pore and toothed fungi and their resupinate allies*

Order	Family
THELEPHORALES	Bankeraceae
	Thelephoraceae
GOMPHUS ALLIANCE	Gomphaceae
	Ramariaceae
POLYPORALES	Schizophyllaceae[w]
	Polyporaceae[w]
	Corticiaceae[w]
	Podoscyphaceae
	Stereaceae[w]
	Punctulariaceae[w]
	Auriscalpiaceae[w] }
	Lentinellaceae[w] }
	Clavicoronaceae[w] }
	Hericiaceae[w] }
	Amylariaceae[w] }
	Bondarzewiaceae[w]
	Echinodontiaceae[w]
	Fistulinaceae[w]
	Ganodermataceae[w]
	Sparassidaceae
	Aphelariaceae
	Clavulinaceae
	Pterulaceae } Clavarioid alliance
	Clavariaceae }
HYMENOCHAETALES	Hymenochaetaceae[a,w]
	Asterostromatoideae
	Hymenochaetoideae
	Vararioideae
	Lachnocladiaceae

[a] The subfamilies within the Hymenochaetaceae probably deserve familial rank. [w] Families composed primarily of wood-rotters. Those families with a very close relationship are braced.

or a resupinate fruit body but all characterised by a uniform spore morphology; and finally, the true Polyporales.

The Thelephoraceae in a very much more restricted sense has been recognised within the system at the level of an order. One member, *Thelephora terrestris*, is a proven and important mycorrhizal fungus, particularly with conifers on acid heaths, but at the moment the ecological picture for the other members is rather cloudy. Undoubtedly many of the resupinate members are active litter decomposers if their habitat preferences are considered alone. The family and order include fungi with a whole series of hymenophoral configurations: hydnoid, agaricoid, cantharelloid, and resupinate. Another segregate order is the Cantharellales which is moulded around the two important genera *Cantharellus* and *Craterellus* and a few segregate genera. The members are either humicolous or mycorrhizal. The ability of these fungi to form mycorrhiza must be examined further.

Hydnum repandum is undoubtedly related to the chanterelles, and there is some evidence to suggest that members of the clavarioid genus *Clavariadelphus* are also related, closing the gap between pileate species and those with a hymenium distributed uniformly over a stalked fruit body, i.e. amphigenous. Several fungi formerly placed in *Cantharellus*, because of differences in chemistry, basidiospore morphology, etc., are now found in quite different orders (*Polyozellus* in Thelephorales and *Gomphus* in Polyporales). These are terricolous, woodland, humicolous or mycorrhizal, fungi. *Gomphus* is very close in anatomy and micromorphology to *Ramaria*, a large genus of branched clavarioid fungi which, like *Gomphus*, are characteristic members of the woodland flora, particularly well represented in Australasia and western USA. Because of their clavarioid nature and the absence of morphological forms showing a transition from one to the other, Corner (1970) placed the clavarioid species, I think correctly, in the family Ramariaceae. With *Gomphus*, this group might be better considered as an independent order, possibly representing one of our most primitive groups of Hymenomycetes.

For many years species of *Ramaria* were classified in *Clavaria* along with all the other coral and fairy-club fungi, but today's systematists, placing considerable emphasis on basidiospore morphology and on the anatomy and development of the fruit body, have changed all this. Thus, not only has the old genus *Clavaria* been split into several genera, but also the segregate genera have been placed, perhaps surprisingly, in several different families. The branched and simple club-shaped fruit bodies have undoubtedly evolved several times and probably represent remnants of a major ancient line of larger fungi. Clavulinaceae has been erected for a small group with large two-spored basidia and dull colours, etc.; Aphelariaceae has been erected for a group of four genera with family connections possibly with the stereoid fungi (*Stereopsis*) and with *Tremellodendron* (Tremellales); and Physalacriaceae has been erected for species that some claim to be reduced agarics (Singer, 1975).

Table 3. *Hymenophore configuration exhibited in families of the Polyporales and their allies*

Families \ Hymenophore configuration	Lamellate	Clavarioid to club-shaped fruit body	Smooth to rugulose resupinate surface	Hydnoid	Tubulate	Poorly developed wrinkles or veins
GOMPHUS ALLIANCE						
{ Gomphaceae			+	+		+
{ Ramariaceae		+				
THELEPHORALES						
Bankeraceae				+		
Thelephoraceae	aphyllophoroid: *Lenzitopsis* agaricoid: *Horakia*	*Thelephora*	*Tomentella*	*Hydnellum*	*Boletopsis*	*Polyozellus*
CANTHARELLALES						
Cantharellaceae			+			+
Clavariadelphaceae		+				
Hydnaceae *sensu stricto*				+		
POLYPORALES						
Schizophyllaceae	+ (not true gills)					
Polyporaceae					+	
Scutigeraceae					+	
Corticiaceae		+	+	+	+	+
Podoscyphaceae			+			
Stereaceae			+			
Punctulariaceae			+			
{ Auriscalpiaceae				+		
{ Lentinellaceae	+					

{ Clavicoronaceae		+				+
{ Hericiaceae			+	+		
{ Amylariaceae		+				
{ Bondarzewiaceae					+	
{ Echinodontiaceae			+	+		
Fistulinaceae					+	
Ganodermataceae					+	
Sparassidaceae			+			
Aphelariaceae		+	+			
Clavulinaceae		+				
{ Pterulaceae		+				
{ Clavariaceae		+				
Favolaschiaceae			+			+
HYMENOCHAETALES						
Hymenochaetaceae	+		+	+	+	+
Lachnocladiaceae		+	+			
BOLETALES						
Coniophoraceae	*Paxillus* – see Table 1		+	+	+	+
TRICHOLOMATALES						
?Physalacriaceae		+				+

The genera are listed, as an example, in one order only. Families bracketed together are often united into single families; Favolaschiaceae is a recently defined family covering small pileate fungi with wrinkles and veins beneath the pileus (see Singer, 1974); Scutigeraceae is often included in the Polyporaceae.

Note: Aphyllophorales is a straight synonym of Polyporales when the latter is taken to cover all non-agaric holobasidial basidiomycetes.

+ indicates that at least one genus within the family possesses the type of hymenial pattern. (Based partly on Donk, 1964.)

E. J. H. Corner believes that the last are remnants of a former larger group of homobasidiomycetes.

The Clavariadelphaceae often has large fruit bodies and in some ways it resembles a group that might be considered ancestral to the Cantharellales. However, this still leaves the nucleus of the old group of fairy clubs. These are now arranged, in three families: Pterulaceae with dimitic fruit bodies, Lachnocladiaceae, and the residual Clavariaceae which houses the majority of our more familiar British species. The first family is the result of nomenclatorial juggling with existing groups, as it had been recognised as a subfamily by Donk (1964) and as a series by Corner (1950). The Lachnocladiaceae as noted below has become a unit within the Hymenochaetales based on the micro-structure of its constituent members. The residue of fairy-club fungi after this wholesale dismembering becomes very homogeneous, with indications of relationships to some members of the Tremellales.

What can be offered as to our knowledge of the ecology of these clavarioid groups? *Typhula* is certainly known to pathologists because many of its species cause diseases of vascular plants, especially dicotyledonous species. Several appear to be particularly prevalent in areas of the world where there are very marked seasonal variations in climate. *Multiclavula* has exploited symbiosis, apparently linking closely to algae and forming basidiolichens, a feature found also in some tricholomatoid agarics (*Omphalina*, e.g. *O. hudsoniana*) and some resupinate fungi (*Dictyonema*). Generally, clavariaceous fungi are woodland or grassland inhabitants, probably playing the rôle of litter decomposers, although this is assumed only from qualitative observations. Certainly species of *Macrotyphula* can be found swarming on rotting leaves and small twigs, and *Clavulina* species like *Ramaria* species are frequently found growing in tight clusters springing from white or pale-coloured mycelia which appear to bind bleached leaf litter; *Ramaria stricta* is the only truly clavarioid lignicolous species in Britain.

Resupinate fungi and their toothed allies

The fairy clubs with their outer (amphigenous) hymenium traditionally have always been linked with the cauliflower-like *Sparassis crispa*. This link is based purely on superficial similarities as *Sparassis* has quite a different hymenial structure, the fertile portion being distributed on only one surface of the strap-like portions of the fruit body, so resembling a whole group of fungi termed resupinates because they lack a pileus, and the hymenium is purely a wrinkled, smooth or warted skin. Apparently species of *Sparassis* are wood rotters and are frequently found growing to enormous sizes; they are deeply rooted and often attached to tree roots. The genus is placed in the Sparassidaceae. In early classifications *Sparassis* sat between the 'Clavariaceae' and the 'Thelephoraceae', but the latter would now be split dramatically into several unrelated units: Stereaceae, Punctulariaceae, Podoscyphaceae and Corticiaceae, and into those resupinate forms similar to *Thelephora terrestris*, i.e. *Tomentella*.

The Podoscyphaceae is largely composed of stipitate 'Stereums' of tropical to subtropical areas of the world and, in keeping with many other groups from these same areas, little is really known about their ecological requirements. The ecology of the Punctulariaceae, with the hymenium distributed over isolated flat-topped tuberculae with cottony (byssoid) hyphae in between, is better known. The Punctulariaceae is an extremely small family of wood-rotting saprophytes (Talbot, 1958).

In contrast, a considerable amount of ecological information has accumulated over the years on several members of the Stereaceae, especially the north-temperate species. This family is retained in a very restricted sense for those fungi similar in structure to our five, common and familiar, British species: *Stereum gausapatum, S. hirsutum, S. rameale, S. rugosum* and *S. sanguinolentum.* The first species is a weak parasite of fagaceous trees, usually forming a white pocket-rot, and there is evidence that *S. rugosum* can sometimes play the same rôle; *S. sanguinolentum* grows commonly on coniferous wood, and *S. rugosum* is a particular feature of the basal stock of coppiced *Corylus. S. hirsutum* is usually considered to be a wood rotter and, if numbers of fruit bodies found are a measure of the activity of this species, it can only be concluded that this widespread and extremely common organism plays an important rôle in the woodland economy. Thus, in this one genus, even in its restricted sense, it is possible to move from a basically saprophytic life style to the parasitic life form; the latter is probably a derived character.

The Corticiaceae, closely related to the more organised Stereaceae, is a vast assemblage of fungi brought together by virtue of their possession of a resupinate fruit body with the hymenium distributed over warts, wrinkles, or short spines, or consisting of a smooth surface. The family, although fairly natural as now defined, probably still has a few anomalous members, reflecting the presence of species at various levels of evolution. The family encompasses typical wood-rotting organisms, but perhaps the constituent species have not demanded much attention in the past because of the difficulty of their identification and what appears to be their continually changing taxonomy. *Hyphoderma sambuci, Phanerochaete velutina*, and the hydnoid *Cristella farinacea* are typical lignicolous members of this group, whilst others fruit freely on leaves, e.g. *Aphinema byssoides*, peat-hags in spruce-woods, e.g. *Tylosperma fibrillosum*, and on soil, e.g. the often poroid *Cristella candidissima.* What we have failed to appreciate, however, is that our knowledge of the ecology of these fungi has been based entirely on where the fruit body finally forms and we have not recognised where the fungus is operational in the ecosystem (Watling, 1977). There are considerable amounts of data to suggest that some, if not a very great number, are potentially mycorrhizal fungi. Evidence is emerging from work carried out by Eriksson & Ryvarden and their collaborators (1973–6) that the resupinate fungi, now that they can be more easily recognised, do have very different and definite ecological requirements. Some species occur only in base-rich, mesophytic woodlands whilst others are only to be found in woodlands on acidic sandy soils, etc.

Donk (1964) has placed the genus *Tulasnella* in the Corticiaceae whilst other authors accommodate members of this same genus in the Tulasnellaceae and even the Tulasnellales. The basidial development in many constituent members is so very different from that of other members of the Polyporales that the last approach is more acceptable, especially as correlated characters are also to be found. However, in morphology, although close to the first, they are intermediate between both the Tremellales and the Polyporales. Members of the Tulasnellales are saprophytic or facultative parasites often of plant tissue in contact with soil, although the ecological significance of many members is only now becoming apparent. Many members, e.g. *Ceratobasidium, Tulasnella* and *Thanatephorus*, are important agents of endotrophic mycorrhizas, particularly with terrestrial orchids, although some species form web blights of tropical crop plants.

One other family, the Coniophoraceae, has the same type of resupinate fruit body and hymenium configuration as in the Corticiaceae, and includes two economically very important members, *Serpula lacrimans* – the dry-rot fungus – and *Coniophora puteana* – the wet-rot or 'cellar' fungus. Both possess thick-walled, strongly pigmented basidiospores, and because of the structure of these propagules this family has been related to the boletoid Paxillaceae. Few members of the latter family are involved in biodeterioration, and then only minimally, e.g. *Paxillus panuoides*, whereas both *Serpula* and *Coniophora* account annually for damage costing thousands of pounds sterling. The very mention of them, because of their ability to rot timbers so rapidly, strikes many a householder with fear (Dickinson: Chapter 10).

Amongst the remaining families in the Polyporales are a few, small, neatly defined families, some with very specific ecological requirements. Three are easily recognised: Echinodontiaceae, Fistulinaceae and Ganodermataceae, the first with toothed hymenophore and amyloid spores, and the last with poroid hymenophore and complex-walled basidiospores with channels running through the exospore. All three families are confined entirely to dead wood or are parasites of living trees, probably entering their host through broken branches and similar wounds. *Echinodontium* is not found in Britain, although *Fistulina hepatica*, forming an extensive heart rot of *Quercus*, is widespread particularly in southern parts of Britain. It is also found on *Castanea* and, in Australia, it attacks *Eucalyptus*, a rather big jump in host preference from Fagaceae to Myrtaceae. The family Fistulinaceae is characterised by free tubes. *Ganoderma adspersum* is a threat to old weakened *Fagus* trees and is common throughout the country; members of the Ganodermataceae are world-wide. *Echinodontium* (Echinodontiaceae) has been extended more recently to include two steroid wood rotters (*Stereum sulcatum* and *S. taxodii*), since it was concluded that only the configuration of the hymenophore was at variance with the family diagnosis and the species were parallel on all microscopic grounds.

Members of the family Hericiaceae are usually large wood-rotting hedgehog fungi. Some constituent taxa grow on weakened hardwood trees and other taxa on softwoods but usually as wound parasites. The genus *Hericium* is probably more a feature of temperate countries, particularly North America, than of the southern hemisphere. There are no terricolous species although in basidiospore morphology the group can be linked to the Clavicoronaceae which is a small group of clavarioid fungi breaking down woody substrates. The Hericiaceae and Clavicoronaceae are characterised by amyloid basidiospores. Two other small families are characterised by the same feature, the polyporoid Bondarzewiaceae and the clavarioid Amylariaceae. The latter was placed with *Hericium* in a single family by Donk (1964), but it was separated by Corner (1970) because in *Amylaria* the club-shaped fruit body may arise from a fertile resupinate *Corticium*-like layer. A relationship between the Bondarzewiaceae and the Russulaceae has been suggested (Singer, 1975), but this would not appear to be supported by microscopic characters.

Members of the Auriscalpiaceae also resemble *Hericium* in basidiospore morphology, but the common European member parallels *Strobilurus* in the agarics by inhabiting cones, fructification occurring well before the cone shows any signs of decay. However, the two other species, *Auriscalpium fimbriato-incisum* found in south-east Asia and *A. umbella* on *Nothofagus* in New Zealand, although terrestrial grow on root stocks not cones. The part played by these species in the colonisation of cones is still unknown. The Lentinellaceae was formerly contained within the Auriscalpiaceae but differs in that its members possess a lamellate hymenophore, i.e. they are agaricoid; indeed, although unrelated, species of *Lentinellus* were at one time classified amongst the agarics together with the pleurotoid *Lentinus* and *Panus*, indicating to some extent their lignicolous habitat preferences.

The re-introduction of the pleurotoid life form into the discussion indicates that as with the clavarioid fruit body, the laterally stipitate, lignicolous agaric has evolved several times. The genus *Pleurotus*, once a large genus, is now restricted to a very homogeneous group and constitutes a link between the polyporoid and agaricoid hymenial configurations. There is undoubtedly a fairly close relationship between the true polypores (Polyporaceae) and *Pleurotus, Lentinus* and *Panus*, if the latter is indeed distinct from *Lentinus.* It is not necessary in the present author's opinion to restrict the family Polyporaceae to a group incorporating the few polypores related to *Polyporus squamosus* and the agaricoid group mentioned above together with *Schizophyllum* as outlined by Singer (1975). *Schizophyllum commune* is isolated amongst both the Polyporales and agarics mainly because of its unique developmental pattern. It has been placed, because of this characteristic, in the Schizophyllaceae in association with a few less familiar fungi, e.g. *Plicaturopsis. S. commune* is, however, also 'off-beat' in its ecology, having been isolated from human toe-nails, sputum, cerebral fluid and mouth ulcers, from various cast-out organic materials ranging from tea leaves to dyes, and from various animal and plant materials including both natural and manufactured wood products; it is also a mycoparasite (Watling & Sweeney, 1974). Miller & Manning (1976) found that out of twenty-three common pleurotoid fungi only two of the species produced brown rots; the two species both belong to the genus *Lentinus.* All the other fungi tested were capable of delignifying woody substrates.

True polypores

The true polypores have attracted the attention of many mycologists in the past because of their importance as wood rotters. They have been dissected taxonomically, and subjected to sophisticated cultural studies and to computer analysis of specific characters. Stalpers (1978) recently produced an excellent key to a very large group of polypores and related fungi in culture; this is a continuation of an approach introduced by Davidson, Campbell & Blaisdell (1938) and extended by Nobles (1948, 1965). Nobles (1973) has been able, based on her long experience of this group of fungi, to plot patterns of correlation amongst members of this group. One of the paths of enquiry has led her to superimpose the ecology of these groups of organisms onto the framework of a modern classification, thereby relating their taxonomy both to basic characters, such as white rot versus brown rot, and to cultural characters, such as presence or absence of clamp connections. Obviously one cannot deal in depth with all the results and their possible ramifications which are discussed in more specialised literature, but some generalisations can be made. Thus there appears to exist a primitive group of polypores, usually with bipolar mating patterns, which induce brown rots and exhibit a preference for gymnosperms. It would appear that polypores producing brown rots are rare in the tropics, the majority of species in this ecological grouping being confined to the northern coniferous zones. This group contrasts with a more advanced assemblage with tetrapolar mating patterns, an ability to induce white rots and with a host preference for frondose trees. In both categories, clamp connections may be present or absent; the latter in the polypores are probably a derived character. In the first group, two secondary groups have also been recognised and are probably derived from the primitive group. In group *a*, interfertility is governed by alleles at one locus only, an extracellular oxidase system is present and clamp connections are formed only on hyphae away from the advancing growing front, e.g. *Phlebia radiata* and its allies; in group *b*, interfertility is governed by alleles at two loci, extracellular oxidase systems are lacking and clamp connections are retained, e.g. *Tyromyces stipticus*.

During all such exercises the significance of the results must be interpreted carefully. For example, *Tyromyces lacteus* would appear to exist in two forms, each with a different spectrum of enzyme systems; these forms are accompanied by small micro- and macroscopic character differences which might indicate that different but closely related taxa were really involved. However, it is only by looking at the picture as a whole that both a full understanding of the family can be achieved, and unsuspected links with other poroid families can be located.

One of the major steps forward in the classification of the polypores was the introduction of the mitic system by Corner (1930, 1932*a,b*) and its subsequent expansion. This has allowed not only relationships to be identified within the morass of poroid fungi but also permitted parallel patterns of structure to be located within supposedly unrelated groups. Apparently, the dimitic and monomitic hyphal patterns are found in many groups of fungi, whereas the trimitic system is unique to the polypores. The possession of such a hyphal pattern has apparently allowed some polypores to exploit with advantage ecological sites not open to soft, more vulnerable, taxa (Bondarstev, 1963). Evidently the trimitic system has evolved on several occasions during the evolution of the polypores. However, care must be taken to relate such observations to correlating characters.

Although usually associated with wood rots, some polypores are humicolous, often forming large sclerotia which act as a resting stage, e.g. *Polyporus mylittae.* These species are frequently associated with arid conditions and/or with vegetation that is periodically burnt. However, one usually associates the Polyporaceae with such fungi as: *Fomes fomentarius*, which has a large perennial fruit body and which in northern latitudes frequently occurs on *Betula*, although in southern areas of Britain it grows on other hosts, especially *Fagus* (Watling, 1978*a*); *Heterobasidion annosum*, an important root-rot organism of plantation and amenity trees; *Laetiporus sulphureus* with its colourful fruit bodies springing from dead or dying trees of *Quercus, Taxus* or even *Pinus*; *Piptoporus betulinus*, a fungus rarely absent from European *Betula* woods; *Meripilus giganteus*, which produces huge multi-pileate fruit bodies often at the base of *Fagus*; *Lenzites betulina* on *Betula* and *Daedalea quercina* on *Quercus* with lamellate and labyrinthiform hymenophores respectively; and finally *Coriolus versicolor* now placed in the genus *Trametes* and a common inhabitant of dead branches, twigs, stumps and standing trunks (Rayner & Todd: Chapter 6). The last is frequently isolated from softwood window and door frames although in the field the same species is rarely found on coniferous wood (D. Aston, personal communication). All these fungi are efficient wood rotters and a complete list of economically important species would be very lengthy.

It can be seen that *Fomes*, delimited at the beginning of the century as a genus with a perennial fruit body, and *Polyporus*, with an annual fruit body, are now very restricted genera, and that many new ones have been defined (Pegler, 1973). Some pore fungi, e.g. *Schizopora*, are undoubtedly closely related to members of the Corticiaceae. Although many polypores characterise a particular tree host or group of hosts, they do not appear to be confined to these hosts. Thus, species characteristic of gymnosperm wood, e.g. *Heterobasidion annosum*, can also be found on *Sorbus, Crataegus, Fagus* and even *Erica* (R. Watling, unpublished), and species characteristic of hardwood timber, e.g. *Ganoderma applanatum*, can sometimes be found on coniferous debris.

A major group of wood rotters are contained in the final order, the Hymenochaetales. As early as 1964, Donk defined the group at family rank as the Hymenochaetaceae but recognised that in the future it might be considered worthy of higher status. He considered that, in addition to the anatomical similarities, the group was natural because every species investigated produced extracellular oxidases in culture and in the field was associated with white rots. The Hymenochaetales, as now defined, includes a clavarioid element (Lachnocladiaceae) as well as resupinate forms (*Hymenochaete*) and poroid forms (*Phellinus*). Patouillard (1900) in his early studies recognised two groups corresponding to the poroid and non-poroid forms above, and Corner (1950) accepted the grouping under the name 'Xanthochroic series'. This latter term reflects the fact that the elements of the fruit body are pigmented from the start and darken even further with the application of alkaline solutions, a unifying character of all the members.

The family Hymenochaetaceae has three subfamilies: the Vararioideae (*Vararia*) with a totally resupinate fruit body and possessing sterile dichohyphidia in the hymenium; the Asterostromatoideae (*Asterostroma*) with resupinate fruit bodies ornamented with stellate structures (asterosetae), sterile fruit bodies of which are found occasionally on house timbers; and the Hymenochaetoideae which ranges from corticioid genera (*Hymenochaete*), typical wood rotters in the floras of base-rich woodlands, to poroid species (*Inonotus* and *Xanthoporia* with or without a well-developed pileus respectively). As in the genus *Stereum*, a range is exhibited within the family from saprophytes to weak but economically important parasites. A member of this family, *Coltricia perennis*, appears to frequent the site of old bonfires or newly disturbed ground in coniferous woodland, but any generalisation on the ecology of such terrestrial larger fungi is frequently based only on observations of where the fungus fruits; little is known yet of the true part that these and other larger fungi play in the habitat.

Table 4. *The Gasteromycetes or stomach fungi – the orders currently accepted in the heterogeneous group often termed the Gasteromycetales*

PHALLALES[l], SCLERODERMATALES[m], LYCOPERDALES, GEASTRALES[l], TULOSTOMATALES, NIDULARIALES[l], HYMENOGASTRALES[m], MELANOGASTRALES[m], GAUTIERIALES[m]

[m] Major mycorrhiza formers and [l] major litter decomposers.

Gasteromycetes or stomach fungi – a text-book nightmare!

In all text books on fungi, the Gasteromycetes or stomach fungi are left to the end as an embarrassing appendix, mainly because of the difficulty of placing the members in an all-embracing scheme. The group is summarised in Table 4. There is no doubt in my mind that some, especially those formerly placed in the Secotiaceae, are related to, indeed better placed in, existing families of agarics (Table 5). However, for others it is still somewhat difficult even to suggest a common origin, if indeed there ever was one (Watling, 1978*b*). What remains abundantly clear is that the Gasteromycetes is a heterogeneous group, and although a single order, the Gasteromycetales, can house them, this is simply a matter of convenience and neither reflects the true relationships of the members nor places sufficient emphasis on the exploitation of the various spore-dispersal mechanisms found within the group. The inability to eject their basidiospores forcibly from the single-celled basidia is the common feature uniting the group because the members are certainly not brought together on a common morphology. The possession of this basidial type (apobasidium) is not unique to the Gasteromycetes and has evolved in other groups such as the Auriculariales and Polyporales (Gareth Jones: Chapter 11). There is evidence, because of the presence of intermediates, that loss of the ejection mechanism has occurred on several occasions in the agaricoid fungi (Thiers & Smith, 1969; Watling, 1976).

The Phallales with both epigeous and hypogeous members is heavily dependent on invertebrate activities for spore dispersal. Perhaps this is why, as in the insect-pollinated angiosperms, such extreme shapes, colours and smells are exhibited by the constituent species, a feature not found in any other single group of fungi. Undoubtedly many members are active wood rotters, e.g. *Phallus impudicus*, but unfortunately we know very little of the ecology of this interesting and mainly tropical group.

Table 5. *Agaric–gasteroid links*. Genera of agarics most closely related to the gasteromycete genus are given in brackets

	Agaricaceae	*Gyrophragmium* (*Agaricus*) *Endoptychum* p.p. *Smithiogaster*
	Bolbitiaceae	*Cyttarophyllum* (*Agrocybe*) *Gastrocybe* (*Bolbitius*)
	Coprinaceae	*Longula* *Montagnea* (*Coprinus*) *Panaeolopsis* (*Panaeolus*)
	Cortinariaceae	*Nivatogastrium* (*Pholiota*) *Setchelliogaster* (*Descolea*) *Thaxterogaster* (*Cortinarius*)
AGARICALES		
	Strophariaceae	*Cyttarophyllopsis* *Clavogaster* *Weraroa* (*Psilocybe* and *Stropharia*) *Galeropsis* *Tympanella*
	Lepiotaceae	*Endolepiotula* *Notholepiota* *Secotium* (*Lepiota* and *Macrolepiota*) *Phyllogaster* *Neosecotium* *Endoptychum* p.p.
TRICHOLO-MATALES	Tricholomataceae	*Hydnangium* (?*Laccaria*) (Hydnangiaceae)
AMANITALES	Amanitaceae	*Torrendia* (*Amanita*)
PLUTEALES	Pluteaceae	*Brauniella* (*Volvariella*) (Brauniellaceae)
	Entolomataceae	*Rhodogaster* (*Rhodophyllus* s. Singer, Moser, etc.) *Richoniella*
RUSSULALES	Russulaceae	Astrogastraceae, e.g. *Macowanites* (*Russula*) *Arcangeliella* (*Lactarius*) Elasmomycetaceae, e.g. *Elasmomyces* (*Russula*) *Zelleromyces* (*Lactarius*)
BOLETALES	Boletaceae	*Gastroboletus* (*Boletus sensu lato*)
	Gomphidiaceae	*Brauniellula* (*Chroogomphus*) *Gomphogaster* (*Gomphidius*)
	Paxillaceae	*Austrogaster*, *Paxillogaster*, *Singeromyces*, *Gymnopaxillus* (*Paxillus*)
	(Coniophoraceae	*Chrysoconia*)
	Stephanosporaceae[a]	*Stephanospora* (*Lindtneria*)

[a] The only polypore–gasteroid connection so far recorded.

The Sclerodermatales includes many mycorrhizal members, and in addition is characteristic of highly mineral soils, in contrast to the superficially similar Lycoperdales whose members generally characterise organic soils. Both groups disperse their spores by mechanical means from a central store; the true puff-balls of the genus *Lycoperdon* (Lycoperdales) and the earth-stars, *Geastrum* (Geastrales), do so by a bellows effect. Drops of water from surrounding herbaceous or woody material fall onto the inner peridium, which on acting as a diaphragm moves puffing out a cloud of basidiospores. Other puff-ball genera, e.g. *Vascellum* and *Calvatia*, and earth-balls (*Scleroderma*, Sclerodermatales) rely on erosion of the apical part of the fruit body to expose the spores, and disturbance to disperse them. *Pisolithus tinctorius*, another member of the Sclerodermatales, provides evidence of the effectiveness of this dispersal mode; spores of this species are scattered everywhere. No wonder it is being heralded as the supreme mycorrhizal former (Marx & Bryan, 1970), for it is not only a very efficient spore producer, but also 'promiscuous' in its choice of hosts. The Sclerodermatales appears to me to be a very ancient group of fungi. *Tulostoma*, a small genus of stalked puff-balls, is placed in the Tulostomatales. This order differs from the puff-balls in development, basidial organisation and morphology, particularly in the significant agaricoid stipe which raises the spore head aloft.

The intriguing 'bird's nest' fungi (Nidulariales) need little comment as they have been dealt with adequately by Brodie (1975). They are mainly litter decomposers and are often found on herbaceous material, leaves and twigs, dung, and wood-chips, either on the forest floor, or in artificial habitats on compounded wood, including chipboard, and similar building materials.

Finally we are left with the false truffles which are distributed in the Gautieriales, Hymenogastrales and Melanogastrales. Just as some members of the puff-ball group were formerly lumped in the genus *Secotium* but can be tied now more satisfactorily to orders and families of agarics, so also some of the hypogeous basidiomycetes can be linked to the boletes and some to the Russulaceae (Table 5). In a similar way, these false truffles are also mycorrhizal with angiosperms and gymnosperms.

The residue of hypogeous fungi, even after the removal of these entities, is probably still far from homogeneous, being drawn together solely by their specialisation for spore dispersal by animal agencies, rather than by anatomical and developmental similarities. Fogel & Trappe (1978) have admirably dealt not only with the relationships between hypogeous fungi and individual rodent species but also with the possible evolution of this relationship and the specialisations of the animals for this food requirement. Normally the fruit bodies of hypogeous fungi are found at the junction of duff and mineral soil; the duff is often dry and pale-coloured because of the permeating mycelium, and, at the junction of the litter and the soil, a bleached zone is observed, suggesting mineral mobilisation. Whether all hypogeous fungi are mycorrhizal needs to be examined in the future.

Conclusions

The ecological characteristics of most of the major groups of the basidiomycetes, more correctly termed the Basidiomycotina, are intercollated within what might be considered an ultra-modern classification. Good correlation has been found in many examples, but a certain amount of data apparently does not as yet fit into the overall pattern. As it is believed that the classification adopted herein is more natural than previous attempts, a baseline can now be offered from which convergence and divergence might be studied.

The specialised mycorrhizal basidiomycetes would appear to be physiologically rather different from most litter or wood decomposers, but even in this first group there is a whole spectrum of organisms ranging from the most specialised to the least specialised, a phenomenon observable both intragenerically and intraspecifically. Thus, in such so-called mycorrhizal fungi as *Paxillus involutus* and *Boletus subtomentosus*, Scandinavian workers (e.g. Lundeberg, 1970) have shown that some strains may be active litter decomposers whilst others morphologically similar are undoubtedly dependent on mycorrhizal associations. However, we have been led to believe that all mycorrhizal relationships are very specific, but records now indicate that a very large number of basidiomycete relationships are less specific than many of the relationships found among wood or litter decomposers. For example *Lactarius rufus*, normally found with *Pinus sylvestris* in Britain, is perfectly at home in the field and in culture with *Picea sitchensis* as well as with *Pinus*; it is only rarely found in the British Isles with arborescent *Betula* species. In Fennoscandia, on the other hand, as one reaches the northern limits of *Pinus, L. rufus* apparently grows with *Betula tortuosa* and in the tundra and palsa bogs with *B. nana*. Thus, under certain conditions, perhaps when competition from more aggressive and possibly more specialised mycorrhiza formers is reduced, *L. rufus* can extend its range into more inhospitable environments.

Some genera, e.g. the bolete genus *Suillus*, are obligatorily mycorrhizal and therefore very specialised, but there is every reason, as indicated above, to suggest that at least some of these fungi can also mobilise nutrients which will assist other fungi in their rôle as wood or litter decomposers. Specialised mycorrhizal fungi are found particularly in the boreal coniferous woodlands and the *Nothofagus* forests of South America, and genera such as *Suillus* have probably advanced in unison as their host genus evolved and expanded its range.

Among litter decomposers one also finds a broad range of specialist ability, some species growing on leaves of only a single tree species whilst others are apparently equally at home colonising either frondose or coniferous debris of various species, or debris of both coniferous and frondose trees. This spectrum of activity would appear to apply also to basidiomycete decomposers in the tropics and subtropics, although in these communities a greater link between litter decomposers and mycorrhizal fungi can be demonstrated, with nutrients probably being quickly removed from the A horizons and made available to the root systems (Frankland: Chapter 14). In these cases, ecological specialisation is at a minimum compared with *Lactarius rufus* and with the majority of the mycorrhizal agarics of the boreal forests all of which appear to partake in highly evolved relationships. This parallels the rather restricted substrate preferences exhibited by many litter and wood decomposers in similar latitudes. It might be suggested therefore that the latter are as derived in character as the boreal mycorrhizal species.

It might even be suggested that the phenomenon of mycorrhizal association is much more fundamental than would at first be supposed. Certainly symbiotic relationships are necessary for our understanding of multicellular organisms (Margulis, 1970) if not of the ability of plants to colonise land (Pirozynski & Malloch, 1975). Symbiosis is therefore not an outrageous suggestion for the ancestral activity of fungi, wood decomposition and related phenomena being derived in character (see Smith, 1979). VA-mycorrhizal fungi are found in rhizomes of Devonian plants and one must look to geological deposits for further clues and possible answers (Pirozynski, 1976). The vast deposits of coal in geologic time indicate a low rate of decomposition by basidiomycetes and soft-rot fungi in these early periods even if they were present in any number; basidiomycetes were apparently present as clamp-connected hyphae and have been found in rocks of this epoch (Dennis, 1970). Therefore, some evidence exists for symbiotic relationships but little for extensive rotting even when very probably not all the coal laid down was totally submerged. There are few decomposers of pteridophytes today; perhaps the dramatic change in chemistry of the substrate during the Cretaceous period with the evolution of the angiosperms allowed the wood-rotting basidiomycetes to expand explosively.

It is a disturbing thought that the species concept adopted for mycorrhizal basidiomycetes may be much narrower than that utilised in the taxonomy of the wood rotters and litter decomposers; this makes comparisons between the two ecological groups difficult. Only time will tell.

A further question remains unanswered, a question which colours the whole discussion, and which, I am the first to admit, could completely destroy our ideas. Has not the method of habitat exploitation of particular taxa influenced the fragmentation of the old Friesian framework in such a way that a positive correlation not only should be expected but should follow automatically? Perhaps when more studies have been completed we shall be able to ascertain whether the tail has been wagging the dog or not!

Since this account was prepared a member of the recently recognised family Favolaschiaceae (*Favolaschia dybowskyana*) has been demonstrated to be mycorrhizal with orchids (Jonsson & Nyland, 1979). Members of the genus are generally considered to be wood rotters or litter decomposers. However, it must also be borne in mind that this phenomenon of a single species in a genus being mycorrhizal is found in other groups of the Aphyllophorales, e.g. a *Clavaria* species (affinis *argillacea*) with ericaceous shrubs (Seviour, Willing & Chilvers, 1973).

References

Ainsworth, G. C. (1966). A general purpose classification of fungi. *Bibliography of Systematic Mycology*, Part 1, 1–4.

Arpin, N. & Kühner, R. (1977). Les grandes lignes de la classification des Boletales. *Bulletin Mensuel de la Société Linnéenne de Lyon*, **46,** 181–208.

Beaumont, P. C. & Edwards, R. L. (1971). Constituents of the Higher Fungi. XI. Boviquinone-3, (2,5-dihydroxy-3-farnesyl-1,4-benzoquinone), diboviquinone-3,4, methylenediboviquinone-3,3, and xerocomic acid from *Gomphidius rutilus* Fr. and diboviquinone-4,4 from *Boletus* (*Suillus*) *bovinus* (Linn. ex Fr.) Kuntze. *Journal of the Chemical Society, C*, 2582–5.

Bondarstev, M. A. (1963). On the anatomical criterion in the taxonomy of Aphyllophorales. *Botanicheskiĭ zhurnal SSSR*, **48,** 362–72.

Brodie, H. J. (1975). *The Bird's Nest Fungi.* Toronto University Press.

Corner, E. J. H. (1930). On the identification of the brown-root fungus. *Gardens' Bulletin, Straits Settlements*, **5,** 317–50.

Corner, E. J. H. (1932*a*). A fruit-body of *Polystictus xanthopus* Fr. *Annals of Botany*, **46,** 71–111.

Corner, E. J. H. (1932*b*). A *Fomes* with two systems of hyphae. *Transactions of the British Mycological Society*, **17**, 51–81.
Corner, E. J. H. (1950). *A Monograph of Clavaria and Allied Genera.* Annals of Botany Memoirs, No. 1. Oxford University Press.
Corner, E. J. H. (1970). Supplement to 'A Monograph of *Clavaria* and Allied Genera'. *Nova Hedwigia*, **33**, 1–209.
Davidson, R. W., Campbell, W. A. & Blaisdell, D. J. (1938). Differentiation of wood-decaying fungi by their reactions on gallic and tannic acid medium. *Journal of Agricultural Research*, **57**, 683–95.
Dennis, R. L. (1970). A Middle Pennsylvanian Basidiomycete mycelium with clamp-connections. *Mycologia*, **62**, 578–84.
Dennis, R. W. G., Orton, P. D. & Hora, F. B. (1960). New check list of British agarics and boleti. *Supplement to the Transactions of the British Mycological Society*, **43**, 1–225.
Donk, M. A. (1964). A conspectus of the families of Aphyllophorales. *Persoonia*, **3**, 199–324.
Donk, M. A. (1966). Check List of European hymenomycetous Heterobasidiae. *Persoonia*, **4**, 145–335.
Edwards, R. L. (1977). Pulvinic acid and derivatives from *Boletus* species. *Abstracts of the Second International Mycological Congress, Tampa*, **1**, 160.
Edwards, R. L. & Elsworthy, G. C. (1967). Variegatic acid, a new tetronic acid responsible for the blueing reaction in the fungus *Suillus* (*Boletus*) *variegatus*. *Chemical Communications*, **135**, 373–4.
Eriksson, J. & Ryvarden, L. (1973–6). *The Corticiaceae of North Europe*. Oslo: Fungiflora.
Fogel, R. & Trappe, J. M. (1978). Fungus consumption (mycophagy) by small animals. *Northwest Science*, **52**, 1–31.
Heath, I. B. (1978). *Nuclear Division in the Fungi.* London & New York: Academic Press.
Heath, I. B. & Heath, M. C. (1976). Ultra-structure of mitosis in the cow-pea rust fungus *Uromyces phaseoli* var. *vignae. Journal of Cell Biology*, **70**, 592–607.
Heim, R. (1973). The interrelationships between the Agaricales and Gasteromycetes. In *Evolution in the Higher Fungi* ed. R. Petersen, pp. 505–34. Knoxville: University of Tennessee Press.
Hering, T. F. (1972). Fungal associations in broad-leaved woodlands in northeast England. *Mycopathologia et Mycologia Applicata*, **48**, 15–21.
Jonsson, L. & Nyland, J. E. (1979). *Favolaschia dybowskyana* (Singer) Singer, a new orchid mycorrhizal fungus from tropical Africa. *New Phytologist*, **83**, 121–8.
Kreisel, H. (1969). *Grundzüge eines naturlichen Systems der Pilze.* Jena: G. Fischer-Verlag.
Kubai, D. F. (1978). Mitosis and fungal phylogeny. In *Nuclear Division in the Fungi*, ed. I. B. Heath, pp. 177–230. London & New York: Academic Press.
Kühner, R. (1977). Les grandes lignes de la classification des Agaricales, Asterosporales, et Boletales. *Bulletin Mensuel de la Société Linnéenne de Lyon*, **46**, 81–108.
Kühner, R. (1978–80). Les grandes lignes de la classification des Agaricales, Plutéales, Tricholomatales. *Bulletin Mensuel de la Société Linnéenne de Lyon*, **47**, 91–164, 234–303, 325–88, 421–84, 517–80; **48**, 17–48, 81–112, 145–76, 201–48, 273–304, 333–64, 393–440, 465–512, 537–68, 609–40; **49**, 73–120, 153–200, 225–72, 297–344, 411–46.
Lundeberg, G. (1970). Utilisation of various nitrogen sources, in particular bound soil nitrogen, by mycorrhizal fungi. *Studia Forestalia Suecica*, No. 79, 1–95.

Margulis, L. (1970). *Origin of Eukaryotic Cells*. New Haven and London: Yale University Press.
Marx, D. H. & Bryan, W. C. (1970). Pure culture synthesis of ectomycorrhizae by *Thelephora terrestris* and *Pisolithus tinctorius* on different conifer hosts. *Canadian Journal of Botany*, **48**, 639–43.
Miller, O. K. & Manning, D. L. (1976). Distribution of the lignicolous Tricholomataceae in the southern Appalachians. In *Distributional History of the Biota of the Southern Appalachians*, Part 4, *Algae and Fungi*, ed. B. C. Parker & M. K. Roane, pp. 307–44. Charlottesville: University Press of Virginia.
Moore, R. T. (1972). Ustomycota, a new division of higher fungi. *Antonie van Leeuwenhoek, Journal of Microbiology and Serology*, **38**, 567–84.
Moore, R. T. (1978). Taxonomic significance of septal ultra-structure with particular reference to the jelly-fungi. *Mycologia*, **20**, 1007–24.
Nobles, M. (1948). Studies in forest pathology. IV. Identification of cultures of wood-rotting fungi. *Canadian Journal of Research*, **C26**, 281–431.
Nobles, M. (1965). Identification of cultures of wood-inhabiting Hymenomycetes. *Canadian Journal of Botany*, **43**, 1097–139.
Nobles, M. (1973). Cultural characters as a guide to the taxonomy of the Polyporaceae. In *Evolution in the Higher Fungi*, ed. R. Petersen, pp. 169–96. Knoxville: University of Tennessee Press.
Oberwinkler, F. (1977). Das neue System der Basidiomyceten. In *Beiträge zur Biologie der niederen Pflanzen*, ed. W. Frey, H. Hurka & F. Oberwinkler, pp. 59–105. Stuttgart & New York: G. Fischer-Verlag.
Oberwinkler, F. (1978). Was ist ein Basidiomycet? *Zeitschrift für Mykologie*, **44**, 13–29.
Patouillard, N. (1900). *Essai Taxonomique sur les Familles et les Genres des Hyménomycètes*. Lons-Les Saunier: L. Declume.
Pegler, D. N. (1973). The Polypores. *Supplement to the Bulletin of the British Mycological Society*, **7**, 1–43.
Pirozynski, K. A. (1976). Fossil fungi. *Annual Review of Phytopathology*, **14**, 237–46.
Pirozynski, K. A. & Malloch, D. W. (1975). The origin of land plants: a matter of mycotrophism. *Biosystems*, **6**, 153–64.
Seviour, R. T., Willing, R. R. & Chilvers, G. A. (1973). Basidiocarps associated with ericoid mycorrhiza. *New Phytologist*, **72**, 381–4.
Singer, R. (1951). The Agaricales in modern taxonomy. *Lilloa*, **22**, 1–832.
Singer, R. (1962). *The Agaricales in Modern Taxonomy*, 2nd edn. Weinheim: J. Cramer.
Singer, R. (1974). A monograph of *Favolaschia*. *Supplement to Nova Hedwigia*, **50**. Weinheim: J. Cramer.
Singer, R. (1975). *The Agaricales in Modern Taxonomy*, 3rd edn. Vaduz: J. Cramer.
Singer, R. (1978). Notes on bolete taxonomy, II. *Persoonia*, **9**, 421–38.
Smith, D. C. (1979). Discussion on: the cell as a habitat (organiser, D. C. Smith). *Proceedings of the Royal Society of London*, **B204**, 113–286.
Stalpers, J. A. (1978). *Identification of wood-inhabiting Aphyllophorales in pure culture*. Studies in Mycology, No. 16. Baarn: Centraalbureau voor Schimmelcultures, Institute of the Royal Netherlands Academy of Arts and Sciences.
Talbot, P. (1958). Studies on some South African resupinate Hymenomycetes. II. *Bothalia*, **7**, 131–87.
Thiers, H. T. & Smith, A. H. (1969). Hypogeous cortinarii. *Mycologia*, **61**, 526–36.
Watling, R. (1976). Observations on the Bolbitiaceae. 13. A xeromorphic member of the family. *Kew Bulletin*, **31**, 587–91.

Watling, R. (1977). Relationships between the development of higher plant and fungal communities. *Abstracts of the Second International Mycological Congress, Tampa*, **2,** 718.
Watling, R. (1978*a*). The distribution of larger fungi in Yorkshire. *Naturalist*, **103,** 39–57.
Watling, R. (1978*b*). From infancy to adolescence. Advances in the study of higher fungi. *Supplement to the Transactions of the Botanical Society of Edinburgh*, **42,** 61–73.
Watling, R. & Sweeney, J. (1974). Observations on *Schizophyllum commune* Fries. *Sabouraudia*, **12,** 214–26.

2 Production, release and dispersal of basidiospores*

C. L. KRAMER
Division of Biology, Kansas State University, Manhattan, Kansas 66506, USA

Introduction

Attempts to define the word 'spore' usually result in a reference to 'microscopic reproduction units', with descriptions of their various attributes and functions. Of most concern here, are the two primary functions of spores: dispersal and survival. As Gregory (1966) pointed out in an excellent paper on the fungal spore, 'the requirements for these two functions are somewhat irreconcilable, and a spore is seldom good at both, so any one spore type in a species of fungus is usually specialised towards one function or the other'. In many of the 'lower fungi' and ascomycetes, dispersal is primarily the function of the asexual spore, while the sexual spore generally serves the survival rôle. What about the basidiomycete decomposing-fungi, the subject of this symposium volume? In most cases, they have but one spore form, the basidiospore.

Since their life histories usually lack two spore forms, basidiomycetes are dependent on the sexually produced basidiospore, at least in part, for both dispersal and survival. As already pointed out, the characteristics necessary for efficient performance of both functions generally are not compatible; it seems that in basidiomycetes, and in the Hymenomycetes in particular, spore forms better suited to dispersal than survival have evolved, and these fungi are dependent on the development of other characteristics for survival as successful organisms.

Spores that function primarily in the survival of the species during periods unfavourable for growth tend to be thick-walled, comparatively large with considerable stored food material, and often require some

* Contribution No. 79-276-A, Division of Biology, Kansas Agricultural Experiment Station, Kansas State University.

factor to break a dormant condition. On the other hand, spores that function primarily for dispersal tend to be thinner-walled, smaller, and with less survival ability during extended dormant periods when conditions may not favour the development of new generations. Such basidiospores possess walls mostly in the 5–7 μm range, they are elongate or somewhat irregular in shape, and in the Hymenomycetes have an apiculus at the point of attachment.

The percentage germination of basidiospores is often low. In a study we undertook to compare germination of spores collected from the atmosphere (Kramer & Pady, 1968), less than 6% of the basidiospores collected produced germ tubes (Fig. 1) compared with approximately 40% of the *Cladosporium* conidia and ascospores and 80% of the *Alternaria* conidia.

One characteristic that could perhaps compensate for poor survival of basidiospores, especially of those species that we are concerned with in this volume, is the development of a perennial mycelium eliminating the necessity for chance establishment of new, sexually compatible, combined mycelial systems each year. The possibility of unions of sexually compatible strains is further increased in these organisms by the development of multi-allele systems for sexual compatibility (Rayner: Chapter 7). As perennial systems, the mycelia become extensive and capable of supporting large numbers of fruit bodies over comparatively long periods of time. This is evident in the 'fairy ring' fungi and in many of the wood decomposers. In many other cases, for example *Laetiporus sulphureus* and *Ganoderma applanatum*, there are fewer individual fruit bodies, but their size is comparatively very large. In some species, even

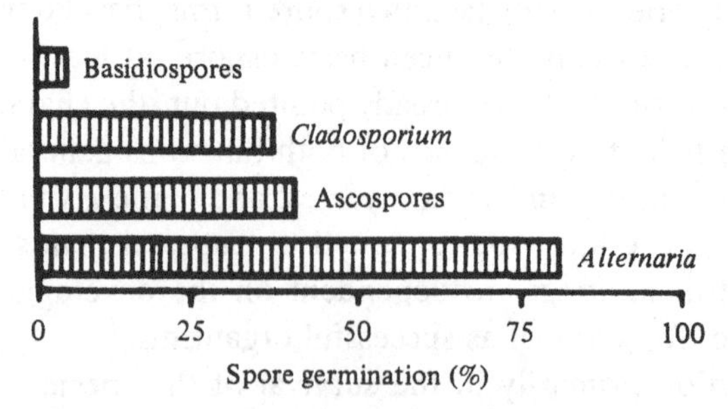

Fig. 1. Percentage of basidiospores, ascospores, and *Cladosporium* and *Alternaria* conidia, collected from the atmosphere on adhesive-coated slides, which germinated to produce germ tubes when exposed slides were sprayed with distilled water and placed in a moist chamber for 24 h.

the fruit bodies themselves may be perennial, so that the energy which otherwise would be needed for the development of new fruit bodies could be used in increased spore production.

All these characters tend to broaden the range of conditions under which sporulation can occur; they extend the time of sporulation, and increase the numbers of spores produced. The combination of such characteristics undoubtedly is sufficient to offset the low survival rate of basidiospores.

Spore production

As indicated, the Hymenomycetes have acquired the ability to produce basidiospores in fantastically large numbers. However, precise knowledge of the number of spores that fungi are capable of producing is meagre. Several researchers, including Buller (1909), have made estimates, but few have done precise quantitative measurements of the total spores that a single fruit body, or a unit area of sporulating tissue, is capable of producing during a given time and under specific environmental conditions. Buller (1909), after collecting spores in a spore deposit from a specimen of *Agaricus campestris* with a cap 8 cm in diameter, estimated that 1.8×10^9 spores were released by that specimen. In a fruit body of *Langermannia gigantea* (syn. *Calvatia gigantea*) that measured $40 \times 28 \times 20$ cm, Buller estimated by determining the spore numbers per unit weight that there were 7×10^{12} spores. White (1919) estimated that a specimen of *Ganoderma applanatum* produced some 30×10^6 spores a day for an estimated six-month growing season, and about 5.5×10^{12} spores during one season.

We became interested several years ago in trying to obtain information on basidiospore production in various Hymenomycetes. In a preliminary study, we used a liquid impinger to collect the total volume of basidiospores as they were released from a specimen of *Agaricus campestris* (Kramer, 1974). The diameter of the cap was approximately 8 cm, about the same as Buller's. During the seven days of sporulation of our specimen, 3.7×10^9 basidiospores were released, compared with Buller's 1.8×10^9. Fig. 2 shows the daily total spore production over the seven-day sporulation. We later developed a collector, the Dodecapus (Rockett, Kramer & Collins, 1974), that automatically collected the spores released each hour in separate liquid impingers. As before, this collector operated continuously with baffles to direct the airflow to the sampler; the total volume of spores was collected as the spores were released from the fruit body. However, because air currents in the field could carry some of the spores away from the collector, we did most of our studies with this device in controlled-environment chambers (Rockett & Kramer, 1974*b,c*). We developed a technique to produce fruit bodies of polypore fungi in the greenhouse on excised portions of wood (Rockett & Kramer, 1974*a*), and then transferred the fruit bodies to the chambers when sporulation began.

In these studies (Rockett & Kramer, 1974*b,c*), we attempted to standardise the measurement of spore discharge by converting the numbers of spores released to numbers per square centimetre of exposed hymenial surface. However, that is not the true area of a polypore's hymenial surface, but rather the under surface of the fruit body with openings to the hymenial tubes. However, in the jelly fungi (Tremellales) that we studied, the basidial layer covers the exterior surface of the fruit body. Table 1 summarises the results from studies of several species of polypore and jelly fungi (Rockett & Kramer, 1974*b,c*). *Coriolus pubescens* produced basidiospores at the phenomenal rate of 11.8×10^6 cm^{-2} of exposed hymenial surface each hour, and, during the one hour of peak release, 21.8×10^6 cm^{-2} were released. This specimen, which had only 7.7 cm^2 of exposed hymenial

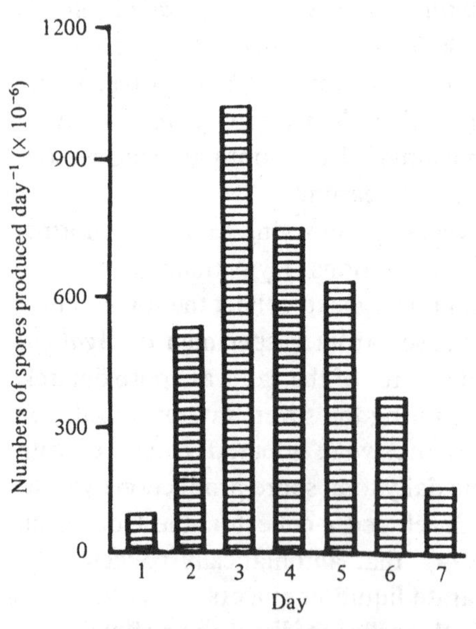

Fig. 2. Total numbers of basidiospores produced and released daily during a 7-day sporulation by a single specimen of *Agaricus campestris* with a cap 8 cm in diameter.

Table 1. *A summary of spore production studies*

Species	Area of exposed hymenial surface (cm^2)	Mean number of spores released per specimen in 24 h	Mean number of spores released (cm^{-2} h^{-1})	Spores released cm^{-2} during hour of peak release
Coriolus pubescens	7.7	2 175 000 000	11 769 500	21 800 000
Piptoporus betulinus	20.6	1 967 000 000	3 978 500	31 000 000
Gloeophyllum sepiarium	7.1	421 000 000	2 470 600	3 000 000
Heterobasidion annosum	11.6	63 440 000	227 900	586 000
Fomes fomentarius	3.9	34 580 000	369 500	794 000
Hirneola auricula-judae	11.3[a]	2 316 000	4 900	200 000
Exidia glandulosa	89.6[a]	196 320 000	6 500	200 000

[a] Since the hymenium covers the entire under surface of the fruit body in these jelly fungi, the figures are of the actual area of hymenium that produced the spores.

surface, produced an average of nearly 2.2×10^9 spores per 24 h. The greatest number of spores released in one hour per square centimetre of exposed hymenium by any of the fungi studied was 31×10^6 by *Piptoporus betulinus*.

Spore production rates will undoubtedly vary with the depth of individual hymenial tubes in different specimens. Also, of course, the state of the fruit body and environmental conditions may drastically affect spore production rates. The studies discussed here were done in optimum conditions, and the specimens and substrates were thoroughly soaked before the spores were collected.

The two species of jelly fungi, *Hirneola auricula-judae* and *Exidia glandulosa*, have a flat hymenial surface that covers all or part of the fruit body, so the area of the hymenial surface given for each of them in Table 1 is the area of actual hymenium, and direct comparisons cannot be made between the volume of spores produced by these fungi and the polypore fungi. However, the spore-production counts for the jelly fungi provide a clear indication of the numbers of spores that can be produced by a unit area of actual hymenial surface.

Spore release and dispersal

Although we have little direct evidence, it seems obvious that fluctuations in the actual formation or production of spores may be responsible, in some cases at least, for variations or rhythms in the release of spores. I think it is especially true in the basidiomycete decomposers with which we are concerned here. In certain of the ascomycetes that we and others have studied (Kramer & Pady, 1970), however, the release or discharge of ascospores depends on specific stimuli to initiate the process. For example, light is necessary to stimulate spore discharge in *Hypoxylon investiens*, while in *Hypoxylon truncatum* light inhibits spore discharge. Although similar situations do not seem to occur in the Hymenomycetes, basidiospore discharge or release, not being continuous at a constant rate, exhibits various fluctuations and rhythms.

Escape of spores from the fruit body

The first stage at which spore release may fluctuate is as spores emerge from the fruit body, when they tend to escape in clouds or plumes of various sizes and durations (Buller, 1922; Haard, 1967). Also, in some of the larger fruit bodies, small plumes may arise from different areas of the fruit body at different time intervals. The plumes of spores can easily be seen at night by directing a beam of parallel light rays across the under side of a fruit body. The spores appear as reflections in the light as they emerge and are carried away in air currents. To observe spore release in the field, Haard (1967) used a spot-type microscope lamp with power from a gasoline electrical generator.

Irregularity in the release of spore plumes from fruit bodies, and the fact that they may be carried in different directions by air currents, presents several problems in studying circadian rhythms of spore release. Also, the volume of spores produced by Hymenomycetes demands precautions against overloading the spore collector. In our studies of the circadian patterns of spore release in Hymenomycetes in the field, we (Haard & Kramer, 1970) usually set up several Kramer–Collins spore samplers (Kramer & Pady, 1966) around single fruit bodies. Although at times there were considerable differences in the amount of spores trapped by different samplers, circadian patterns with night-time maxima and day-time minima were nearly always evident. When the counts from each sampler were totalled, circadian patterns became quite distinct. McCracken (1971) designed an electronic timer that would accurately operate the sampler for seconds rather than minutes. This enabled him to set a single sampler quite close to a fruit body without collecting too heavy a deposit of spores.

Circadian rhythms

Both ascomycetes and Hymenomycetes exhibit distinct circa-

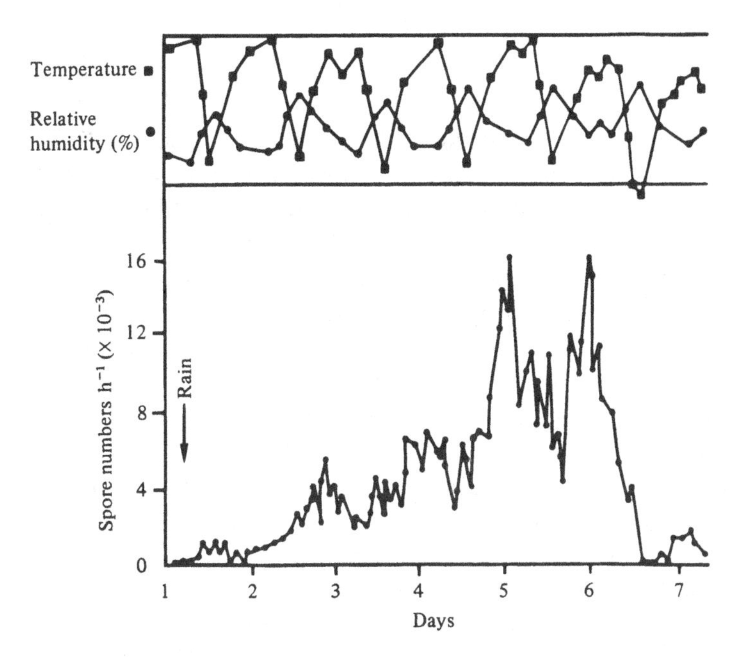

Fig. 3. Circadian pattern of basidiospore release from a single specimen of *Cortinarius* in the field, over the 7 days that sporulation occurred.

dian rhythms in spore release with maxima at night and minima during the day. Consequently, their spores are often referred to as 'night spora', as opposed to the conidia of most Hyphomycetes which are released during the day-time, and are referred to as 'day spora'.

Circadian patterns of spore release in the Hymenomycetes seem to be most closely correlated with fluctuations in relative humidity. In field studies of a number of fleshy fungi, Haard & Kramer (1970) and McCracken (1978) found that the low point in spore release occurred around noon, and that by 1800 hours spore numbers were already increasing rapidly towards a maximum level, correlated with rising relative humidity. A study of a species of *Cortinarius* by Haard & Kramer (1970) will serve as an example of spore release by a fleshy basidiomycete. In this study, the sampler was placed with the intake tube positioned just over the specimen, so that the spores were picked up in the air-flow as they escaped from the fruit body. As can be seen from Fig. 3, there was considerable variation from hour to hour, but night-time maxima were quite distinct from day-time minima in spore release. Increase in spore release was correlated with rise in relative humidity.

As an example of circadian patterns of spore release in a 'woody' or coriaceous hymenomycete fruit body, we can cite the study of *Polyporus hirsutus* (= *Coriolus hirsutus*) by Rockett & Kramer (1974*b*) (Fig. 4). Spore data shown in Fig. 4 were obtained with the Dodecapus spore collector and counts represent the total number of spores released each hour during that five-day study. In fruit bodies of this type, the volume of spores released during peak periods may vary depending on moisture conditions. Spore numbers are high immediately after rain, but gradually decrease as the substrate and fruit bodies begin to dry. The effects of drying conditions can be seen with an overall decrease in spore production and release by the fifth day (Fig. 4).

McCracken (1972), working with *Pleurotus ostreatus* under controlled conditions, found that with constant temperature, light, and relative humidity, there were no circadian patterns of spore release. Also, alternation of light and dark with both constant temperature and relative humidity produced no circadian pattern. However, at constant temperature (17 °C) and alternating relative humidity (30 and 90%), distinct circadian patterns became evident. Bohaychuk & Whitney (1973) reported similar results with *Coltricia tomentosus* under controlled conditions.

It is interesting to compare a 'typical' circadian pattern of spore

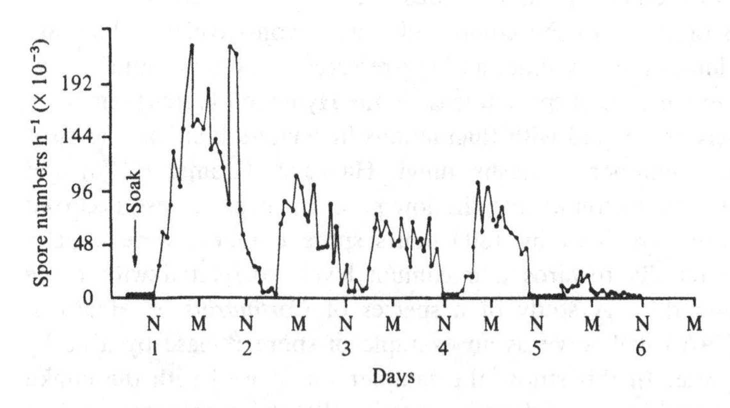

Fig. 4. Circadian pattern of basidiospore release in a single specimen of *Coriolus hirsutus* in a controlled-environment chamber, following artificial soaking of the specimen and wood substrate. N = noon; M = midnight.

release in the Hymenomycetes with that of an ascomycete like *Hypoxylon*. In Fig. 5 I have tried to show graphically several basic differences. First, many ascomycetes seem to be directly affected by light with few, if any, spores being released during the light period. On the other hand, spore release in the Hymenomycetes does not seem to be affected by light, but instead an increase in spore release is correlated with rise in relative humidity. Consistent with that is the fact that Hymenomycetes release spores during the day-time, sometimes in comparatively high numbers. The differences between minimum and maximum levels of spore release become less during wet periods when differences between daily minimum and maximum relative humidities are also less and the level of spore release is much higher.

Hymenomycetes are able to continue spore release during dry periods because the fruit bodies are larger and often 'woody' or coriaceous and probably have a more extensive mycelial support system than those of ascomycetes. Even the larger fleshy fruit bodies are able to sustain spore release for several days during dry conditions. However, in ascomycetes spore release declines soon after rain and ceases during dry periods.

In both ascomycetes and Hymenomycetes, spore release increases to a peak then declines sharply. In some species of *Hypoxylon* (e.g.

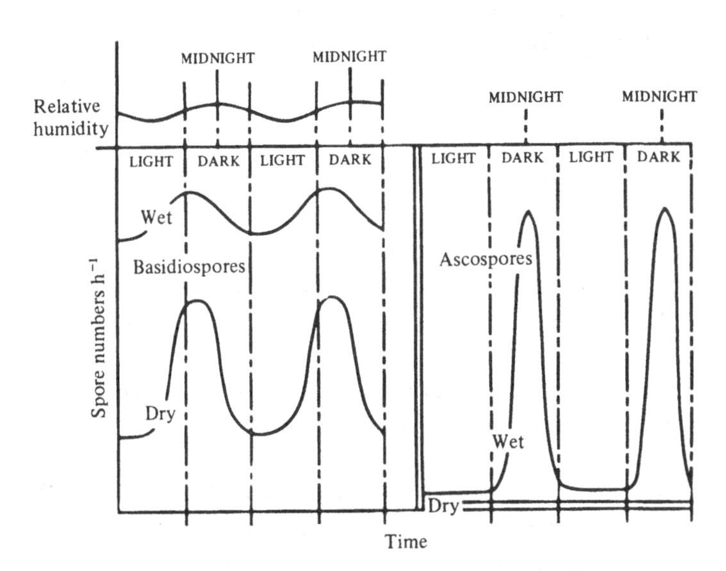

Fig. 5. Generalised circadian patterns of ascospore and basidiospore release during favourable (wet) and unfavourable conditions for growth and sporulation.

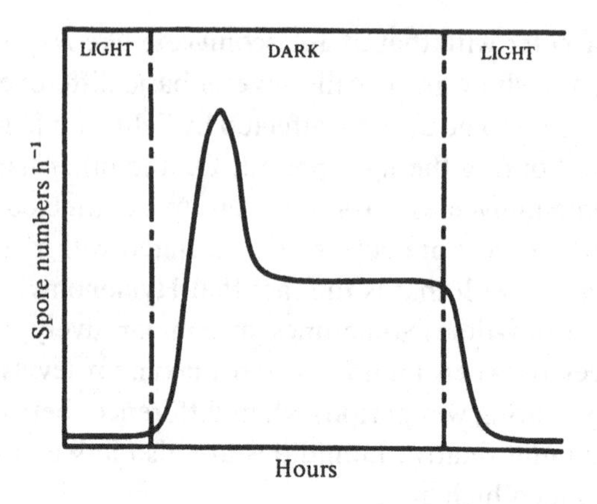

Fig. 6. Spore release in the ascomycete *Hypoxylon truncatum* as affected by light conditions in a controlled-environment chamber. The dark period was 24 h in duration.

H. truncatum) that we have studied (Kramer & Pady, 1970), spore release declined to about midway between maximum and minimum levels and, if light conditions were then held constant, the rate of spore release levelled-off and remained at that level until light conditions were changed (Fig. 6). This indicates that spore production is continuous during dark and light periods, resulting in an accumulation of 'mature' ascospores during light periods. With an ensuing dark period, accumulated ascospores are released faster than new ones are being produced. The resulting temporary peak lasts until the accumulated spores have been discharged. Then the rate of spore release declines and equals spore production. How general this situation is in the ascomycetes, and whether it also occurs in the Hymenomycetes, we do not know However, the rate of spore release by the Hymenomycetes increases to a maximum and then declines while the relative humidity remains at a high percentage. This may indicate that a somewhat similar situation exists in the Hymenomycetes with spore production continuing at a rate greater than that of spore release during periods of low relative humidity.

With regard to the effects of light, Bohaychuk & Whitney (1973) indicated that spore discharge rates in *Coltricia tomentosus* generally were higher during continuous darkness when temperature and relative humidity were optimum, but that the effect could not be totally separated from relative humidity. Grosclaude (1969), working with *Chondro-*

stereum purpureum, mentioned an increase in spore discharge after a light stimulus 'with some delay'. I think that the increases in spore release probably correlated with increases in relative humidity and not with light.

Temperature also has been shown to affect spore release. McCracken (1972) demonstrated a fluctuation in spore release in *Pleurotus ostreatus* by alternating temperatures between 5 and 25 °C and holding relative humidity constant. In other studies, McCracken (1970) found that in *Hericium erinaceum* at 85–95% relative humidity spore release increased as temperatures were raised from 0 to 24–27 °C but ceased at 31–33 °C. Bohaychuk & Whitney's (1973) findings for *Coltricia tomentosus* were similar with alternating temperatures under controlled conditions. However, under natural conditions in the field, temperatures probably affect the average rate of spore production but do little towards establishing periodicity in spore release. The average daily rate of spore release (and production) may decrease with a decrease in mean daily temperature, but maxima would still occur at night when relative humidity is highest and temperatures are lowest. There are times, however, when temperatures become low enough or high enough to cause spore release and spore production to cease. An example of this may be seen in *Hirneola auricula-judae* when a drop from 20 to 2 °C inhibited spore release (Rockett & Kramer, 1974*c*). When the temperature was raised again, sporulation resumed.

Double peak patterns

There are occasional references in the literature to possible double peaks in the release of fungal spores during a 24-h period. In our field studies of *Ganoderma applanatum* (Haard & Kramer, 1970), indications of double peaks prompted us to study that organism under controlled environmental conditions. Our first study was with a fruit body collected in August on the Olympic Peninsula of Washington. An excised portion of wood sent to us was placed on wet peat moss in a plastic tray, soaked with water, and placed in the greenhouse. When the fungus began to sporulate, it was removed to an environmental chamber for study. If relative humidity and temperature were kept constant, maxima in spore releases occurred at 9–11-h intervals (Kramer & Long, 1970). This pattern remained unchanged regardless of light conditions. We obtained similar results with two other specimens from Michigan, but others from Kansas and Colorado failed to show double peaks. It is an interesting phenomenon which we cannot explain. Sreeramulu

(1963) reported collecting basidiospores of this fungus with a Hirst trap some 100 metres from a specimen in Silwood Park, Ascot, Berks., England. At that distance he found that spore numbers began to increase at about 1800 hours and peaked about midnight, then immediately began to decline to a minimum at about 0600 hours. Perhaps the English specimen did not exhibit double-peaked periodicity in spore release, but it shows that when studies of spore release are undertaken, precautions must be made to collect spores at the source as they are released from the fruit body.

'Life span' patterns in spore release

In addition to the distinct circadian rhythms so characteristic of spore release in these fungi, there is also a pattern characteristic of the release of spores over the 'life span' of the fruit body. In the graph of *Cortinarius* (Fig. 3), spore release is shown to have increased slowly over several days as the cap of the mushroom opened and matured, and reached its maximum potential on the fourth and fifth days. Then, with drying conditions (decreased relative humidity) and senescence of the fruit body, spore release declined abruptly. This particular study was made in October when temperatures were rather low. With higher temperatures, peak spore release might have been reached sooner, resulting in a slightly different pattern.

In basidiomycetes with small fruit bodies, the 'life span' may be relatively short, i.e. as little as one day. In some of the smaller species of *Coprinus*, the fruit bodies mature, sporulate, and are expended within 24 hours (Haard & Kramer, 1970). However, these fungi have the property of producing successive crops of fruit bodies each day for many days in suitable environmental conditions.

Fruit body revival patterns

In addition to circadian rhythms and variation in spore release over the 'life-span' of a fruit body, other types of fluctuation in spore release may result from a variety of factors. Some of the small forms of agarics and many of the jelly fungi have attained the ability to extend their spore-producing life by becoming dehydrated during dry periods and then reviving when wetted by rain. They are then able to resume and continue spore production and release until they are dehydrated again.

The ability to revive when rehydrated is important in the taxonomy of the Tricholomataceae, particularly in separating *Marasmius* from *Mycena* and *Collybia*. Gilliam (1975) found that one specimen of *Marasmius rotula* retained the potential for spore release for three weeks. During that time, two large and three smaller episodes of spore release were correlated with rewetting by rain. Fig. 7 shows the effects of heavy rain (5 cm) that began at 1100 hours on day 1. Although a few spores were released immediately, there was no significant rise until 6 h later, then maximum spore release was reached in another 3 h. The moisture from such heavy rain made it possible for the fruit body to release spores continuously for some 28 h. During the night of day 7, a trace of rain fell which caused spore release to begin within one hour, reach a peak within three hours, and then decline rapidly (Fig. 7). Spore production and release during this period was only 2–3% of that after the heavy rain.

The jelly fungi are perhaps the most efficient fungi at rehydrating and resuming spore production and release over extended time periods. We (Rockett & Kramer, 1974*c*) studied several species of *Exidia* and *Hirneola* both in the field and in the laboratory where they were rehydrated by rain or artificially rehydrated, and then spores were collected from single specimens. Many of the fungi were revived repeatedly. Table 2 lists the number of times that several specimens were revived and resumed sporulation during a 191-day period from March 28 to October 4. Fig. 8 is of a study of *Hirneola auricula-judae* in

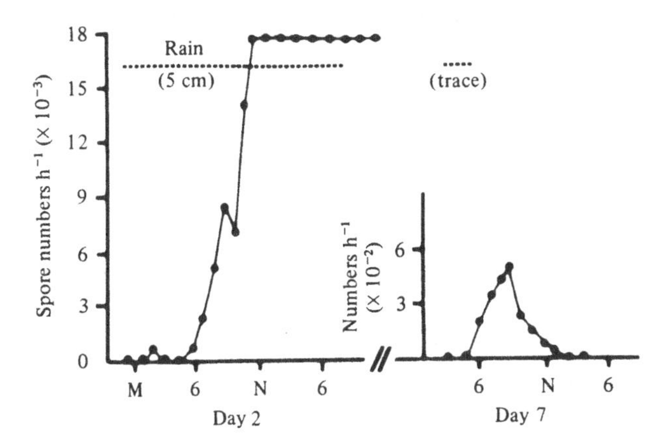

Fig. 7. Basidiospore release by *Marasmius rotula* following revival of the fruit body from heavy rain (5 cm) and a light shower (trace) 6 days later. Prior to the light shower on day 7, the fruit body had become dehydrated and spore release ceased. N = noon; M = midnight. (After Gilliam, 1975.)

Table 2. *Summary of rehydration and sporulation of three species of jelly fungi (Tremellales) in the field from April to October*

Species	Number of times fruit body was rehydrated and sporulated	Total days spore release occurred	Longest period of continuous spore release (days)
Exidia glandulosa	20	97	21
E. recisa			
Specimen 1	10	38	10
Specimen 2	10	33	11
Hirneola auricula-judae			
Specimen 1	5	18	8
Specimen 2	11	39	11
Specimen 3	4	17	7

which each of three periods of rain were followed by rehydration and sporulation of the fruit bodies during July when temperatures were high and dehydration was rapid, so sporulation periods were comparatively short (Rockett & Kramer, 1974*c*). At other times of the year when temperatures are lower and dehydration slower, sporulation may last many days.

In a study of a small specimen (2 mm × 1 cm tall) of *Calocera cornea* with a large surface:volume ratio, rapid dehydration followed artificial soaking even though relative humidity was maintained at 94% in the

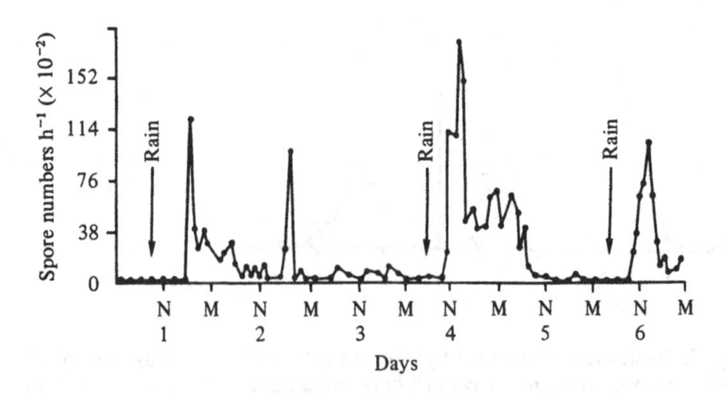

Fig. 8. Three episodes of basidiospore release in *Hirneola auricula-judae* following rehydration by rain. N = noon; M = midnight.

environmental chamber. Spore release began within 2–3 h after soaking and increased rapidly to a peak within 6–8 h. This was followed by a decline over the next 12 h to a low level of spore release (Rockett & Kramer, 1974*c*). In a somewhat larger species, *Dacrymyces palmatus* (1 × 2 cm), sporulation patterns were similar to those of *C. cornea*, except that spore release was maintained at a higher level for a longer period under the same conditions (Rockett & Kramer, 1974*c*). Depending on conditions that might affect dehydration of the fruit body, larger species such as *Hirneola* may be capable of continued spore release at a comparatively high level for many days (Fig. 9).

Since the fruit bodies of these jelly fungi are easily found in the field during the winter, they may be capable of overwintering and, with warmer temperatures and rehydration in the spring, may begin sporulation again. Specimens brought into the laboratory during January and February and rehydrated began to sporulate within a few hours. Other specimens stored for several months at 28 °C usually began to sporulate within a few hours after being moved to room temperatures and rehydrated. Fig. 9 illustrates the effect of cold temperatures in inhibiting sporulation, and the resumption of sporulation when returned to warmer temperatures, by *Hirneola auricula-judae* (Rockett & Kramer, 1974*c*).

In the species of *Marasmius* studied by Gilliam (1975) and in all the jelly fungi we studied (Rockett & Kramer, 1974*c*), there was no

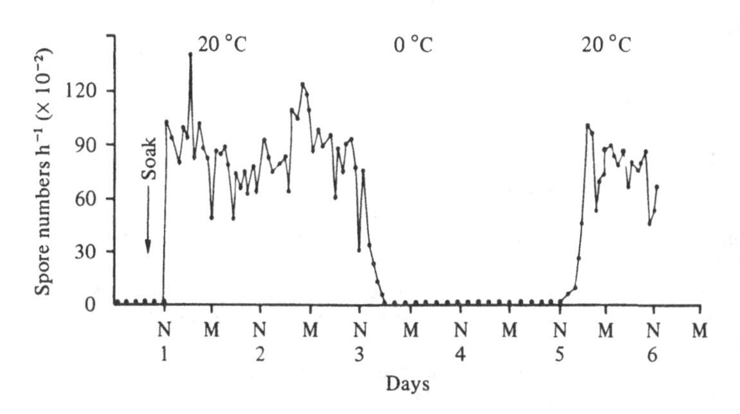

Fig. 9. Basidiospore release in *Hirneola auricula-judae* following artificial soaking to rehydrate the fruit body. Spore release ceased when the temperature was lowered in the controlled-environment chamber. N = noon; M = midnight.

evidence of a circadian pattern of sporulation. Generally, a lag period of a few hours followed rehydration before spore release began. Spore release then increased rapidly to a 'maximum' and continued at an almost constant rate as long as the fruit body remained in the same condition. Hour-to-hour variation was high in some, but never had a distinct circadian pattern. Then, as the fruit bodies began to dehydrate, spore release correspondingly began to decrease.

Conclusions

Evolution in the wood- and litter-decomposing basidiomycetes has been towards the development of a capacity to produce tremendously large numbers of basidiospores, well adapted for air dispersal. Because of the extensive mycelial systems and nature of the fruit bodies, spore production by these fungi is generally maintained for extended periods, during which environmental conditions may not be optimum. In some species, sporulation may be interrupted during long dry periods, but, with rain and revival of the fruit body, spore production and release resume. These and other characteristics seem to do more than compensate for the poor survival of basidiospores in making these fungi successful competitors.

References

Bohaychuk, W. P. & Whitney, R. D. (1973). Environmental factors influencing basidiospore discharge in *Polyporus tomentosus*. *Canadian Journal of Botany*, **51**, 801–15.

Buller, A. H. R. (1909). Spore deposits – the number of spores. In *Researches on Fungi*, vol. 1, pp. 79–88. London: Longmans, Green & Co.

Buller, A. H. R. (1922). Spore-discharge in the Hydneae, Tremellineae, Clavarieae, and Exobasidieae. In *Researches on Fungi*, vol. 2, pp. 149–94. London: Longmans, Green & Co.

Gilliam, M. S. (1975). Periodicity of spore release in *Marasmius rotula*. *The Michigan Botanist*, **14**, 83–90.

Gregory, P. H. (1966). The fungus spore: what it is and what it does. In *The Fungus Spore*, ed. M. F. Madelin, pp. 1–13. London: Butterworths.

Grosclaude, C. (1969). Note sur l'influence de la lumière sur la libération des basidiospores du Stereum purpureum Pers. *Annales de Phytopathologie*, **1**, 107–11.

Haard, R. T. (1967). The periodicity of basidiospore release in the Hymenomycetes. Dissertation, Department of Botany and Plant Pathology, Kansas State University.

Haard, R. T. & Kramer, C. L. (1970). Periodicity of spore discharge in the Hymenomycetes. *Mycologia*, **62**, 1145–69.

Kramer, C. L. (1974). Seasonality of airborne fungi. In *Phenology and Seasonality Modeling*, ed. H. Lieth, pp. 415–24. New York: Springer-Verlag.

Kramer, C. L. & Long, D. L. (1970). An endogenous rhythm of spore discharge in *Ganoderma applanatum*. *Mycologia*, **62**, 1138–44.

Kramer, C. L. & Pady, S. M. (1966). A new 24-hour spore sampler. *Phytopathology*, **56**, 517–20.

Kramer, C. L. & Pady, S. M. (1968). Viability of airborne spores. *Mycologia*, **60**, 448–9.

Kramer, C. L. & Pady, S. M. (1970). Ascospore discharge in *Hypoxylon*. *Mycologia*, **62**, 1170–86.

McCracken, F. I. (1970). Spore production of *Hericium erinaceus*. *Phytopathology*, **60**, 1639–41.

McCracken, F. I. (1971). Modified sampler accurately measures heavy spore production of *Fomes marmoratus*. *Phytopathology*, **61**, 250–1.

McCracken, F. I. (1972). Sporulation of *Pleurotus ostreatus*. *Canadian Journal of Botany*, **50**, 2111–5.

McCracken, F. I. (1978). Spore release of some decay fungi of southern hardwoods. *Canadian Journal of Botany*, **56**, 426–31.

Rockett, T. R. & Kramer, C. L. (1974*a*). A technique to induce sporocarp production and sporulation in lignicolous basidiomycetes. *Mycologia*, **66**, 524–6.

Rockett, T. R. & Kramer, C. L. (1974*b*). Periodicity and total spore production by lignicolous basidiomycetes. *Mycologia*, **66**, 817–29.

Rockett, T. R. & Kramer, C. L. (1974*c*). The biology of sporulation of selected Tremellales. *Mycologia*, **66**, 926–41.

Rockett, T. R., Kramer, C. L. & Collins, T. I. (1974). A new collector to measure total spore production. *Mycologia*, **66**, 526–30.

Sreeramulu, T. (1963). Observations on the periodicity in the air-borne spores of *Ganoderma applanatum*. *Mycologia*, **55**, 371–9.

White, J. H. (1919). On the biology of *Fomes applanatus* (Pers.) Wallr. *Transactions of the Royal Canadian Institute*, **12**, 133–74.

3 The rôle of polysaccharidase enzymes in the decay of wood by basidiomycetes

R. A. P. MONTGOMERY
Department of Botany, Imperial College, University of London, London SW7 2BB, England

Introduction

Wood, although a high-calorie carbon source, suffers from the disadvantage of being highly polymeric (Norkrans, 1967) and thus insoluble. Before it can be utilised by a wood-destroying organism, such as a basidiomycete fungus, it must first be depolymerised. As wood is a very stable polymer, catalysts are needed and, since the wood is outside the organism, they must be extracellular enzymes. This implies that the enzymes must function in an uncontrolled environment which, together with the nature of the substrate, puts several constraints on the enzymes and the way they act. They should be small to facilitate penetration of the substrate, although there must be a minimum functional size; they should remain active for as long as possible, although active turnover, i.e. absorption and production of the enzymes by the organism, would reduce this requirement and effect a control mechanism; they should also be ionically suited to their environment, because if they were highly charged they would either be attracted by the wood and immobilised, or repelled and unable to function.

These requirements should be considered in the context of the microstructure of wood. A hardwood is characterised by vessels serving for fluid conduction, and by fibres for strength. A softwood has tracheids which perform both functions. These cells all have a similar structure, with laminated walls. Most of their strength is in the thick secondary walls, which can account for 76% of the cell-wall volume (Preston, 1974). Kerr & Bailey (1934) termed the laminations of the secondary wall the S_1, S_2 and S_3 layers. The S_2 layer has cellulose microfibrils arranged in a steep helix, whilst in the S_1 and S_3 the microfibrils follow a shallow helix. The general structure of secondarily thickened cells as seen in a transverse section is illustrated in Fig. 4.

Wood – its chemical constitution

The constituents of a wood cell wall are cellulose, hemicelluloses, lignin, extractives, minerals and water, the chemical constituents varying throughout the wall. Cellulose is a homopolymer composed entirely of glucose units whereas the hemicelluloses are heteropolymers composed of a mixture of sugars. The homopolymers of cellulose can have a degree of polymerisation (DP) of up to 10 000 (Norkrans, 1967), i.e. 10 000 glucose units totalling 7 μm in length! These homopolymer chains combine where they lie parallel to form crystallites. At certain points the chains do not align and amorphous zones occur (Fig. 1). One chain may be part of several crystallites. Surrounding the crystallites and infiltrating the amorphous zones are stray cellulose chains and the heteropolymers, the hemicelluloses, which generally do not exceed 200 in their degree of polymerisation. The crystallites form the basis of microfibrils, which measure about 10 × 4 nm in oval cross-section and

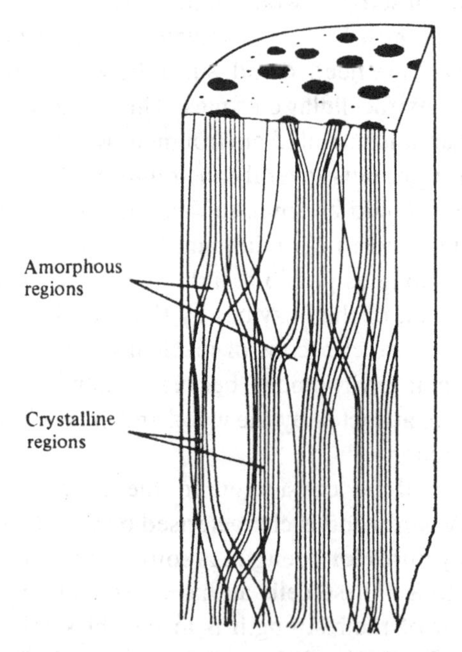

Fig. 1. A two-dimensional representation of the three-dimensional structure of a cellulose microfibril. Microfibrils, which measure about 10 × 4 nm in oval cross-section, lie centrally in a sheath of polysaccharide and water about 100 nm wide. The cellulose chains, which can be up to 7 μm long, may each be part of several crystallites. Hemicelluloses infiltrate the amorphous regions and surround the crystallites. (From Sihtola & Neimo, 1975.)

lie centrally in a sheath of polysaccharide and water abut 100 nm wide. It is the crystallites that make it possible to see the cell-wall layers in the polarising microscope; the long axes of the crystallites lie parallel to the microfibrils and thus are at different angles in the different layers of the cell wall.

There are a variety of bond types in interchain bonding. Mono- and divalent ions can affect cell wall growth and therefore salt (ionic) links may be important (Preston, 1974), but hydrogen bonds are thought to be the most important. Weak chemical extraction will remove most heteropolysaccharides from the microfibrillar sheath, showing them to be loosely bound about the microfibril, as does the little changed X-ray diffraction pattern before and after extraction. Treatment for 1 h at 100 °C with 1.25 M sulphuric acid yields 'rodlets' which give the X-ray diffraction pattern of pure cellulose (Rainby, 1957). It would, therefore, seem likely that the residual hemicellulose is bound in the amorphous regions of the cellulose microfibrils.

Wood – a microenvironment for decay organisms

All cellulosic materials (with the possible exception of reconstituted forms such as cellophane and rayon) have microscopic capillaries, and all have submicroscopic capillaries and spaces. In fact the surface area of one cubic centimetre of wood is of the order of 0.8 ha, comprising the microscopic cell lumena and pits (2×10^3 cm^2) and the submicroscopic inter-microfibrillar spaces (3×10^6 cm^2). The size of the inter-microfibrillar spaces depends on the moisture content of the wall, reaching a maximum at fibre saturation point and ranging from 0.5 to 4 nm as a skewed distribution, with an average of about 1.6 nm (Reese, 1977) (Fig. 2). The open spaces are important for decay organisms. They provide routes for colonisation; they allow ready diffusion of oxygen, and they allow water to flow and so facilitate the diffusion of enzymes and their products to and from decay organisms.

The action of polysaccharidase enzymes

Seillière was probably the first to report on in-vitro cellulase activity in 1906. Since then much work has been carried out, but the actual action of the polysaccharidase enzymes still has not been elucidated. This is mainly due to the great number of complex variables and the difficulties of studying enzyme activity on a solid-phase substrate. Most of the work has been directed towards cellulase, as cellulose is the major constituent of interest, but although less work has been done on hemicellulases the mode of the action appears to be similar.

Work on polysaccharides has centred on ascomycetes and the fungi imperfecti ('moulds' and 'soft rots') especially *Trichoderma*, and it is interesting to contrast the modes of action of their enzymes with those of the basidiomycetes of the brown- and white-rot types. It has been shown (Eriksson, 1978; Eriksson & Pettersson, 1973) that varying numbers of cellulase enzymes are produced for the same substrate. They are of two main types: exo-glucanases and endo-glucanases. The mode of action of each type is shown in Fig. 3.

Exo-glucanases split glucose or cellobiose from the end of the polyglucose chain. This enzyme must become attached to the end of the chain in order to remove glucose units. The enzyme would, therefore, be ineffective against a crystallite, where the ends of the cellulose chains would not be sufficiently exposed. *Endo-glucanases* act randomly within the cellulose chain, breaking chemical bonds and thus creating free ends on which the exo-glucanases can act (Eriksen & Goksøyr, 1977). It has been shown that they act synergistically with exo-glucanases, since without the latter the chains may realign.

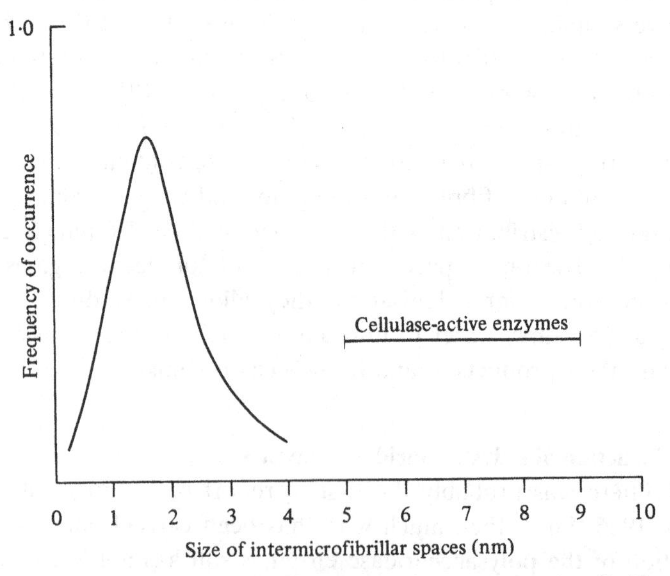

Fig. 2. Comparison of the size range of inter-microfibrillar spaces in the cell walls of *Picea* with that of known cellulase-active enzymes. It is shown that these enzymes are excluded by their size from the microspaces of intact wood cell walls. (After Cowling, 1963.)

Jermyn noted in 1955 that there appeared to be at least two binding centres in the cellulase molecule. It would appear that cellulases 'recognise' sites about five sugar residues in length.

Initially it was found that some culture filtrates containing active cellulases could not attack highly ordered celluloses such as cotton, although they did reduce its strength (Nilsson, 1974*a*). Other culture filtrates could attack the highly ordered celluloses and it was thought that they produced an enzyme responsible for disordering the cellulose, opening it up to attack by other enzymes in the system. The disordering enzyme was termed the C_1 enzyme, while the enzyme attacking the disordered cellulose was termed the C_x enzyme (Reese, Siu & Levinson, 1950). It has proved very difficult to isolate a C_1 fraction devoid of cellulase activity. The C_1 fraction is usually isolated with an active exo-enzyme fraction and it is now widely accepted that C_1 and exo-enzyme activity are synonymous. Synergism is explained by the opening up of more chain ends by the endo-enzyme, enabling increased activity for the exo-enzyme. The current concept is therefore of a system where an endo-enzyme can break bonds on the surface of a crystallite, thus opening it up for the exo-enzymes to act. Reese, the originator of the C_1/C_x concept, however disagrees with the idea of C_1 as an exo-enzyme; he believes that the concept of a 'disordering enzyme' is more consistent with a specialised form of endo-enzyme (Reese, 1977). He has therefore

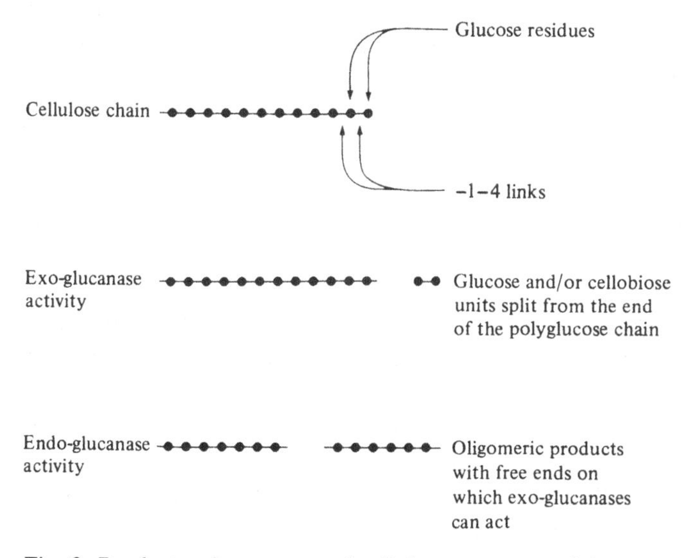

Fig. 3. Products of two types of cellulase enzyme activity.

modified his scheme in favour of the C_1 enzyme being an endo-enzyme acting only on highly ordered substrates. He offers as proof the analogous situation of the enzyme which is responsible for attacking and unravelling collagen but is unable to act on non-ordered (i.e. already unravelled) collagen. He emphasises that a remaining area of ignorance is that of enzyme activity in liquid/solid phases. He also quite rightly points out that substrate conformation is of the utmost importance. In view of the changing meaning of the terms C_1 and C_x it would seem sensible to develop an alternative terminology to describe the cellulase system, such as endo- and exo-enzymes.

Localisation of polysaccharidases as shown by the morphology of cell-wall degradation

Figure 4 shows diagrammatically the three types of wood degradation. The white and brown rots are caused by basidiomycetes and soft rot by fungi imperfecti and ascomycetes (for a more detailed description see Levy: Chapter 9).

From Fig. 4 it can be seen that enzymic attack on the wall is local to the hypha in the case of white and soft rots, but distant from the hypha in the case of brown rot. The postulated mechanisms to explain these differences in types of attack include enzyme size. As discussed, any enzyme larger than 4 nm in any one dimension would be excluded from the wall microspace and thus would be unable to diffuse freely. It is considered that the maximum pore size should be larger than the maximum enzyme dimension for free diffusion to take place (Cowling, 1963; Cowling & Brown, 1969). As no active enzymes below this size limit have been found, it seems safe to reason that enzymes do not penetrate unaltered wall layers. This would explain the free diffusion of enzymes from brown-rot organisms, for, although they are no smaller than enzymes from the other rot types, it seems likely that they penetrate an altered wall.

Polysaccharide degradation by brown-rot basidiomycetes

Recent observations suggest that brown-rot basidiomycetes degrade polysaccharides by a mechanism different from that of other fungi. Highley (1978) found that *Poria placenta* was unable to digest crystalline cellulose unless glucose, starch, holocellulose or mannose was added. Xylan or a water extract of wood was not effective. Nilsson (1974*b*) also found that cotton placed between two blocks of wood infected with brown rot lost strength but little weight. He found that the greatest reduction in degree of polymerisation occurred with a brown-rot basidiomycete that induced little weight loss. This was not shown by soft- or white-rot fungi which rapidly utilised polymers when breakdown started. A 0.4% solution of hydrogen peroxide in 0.2 M ferrous sulphate at pH 4.2 can solubilise cotton fibres at the rate of about 5 mg per 7 days. Koenigs (1972*a*, 1974*b*) showed that wood breakdown by a H_2O_2/Fe^{2+} system was very similar to that of brown rot but unlike that of soft or white rot or of acid hydrolysis. He suggested that the pathway of breakdown is oxidation of the C_6 carbon atom of the pyranose unit,

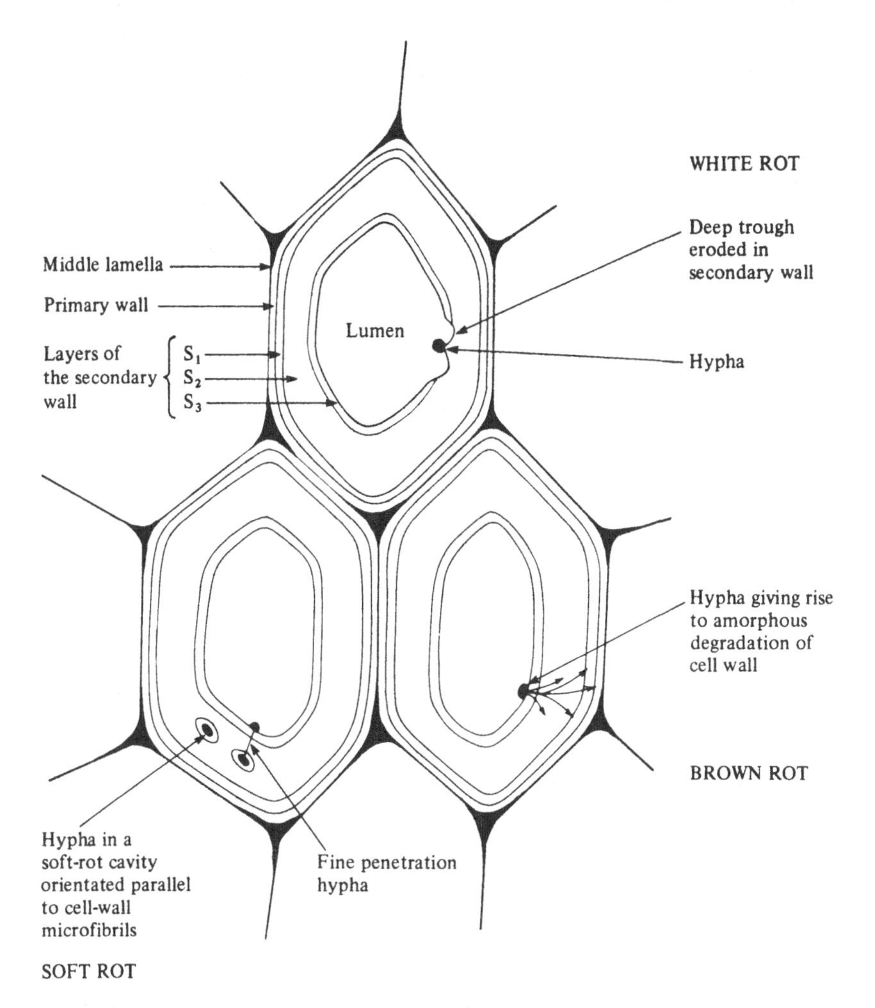

Fig. 4. Diagram of wood cells in transverse section showing patterns of degradation produced by three types of wood decay fungi.

followed by that of the C_2 and C_3 atoms, to an aldehyde and hence to a carboxyl group (Fig. 5). In this way the pyranose ring is broken and so also is the chain. The ends are altered to products which are possibly harder for the fungi to metabolise. As only very few of these reactions would open up the cellulose (about 1%), the oxidation products would be hard to find and would not deter the fungus. The demonstration of a degradation of cellulose in wood by brown-rot basidiomycetes via this mechanism is therefore difficult, but Koenigs (1972*a,b,* 1974*a*) showed that these fungi produce hydrogen peroxide. Hydrogen peroxide acts very slowly, but a H_2O_2/Fe^{2+} system (Fenton's reagent) works much more quickly. It has been estimated that there is enough Fe^{3+} present in decaying wood, but it would have to be enzymically con-

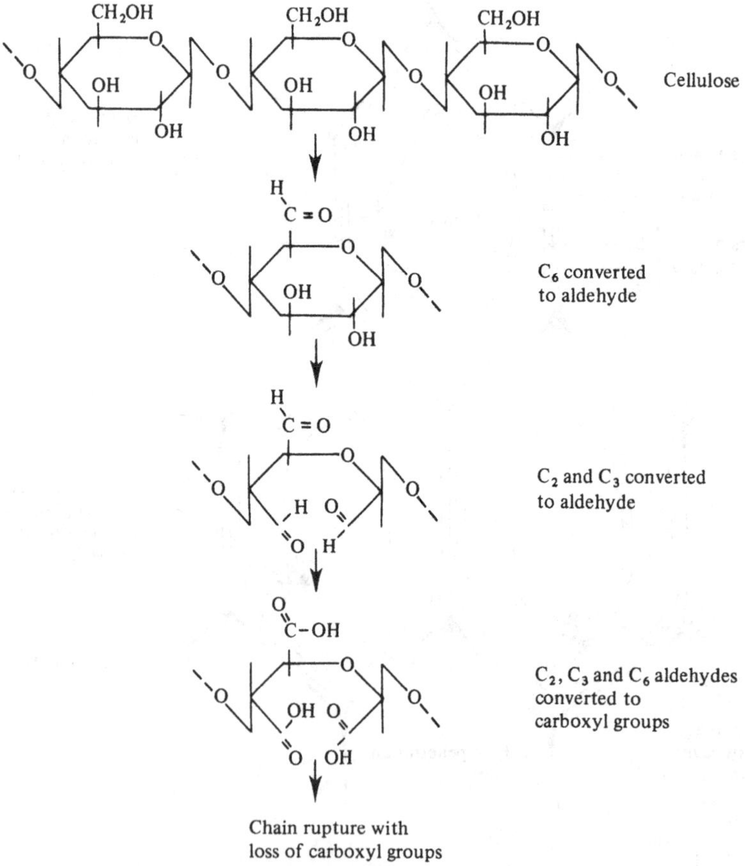

Cellulose

C_6 converted to aldehyde

C_2 and C_3 converted to aldehyde

C_2, C_3 and C_6 aldehydes converted to carboxyl groups

Chain rupture with loss of carboxyl groups

Fig. 5. Proposed mechanism for the breakdown of cellulose by a H_2O_2/Fe^{2+} system.

verted into the active Fe^{2+} form (enzymes capable of this are known in other spheres of biochemistry). The hydrogen peroxide could come from oxidation of sugars to sugar lactones by the fungal enzymes. This produces hydrogen peroxide and a sugar lactone which spontaneously breaks down to an equilibrium with the sugar acid. This would explain the necessity for the presence of other carbohydrate (such as hemicellulose) for the metabolism of crystalline cellulose by *Poria placenta*. Highley (1977, 1978) concluded that there was strong evidence that brown-rot fungi, in contrast to most other groups of fungi, employ a non-protein oxidation mechanism to break down cellulose in wood. He also found that washing the culture solids in a variety of chemical reagents failed to liberate further activity, indicating that immobilised enzymes did not occur with this type of rot.

If a H_2O_2/Fe^{2+} cellulose-degrading system exists in the brown-rot basidiomycetes, a suitable safety mechanism must also have been evolved as hydrogen peroxide is very toxic to living cells. It can form, by homolytic cleavage, free radicals (molecules with an unpaired electron) which are capable of rapidly depolymerising the lipid-forming membranes (Slater, 1972). This action is a chain reaction, the free radical being regenerated at each break, and thus only a small amount of hydrogen peroxide will cause considerable damage to the cell. Three mechanisms of protection are possible: (*a*) production of hydrogen peroxide at a safe distance from the hypha; this would lower the concentration at the cell membrane but would have to be used with (*b*) or (*c*); (*b*) a mechanism to remove and deactivate free radicals; (*c*) a mechanism to convert H_2O_2 to $H_2O + \frac{1}{2}O_2$ without producing free radicals (i.e. by heterolytic cleavage).

In the case of (*b*) or (*c*), and assuming (*a*), we would expect the protective mechanism to be on or about the hyphal cell wall. Mechanism (*c*) would also increase the oxygen concentration local to the hypha and would involve the production of a catalase. These enzymes are usually of a high molecular weight but are very efficient, with a high turnover rate; the number of reactions they can catalyse per minute is high.

Polysaccharide degradation by white-rot basidiomycetes and the soft-rot fungi

In one type of soft rot (Fig. 4), no other retention mechanism is needed as fungal enzymes are totally enclosed within the cavity in the S_2 layer. In a second type of soft rot and in white rot, on the other hand, enzymes are exposed in the cell lumen, but sharp localisation of degradative action is still shown. In these cases there must be some mechanism retaining the enzyme about the hypha. Retention of enzyme about the hypha would be logical, as in a nitrogen-limited system it would tend to minimise protein loss. It also means that the products of enzyme action would be local to the hypha for easy uptake, thereby minimising 'poaching' of nutrients by the secondary moulds. With brown rots the latter benefit is lost by the mode of action of the hydrogen peroxide system, and retention of enzyme about the hypha is therefore a disadvantage.

Indirect evidence of a possible mechanism has been obtained by electron microscopy. In some scanning electron micrographs, debris can be seen surrounding the hypha in soft-rot types 1 and 2. At first it was thought that this was material stripped from the cavity wall during degradation, but it could also be mucilage. Crossley (1979) developed a technique for embedding decayed wood with better preservation of the matrix structures which are lost in dehydration preparations of conventional transmission electron microscopy (TEM). The results showed similarity between embedded mucilage from *Rhinocladiella mansonii* and the matrix surrounding the hyphae of *Chaetomium globosum*, a soft-rot fungus, and of *Coriolus versicolor*, causing white rot. This mucilage matrix was absent from the hyphae of *Coniophora puteana*, a brown-rot fungus.

All the enzymes so far studied have their optimum pH for activity near their isoelectric point (the pH at which the protein has no net charge). Thus, we may conclude that they are only weakly charged in their active state, thereby facilitating ease of movement in a highly charged environment. This also means that ionic bonding is unlikely to be the retaining force. The components of five enzymes from *Sporotrichum pulverulentum*, the conidial state of the white-rot basidiomycete *Phanerochaete chrysosporium*, have been elucidated (Almin, Eriksson & Pettersson, 1975). All but one have carbohydrate attached. In the absence of a stronger attraction, the weak affinity between the carbohydrate fraction of the enzyme and the mucilage could retain the enzymes, while still allowing mobility within the mucilage. In the presence of a stronger binding force, such as enzyme–substrate, the enzyme would be held by the stronger force, i.e. bound to the substrate. R. A. P. Montgomery, N. Green & J. F. Levy (unpublished) tested this theory by taking the solids from liquid cultures of *Coniophora puteana, Chaetomium globosum*, and *Coriolus versicolor*, and assaying them before and after washing with the culture medium. It was consistently shown, by a variety of techniques, that retention of the enzymes took place with the white and soft rots, but was washed out from the brown-rot organism. Highley (1978) also found it impossible to increase the yield of enzyme from the brown-rot fungus *Poria placenta* by washing, indicating that all the enzyme had already been washed out. The above tests were performed with fungi grown on either sawdust or a soluble cellulose (carboxymethyl cellulose). Retention was shown in both cases. From these results it was concluded that the enzymes were bound about the hyphae, and mucilage could provide a binding site, which would be consistent with the ability of the white- and soft-rot fungi to limit the degradation of the wall to a region in relatively close association with the fungal hypha.

Given retention of enzymes by soft and white rots, the cavity shapes need some explaining (see also Levy: Chapter 9). Type 1 soft-rot fungi seem to form cavities by pushing a proboscis hypha forward, possibly by hydrostatic pressure, along the line of least resistance parallel to the microfibrils. They then secrete enzymes to erode a cavity sideways. The characteristically pointed ends of the cavity could be due to the exo-enzyme having longer to attack the chain end nearer the hypha. The bridges of unattacked material would suggest 'pulses' of enzyme activity. This may be because the hypha is unable to synthesise or secrete enzyme under the conditions for proboscis extension.

Type 2 soft-rot and white-rot decay take the form of an eroded groove with a central ridge on which the hypha lies. It is thought (Crossley, 1979) that this is due to surface-tension effects. As the hypha will be growing in a water film, surface tension would clamp it to the wall, thereby excluding enzyme-bearing mucilage from immediately underneath it. Thus, with little enzyme attack directly below the hypha, a ridge would form.

Regulation of enzyme production and activity

Since wood is low in nitrogen, invading organisms must use any available nitrogen efficiently (see Frankland: Chapter 14). Close regulation of enzyme production is likely to be of benefit to the hypha. A means of controlling activity is necessary, as any excess products (i.e. those which the hypha cannot use immediately) could be used by scavengers. Regulation of enzyme production seems to occur by both induction and repression (Mandels & Reese, 1957; Jacob & Monod, 1961; Norkrans, 1967; Conn & Stumpf, 1972). Such control would be expected to regulate numerous enzymes at the same time, hemicellulases being required as well as cellulases because hemicelluloses encompass the cellulose microfibrils and presumably have to be cleared first. During the invasion of wood by fungi there is unlikely to be repression of cellulase or hemicellulase (save perhaps in the ray cells) and thus induction would be the important regulator at this stage. When the fungus encounters a more easily metabolised nutrient source, catabolite repression of the cellulase/hemicellulase system becomes the regulator, with enzymes able to degrade the new substrate being induced. At intermediate levels, both systems may be partially induced/depressed. The concentration of an easily metabolised carbon source at which repression takes place will depend upon the enzyme being affected and on the nature of the repressor, but, in general, the concentration is low and this leads to problems if the fungus is using a natural polymer. Any impurities in the polymer may cause varying degrees of repression of different enzymes, giving varying results with new batches of substrate.

The induction/repression system outlined above has been verified by the author for *Chaetomium globosum, Coriolus versicolor* and *Coniophora puteana* by growing these fungi on cellulose (1%) plus a non-cellulosic carbon source (0.5%), and on the non-cellulosic carbon source alone (1%) (unpublished data). The results showed repression of cellulase on the non-cellulosic carbon source. After two and a half weeks it was assumed that carbon was limiting, but there was still no cellulase activity and thus it was assumed that the cellulase was not constitutive. All the flasks with added cellulose developed cellulase activity in two and a half weeks. If enzyme regulation is by this means alone, one would expect a fairly rapid enzyme turnover, i.e. the old enzyme being taken in and resynthesised, otherwise there would be a large lag before the enzyme complement changed in response to the changing environment, but no figures for the turnover of cellulases appear to have been recorded. Mandels & Reese (1965) also found that rapid metabolism of sugars repressed cellulase production and also caused the disappearance of preformed cellulase, although if production is stopped the disappearance is likely to be due to turnover of enzyme rather than to specific uptake.

Another mechanism for controlling enzyme activity could affect the preformed enzyme, but the mechanism is unlikely to be direct when the enzymes are extracellular. The drop in pH concurrent with metabolism of sugars and cellulose has been suggested (Mandels & Reese, 1965) as a means of feedback control of the enzymes, as some are very pH-sensitive. Thus, when the fungus has enough sugars for its metabolic processes, the rate of cellulose breakdown can slow, limiting the pool of sugars available to scavengers.

Conclusions

From the preceding description of the likely mechanisms by which fungi generate the three morphologically distinct types of wood rot, it can be seen that, while soft-rot organisms are very adaptive and able to degrade a wide variety of woods, it is the basidiomycetes that are the climax flora in normal circumstances. This is probably because of their specialised techniques which deal with the lignin content of their substrate.

Soft-rot fungi grow slowly and only in the S_2 layer of the secondary wall, which is low in lignin, whereas the white- and brown-rot fungi grow faster and utilise all the cell-wall layers. White-rot fungi have a specialised enzyme system for degrading lignin, although the extent to which the fungi can utilise it is still not understood. The mechanism of lignin degradation by these fungi is discussed by Kirk & Fenn (Chapter 4). Brown-rot fungi circumnavigate the need for lignin degradation by using a carbohydrate-degrading system which does not require the removal of lignin in order to function. The result of brown-rot decay is, therefore, the digestion of carbohydrate from within the lignin matrix, leaving a brown mass of lignin.

Acknowledgements. I would like to acknowledge the assistance of Mr N. Green and Mr A. Crossley in the preparation of this chapter and also of Hicksons and Welch for sponsoring this research.

References

Almin, K. E., Eriksson, K.-E. & Pettersson, B. (1975). Extracellular enzyme system utilised by the fungus *Sporotrichum pulverulentum* (*Chrysosporium lignorum*) for the breakdown of cellulose. 2. Activities of the five endo-1,4-β-glucanases towards carboxymethyl cellulose. *European Journal of Biochemistry*, **51**, 207–11.

Conn, E. E. & Stumpf, P. K. (1972). *Outlines of Biochemistry*, 3rd edn. New York: Wiley.

Cowling, E. B. (1963). Structural features of cellulose that influence its susceptibility to enzyme hydrolysis. In *Advances in Enzymic Hydrolysis of Cellulose and Related Materials*, ed. E. T. Reese, pp. 1–32. Oxford: Pergamon Press.

Cowling, E. B. & Brown, W. (1969). Structural features of cellulosic materials in relation to enzymic hydrolysis. In *Cellulases and their Applications*, ed. G. J. Hajny & E. T. Reese, pp. 152–87. Washington, DC: American Chemical Society Press.

Crossley, A. (1979). The use of electron microscopy to compare wood decay mechanisms. PhD thesis, University of London.

Eriksen, J. & Goksøyr, J. (1977). Cellulases from *Chaetomium thermophilum* var. *dissitum*. *European Journal of Biochemistry*, **77**, 445–50.
Eriksson, K.-E. (1978). Enzyme mechanisms involved in cellulose hydrolysis by the rot fungus *Sporotrichum pulverulentum*. *Biotechnology and Bioengineering*, **20**, 317–32.
Eriksson, K.-E. & Pettersson, B. (1973). Extracellular enzyme system utilised by the rot fungus *Sporotrichum pulverulentum* (*Chrysosporium lignorum*) for the breakdown of cellulose. 1. Separation, purification and physico-chemical characterisation of five endo-1,4-β-glucanases. *European Journal of Biochemistry*, **51**, 193–206.
Highley, T. L. (1977). Requirements for cellulose degradation by a brown rot fungus. *Material und Organismen*, **12**, 25–36.
Highley, T. L. (1978). Degradation of cellulose by culture filtrates of *Poria placenta*. *Material und Organismen*, **12**, 161–74.
Jacob, F. & Monod, J. (1961). Genetic regulatory mechanisms in the synthesis of proteins. *Journal of Molecular Biology*, **3**, 318–56.
Jermyn, M. A. (1955). Fungal cellulases. VI. Substrate and inhibitor specificity of the β-glucosidase of *Stachybotrys atra*. *Australian Journal of Biological Sciences*, **8**, 577–602.
Kerr, T. & Bailey, I. W. (1934). The cambium and its derivative tissues. I. Structure, optical properties and chemical composition of the so-called middle lamella. *Journal of the Arnold Arboretum, Harvard University*, **15**, 327–49.
Koenigs, J. W. (1972*a*). Effects of hydrogen peroxide on cellulose and its susceptibility to cellulase. *Material und Organismen*, **7**, 133–47.
Koenigs, J. W. (1972*b*). Production of extracellular hydrogen peroxide and peroxidase by wood-rotting fungi. *Phytopathology*, **62**, 100–10.
Koenigs, J. W. (1974*a*). Hydrogen peroxide and iron: a proposed system for decomposition of wood by brown rot basidiomycetes. *Wood and Fibre*, **6**, 66–79.
Koenigs, J. W. (1974*b*). Production of hydrogen peroxide by wood-decaying fungi in wood and its correlation with weight loss, depolymerisation and pH changes. *Archives of Microbiology*, **99**, 129–45.
Mandels, M. & Reese, E. T. (1957). Induction of cellulase in *Trichoderma viride* as influenced by carbon sources and metals. *Journal of Bacteriology*, **73**, 269–78.
Mandels, M. & Reese, E. T. (1965). Inhibition of cellulase. *Annual Review of Phytopathology*, **3**, 85–102.
Nilsson, T. (1974*a*). Microscopic studies on the degradation of cellophane and various cellulosic fibres by wood-attacking microfungi. *Studia Forestalia Suecica*, No. 117, 1–32.
Nilsson, T. (1974*b*). Comparative study on the cellulolytic activity of white-rot and brown-rot fungi. *Material und Organismen*, **9**, 173–8.
Norkrans, B. (1967). Cellulose and cellulolysis. *Advances in Applied Microbiology*, **9**, 91–129.
Preston, R. D. (1974). *The Physical Biology of Plant Cell Walls*. London: Chapman & Hall.
Rainby, B. G. (1957). The size and shape factor in colloidal systems. *Discoveries of the Faraday Society*, **11**, 158.
Reese, E. T. (1977). Degradation of polymeric carbohydrates by microbial enzymes. In *The Structure, Biosynthesis and Degradation of Wood*, ed. F. A. Loewus & V. C. Runeckles, pp. 311–68. *Recent Advances in Phytochemistry*, **11**, New York & London: Plenum Press.
Reese, E. T., Siu, R. G. H. & Levinson, H. S. (1950). The biological degradation of

soluble cellulose derivatives and its relationship to the mechanism of cellulose hydrolysis. *Journal of Bacteriology*, **59**, 485–97.

Seillière, G. (1906). Sur un cas d'hydrolyse diastasique de la cellulose du coton, après dissolution dans le liqueur de Schweitzer. *Compte rendu des Séances de la Société de Biologie*, **61**, 205–6.

Sihtola, H. & Neimo, L. (1975). The structure and properties of cellulose. In *Enzymatic Hydrolysis of Cellulose*, ed. M. Bailey, T. M. Enari & M. Linko, pp. 9–21. Aulanka, Finland: SITRA.

Slater, T. F. (1972). *Free Radical Mechanisms in Tissue Injury*. London: Pione.

4 Formation and action of the ligninolytic system in basidiomycetes

T. K. KIRK and P. FENN

Forest Products Laboratory, Forest Service, United States Department of Agriculture, Madison, Wisconsin 53705, USA

Introduction

It has been estimated that the annual production of terrestrial biomass on the earth is 100×10^9 tons (Bassham, 1975). Of this, approximately 20×10^9 tons are lignin. Because of its high carbon content, lignin is one of the largest single repositories of the energy trapped annually on earth via photosynthesis. Biodegradation of this ubiquitous and abundant aromatic polymer, therefore, is obviously of enormous ecological significance, and evidence to date suggests that the basidiomycetes play a major rôle in accomplishing this task.

The study of lignin biodegradation has until recently been a neglected area, but recent efforts have disclosed a fascinating and unusual process, even before many of the detailed features have been delineated. This review summarises current understanding of lignin metabolism by basidiomycetes with an emphasis on physiological aspects. Two recent reviews (Kirk, Connors & Zeikus, 1977; Ander & Eriksson, 1978) and a comprehensive book (Kirk, Higuchi & Chang, 1980) provide more complete coverage of the literature.

Although the only microbes shown to metabolise lignin efficiently are basidiomycetes, other fungi are known to degrade it slowly (Lundström, 1973; Eslyn, Kirk & Effland, 1975; Haider & Trojanowski, 1980). But little is known about their rôle in the natural process, and very little is known about the chemical and physiological aspects of their activities. Many fungi, such as *Trichoderma* species, are strong cellulose-degraders, but because they cannot degrade lignin they cannot degrade the cellulose in intact lignified tissues.

A number of reports of bacterial degradation of lignin have appeared, several quite recently (see Haider & Trojanowski, 1980). As with the non-basidiomycete fungi, bacterial attack on lignin seems to be slow. It is relevant to note, too, that although many bacteria can degrade isolated cellulose and hemicelluloses, they are not rapid degraders of wood (Levy: Chapter 9).

Many species – perhaps thousands – of wood- and litter-degrading basidiomycetes metabolise lignin as they decompose lignocelluloses. Metabolism of lignin by the litter-degraders, unfortunately, has received little attention since the early work of Lindeberg (1944). The basidiomycetes that decompose wood have been studied extensively, and are known to cause either of two types of decay: white rot, in which the lignin and polysaccharides are metabolised (Cowling, 1961), and brown rot, in which the lignin is modified (Kirk & Adler, 1970; Kirk, 1975) and the polysaccharides (cellulose and hemicelluloses) metabolised (Cowling, 1961). The recent demonstration by Haider & Trojanowski (1980) of conversion of synthetic [^{14}C]lignins to $^{14}CO_2$ by the brown-rot fungus, *Gloeophyllum trabeum*, in liquid culture raises the question of why lignin is not depleted significantly as wood is decayed by these organisms.

The white-rot fungi have been studied far more than other microbes from the standpoint of lignin biodegradation, and it is their activities which are the theme of this paper. Before these are treated, however, a brief look at the substrate is in order.

Lignin is a generic name for the complex aromatic polymers that are major components of vascular tissues in terrestrial plants. Lignin is the product of the free-radical polymerisation of substituted *p*-hydroxycinnamyl alcohols; the three starting 'monomers' are *p*-coumaryl, coniferyl and sinapyl alcohols (Fig. 1). Peroxidase-catalysed

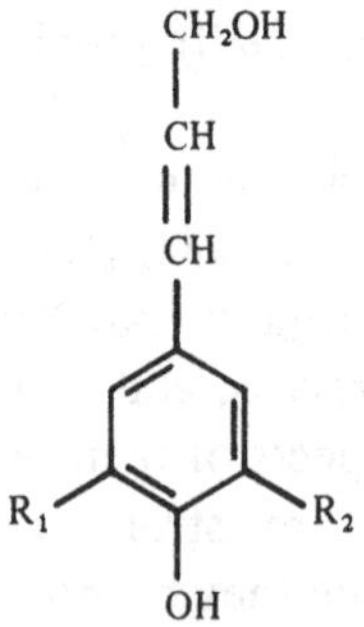

Fig. 1. Lignin precursors: *p*-coumaryl alcohol ($R_1 = R_2 = H$), coniferyl alcohol ($R_1 = OCH_3$, $R_2 = H$), and sinapyl alcohol ($R_1 = R_2 = OCH_3$).

single-electron oxidation of these precursors produces radical species which exist in several mesomeric forms. These couple in an essentially random fashion with each other, and primarily with radicals in the growing lignin polymer (which contains phenolic groups and is itself oxidised by peroxidase), to produce the cross-linked, three-dimensional and polydisperse natural plastic lignin within plant cell walls. A variety of interunit linkages occurs in the polymer. A diagram showing the various interunit linkages in a typical coniferous lignin (formed almost entirely from coniferyl alcohol) is shown in Fig. 2; such lignin is termed a 'guaiacyl' lignin. Angiosperm and some atypical gymnosperm lignins contain high proportions of units derived from sinapyl units (e.g. unit 13 in Fig. 2). Angiosperm tissues contain heteropolymers formed from coniferyl and sinapyl alcohols, as well as near homopolymers of coniferyl alcohol (Wolter, Harkin & Kirk, 1974) and, as discovered recently (Yamasaki, Hata & Higuchi, 1978), of sinapyl alcohol; the latter are termed 'syringyl' lignins. Most lignins contain small amounts of *p*-coumaryl alcohol-derived units (e.g. unit 2 in Fig. 2). The reader is

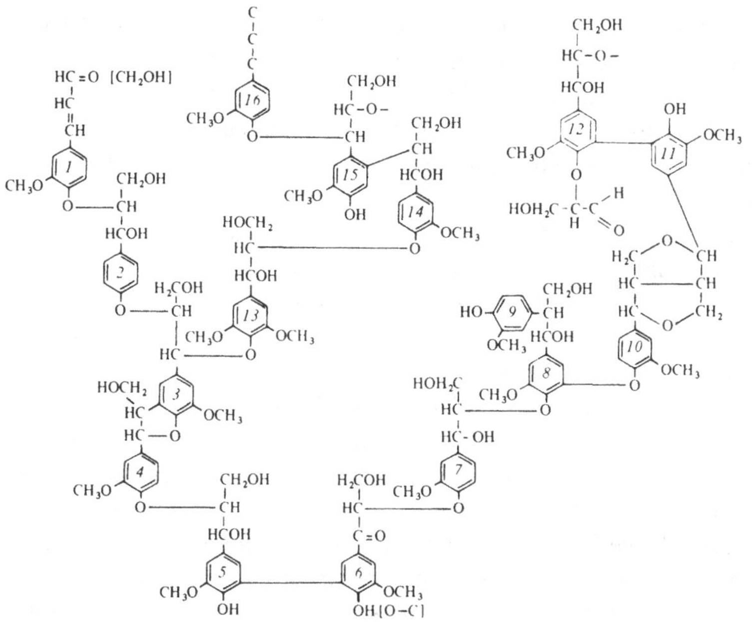

Fig. 2. Prominent structures in a typical coniferous lignin. Interunit linkages between units *1* and *2*, *2* and *3*, *4* and *5*, *6* and *7*, and *7* and *8* are the most abundant types. Other quantitatively important linkage types are between units *3* and *4*, *5* and *6*, and *8* and *9*. (From Adler, 1977.)

referred to Freudenberg (1968), Higuchi (1971), Sarkanen & Ludwig (1971), and Adler (1977) for detailed accounts of lignin formation and structure.

Wood and the woody vascular tissues of herbaceous plants generally contain 20–35% lignin by weight. Together with hemicelluloses, the lignin fills the spaces between cellulose fibrils, in effect coating the cellulose as part of an interpenetrating network with the hemicelluloses. Lignin physically protects the polysaccharides from attack by cellulases and hemicellulases. Occasional covalent bonds between lignin and hemicelluloses are indicated by indirect methods (see e.g. Merewether, Samsuzzaman & Calder, 1972), but their chemical nature and frequency are not yet known, and as a result their significance to lignin metabolism by micro-organisms is not yet approachable experimentally.

Structural changes in the lignin polymer during degradation by basidiomycetes

Understanding of the specific reactions that comprise lignin biodegradation by white-rot fungi is far from complete, but good progress is now being made in several laboratories. Most of what is known has come from chemical and physical characterisation of partially biodegraded lignin isolated and purified from rotted wood. The major conclusions of those findings are presented here; detailed treatment of the data can be found elsewhere (Kirk & Chang, 1975; Chang, Chen & Kirk, 1980).

Elemental and methoxyl analyses and summative C_9-unit formulae for sound and white-rot lignins are given in Table 1. Changes in other properties caused by the fungal degradation are summarised in Table 2. It is apparent from such investigations that biodegradation of the polymer consists primarily of an oxidative attack.

Interpretation of the detailed chemical and physical comparisons of sound and white-rot lignins have led to the conclusion that the oxidative attack by basidiomycetes occurs in both the aliphatic side chains, and in the aromatic nuclei *still bound in the polymer* (Kirk & Chang, 1975). Direct evidence for side-chain oxidation is provided by the presence, in the degraded polymer, of aromatic acid moieties, and by the increase in α-carbonyl content (Hata, 1966; Kirk & Chang, 1975). Evidence for the presence of residues of aromatic ring cleavage in the degraded lignin polymers is indirect: (*a*) the high carboxyl and oxygen contents which cannot be accommodated by side-chain oxidation alone; (*b*) the low methoxyl content and absence of (intact) demethylated aromatic nuclei;

Table 1. *Elemental and methoxyl analyses (%) and C_9-unit formulae for sound and white-rotted spruce lignins*[a]

Lignin sample[b]	C	H	O	OCH_3	C_9-formulae	Mol. wt of C_9-unit
Sound	62.85	6.08	31.07	15.11	$C_9H_{8.66}O_{2.75}(OCH_3)_{0.92}$	189.2
Fungus-degraded	57.97	4.70	37.23	11.33	$C_9H_{7.26}O_{3.95}(OCH_3)_{0.74}$	199.4

[a] From Kirk & Chang (1974).
[b] The sound sample was a milled wood lignin. The fungus-degraded sample was purified from wood which had been decayed to 50% weight loss by *Coriolus versicolor*.

Table 2. *Changes in properties of lignin caused by white-rot basidiomycete attack*

Property	Change[a] Increase	Decrease	Method of analysis[b]	Reference
Carboxyl content	+		C, S	Hata, 1966; Kirk & Chang, 1974, 1975
Carboxyl content	+		S	Hata, 1966; Kirk & Chang, 1974, 1975
Hydroxyl content				
Aliphatic		+	C, S	Hata, 1966; Kirk & Chang, 1974, 1975
Phenolic	+[c]	+	C, S	Hata, 1966; Kirk & Chang, 1974, 1975
Aromatic content		+	C, S	Kirk & Chang, 1975
Yield of low mol. wt aromatic compounds on oxidative chemical degradation		+	C	Higuchi, Kawamura & Kawamura, 1955; Hata, 1966; Kirk & Chang, 1975
Yield of low mol. wt aromatic compounds on hydrolytic degradation		+	C	Hata, 1966; Kirk & Chang, 1975

[a] Purified sound and fungus-degraded lignins were compared.
[b] C = various chemical procedures; S = spectroscopic methods (ultra-violet, infra-red and/or proton magnetic resonance).
[c] Hata (1966) reported an increase in phenolic hydroxyl content and variable results with aliphatic hydroxyl content.

(*c*) the decrease in aromatic content; (*d*) the presence of α, β-unsaturated carboxyl groups which are not aryl-conjugated, and (*e*) the decrease in phenolic hydroxyl content. Data obtained recently (C.-L. Chen, H.-m. Chang & T. K. Kirk, unpublished) provide more direct support for ring cleavage in the polymer. Among the low molecular weight aromatics in extracts of wood partially decayed by *Phanerochaete chrysosporium* were minute quantities of several compounds, each containing one intact aromatic ring attached to an aromatic carbon-derived aliphatic moiety. Fig. 3 shows the tentatively assigned structures of two of these compounds (I and II). Reference to the lignin structure in Fig. 2 shows that compound I could have arisen from units 5–6, and compound II from structures of the type illustrated by units 1–2. These new results show clearly that some aromatic rings are oxidatively cleaved while still attached to a second aromatic ring.

Higuchi (1980) demonstrated recently that degradation of the dimeric lignin model compound dehydrodiconiferyl alcohol (III) by *Fusarium solani* proceeds via compound IV to 5-acetylvanillyl alcohol (V). He proposed that the key degradative step is side-chain oxygenation; a possible pathway is shown in Fig. 4. Shimada (1980) suggested that analogous oxygenation in the side-chains of other substructures as well as in this type would lead to almost complete degradation of the lignin to low molecular weight aromatics.

Such side-chain oxidation reactions can explain the origin of the aromatic acid moieties found in white-rot lignin discussed above. However, it is not possible for these side-chain oxidations alone, even with retention of the oxidised fragments in the polymer, to account for

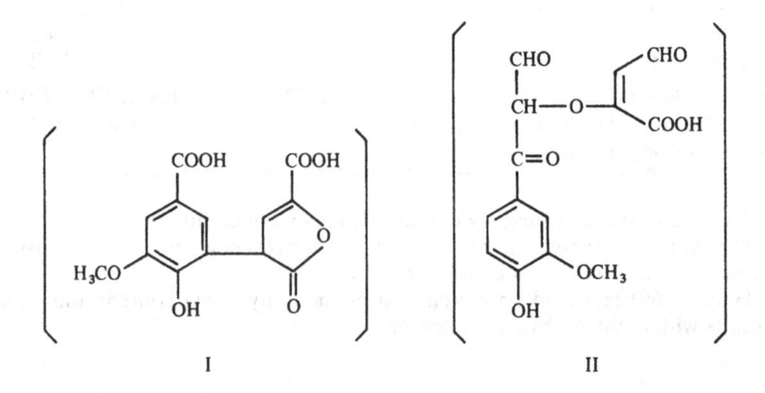

Fig. 3. Tentatively assigned structures of two lignin oxidation products in an extract of spruce wood partially decayed by *Phanerochaete chrysosporium*. Products were separated and characterised by gas chromatography/mass spectrometry. (C.-L. Chen, H.-m. Chang & T. K. Kirk, unpublished.)

all the properties of the white-rot lignin polymer (Kirk & Chang, 1975; Chang *et al.*, 1980). Oxidative cleavages of both side chains and aromatic nuclei are indicated, as hypothesised originally (Kirk & Chang, 1975).

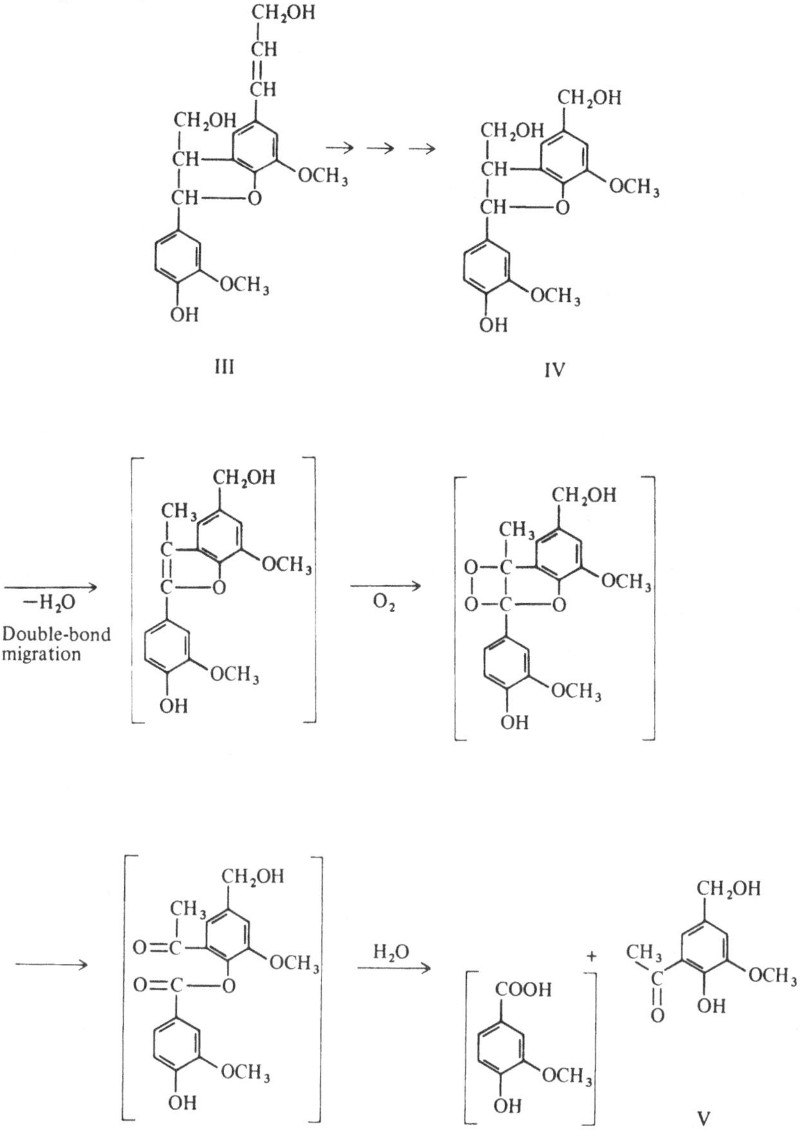

Fig. 4. A possible pathway for degradation of dehydrodiconiferyl alcohol (III) to 5-acetylvanillyl alcohol (V) in cultures of *Fusarium solani*. Compounds in brackets are hypothetical. (Adapted from Higuchi, 1980.)

Physiology of lignin metabolism by basidiomycetes

Experimental strategies

Because of its complex and heterogeneous structure, lignin does not lend itself to specific quantitative determination. Classical chemical and physical methods (Sarkanen & Ludwig, 1971) are not suitable for detailed physiological investigations, and consequently radio-isotopic methods have been developed to provide the rapid qualitative and quantitative assays needed for biodegradation (see Crawford, Robinson & Cheh, 1980). Studies in this laboratory (Kirk *et al.*, 1975) and in that of Haider & Trojanowski (1975) employed synthetic [^{14}C]lignins prepared from [^{14}C]coniferyl alcohol with peroxidase and H_2O_2. Side chain-, ring- and methoxyl-labelled lignins have all been prepared. Crawford *et al.* (1980), Haider & Trojanowski (1975) and Haider & Martin (1980) have employed lignocelluloses labelled by feeding [^{14}C]lignin precursors to growing plants. Both types of preparation have advantages; for physiological studies, the synthetic [^{14}C]lignins have the advantages of having high specific activities and of being well defined, with little chance for non-lignin ^{14}C contamination. With both types of lignin, $^{14}CO_2$ produced by microbial metabolism provides the desired assay.

Almost all of the physiological investigations have been, and are being, conducted with a single fungus in the family Corticiaceae, *Phanerochaete chrysosporium* (taxonomy: Burdsall & Eslyn, 1974). The fungus has several advantages over other white-rot fungi for physiological investigations: (*a*) it produces copious conidia (aleuriospores), which greatly simplify handling; (*b*) its optimum temperature for growth is 39–40 °C, which minimises contamination; (*c*) it grows more rapidly than most white-rot fungi; and (*d*) examined strains produce only barely detectable levels of phenol-oxidising enzyme activity, an advantage when phenolic model compounds are being studied. Gold and coworkers recently described techniques for mutation, colony restriction and replica plating (Gold & Cheng, 1978, 1979), as well as methods for producing sexual fruiting structures (Gold *et al.*, 1980) in *P. chrysosporium*.

Optimisation of culture parameters for lignin metabolism

During the past three to four years, studies have been conducted to define the culture conditions important for metabolism of [^{14}C]lignins to $^{14}CO_2$ by *Phanerochaete chrysosporium* in defined liquid media. Many of the findings have been re-examined using *Coriolus versicolor*; in all cases comparable results were obtained. Detailed descriptions of procedures and results have been published (Kirk, Connors & Zeikus, 1976; Kirk, Schultz *et al.*, 1978; Fenn & Kirk, 1979).

It was discovered early on that fungal growth and metabolism of [^{14}C]lignin were negligible with an isolated spruce lignin or with synthetic lignin as sole carbon addition to an otherwise complete medium. A readily utilised carbon source in addition to the [^{14}C]lignin is required. A variety of carbohydrates, glycerol, ethanol and succinate all suffice for growth and lignin metabolism, and, as expected, they support these activities to differing extents. Most subsequent work has been done with glucose as growth substrate.

The requirement for a growth substrate is curious when the high content of reduced carbon in lignin is considered. Recent studies, described on p. 85, showed that lignin carbon enters the central metabolic pathways of the fungus. Whether the net gain in energy from lignin metabolism is sufficient to support growth, however, is not clear, and it is also not clear whether the process can occur rapidly enough to sustain some minimal metabolic rate. Recent research (Keyser, Kirk & Zeikus, 1978) showed that ligninolytic activity *does not appear in cultures until after primary growth is complete*. This obviously precludes primary growth on lignin, and leaves open the question whether lignin can, theoretically, serve as sole carbon and energy source (see p. 83).

It was also observed early on, in attempts to optimise lignin metabolism, that the concentration of nutrient nitrogen exerts an especially powerful influence on the rate of metabolism. High levels are deleterious. For example, the rate of lignin metabolism during 15 days in cultures containing 2 mM nitrogen was more than twice the rate in cultures containing 20 mM nitrogen at constant pH. A variety of amino acids, as well as NH_4^+, serve as a nitrogen source. Even nitrate supports growth and lignin degradation by *P. chrysosporium* (which is unusual for Hymenomycetes), although growth is slow and the deleterious effect of high nitrogen on lignin degradation is not very pronounced with nitrate. The basis for the suppressive effect of nutrient nitrogen on lignin metabolism is discussed below.

The optimum pH for lignin metabolism by *P. chrysosporium* is near pH 4, whereas the optimum for growth appears to be closer to pH 5. Degradation of lignin above pH 5.5 and below pH 3.5 is less than 20% of that at pH 4.0–4.5. Choice of buffers is also important. Much of the earlier work employed *o*-phthalate, which was chosen in preference to several others, but recently this compound has been found to inhibit degradation by at least 50% at a concentration of 0.01 M (Fenn & Kirk, 1979). Recent work has employed 2,2-dimethylsuccinate at 0.01 M, which gives good control of pH, is not inhibitory to lignin degradation, and is not metabolised. Inhibition by *o*-phthalate is apparently due to a direct effect on the ligninolytic system.

One of the most important culture parameters affecting lignin metabolism is oxygen concentration. Studies conducted with cultures grown under atmospheres containing various concentrations of oxygen showed that the rate and extent was two- to three-fold greater under 100% O_2 than under air (21% O_2), despite the fact that growth is somewhat better in air (Kirk, Schultz *et al.*, 1978). Growth, but not lignin metabolism, occurred under 5% O_2 in nitrogen. Recent studies (S. Bar-Lev and T. K. Kirk, unpublished) showed that 40–60% O_2 is superior to 100% O_2 for lignin metabolism.

All the above results were obtained with shallow stationary cultures. Culture agitation resulting in formation of mycelial pellets greatly suppressed lignin metabolism. If mycelial mats were formed first (in stationary cultures), subsequent agitation, without pellet formation, did not have a significant influence on cultures under 21% O_2. Agitation under 100% O_2 completely inhibited lignin metabolism whether or not pellets were formed, and agitation did not enable cultures under 5% O_2 to metabolise lignin. It is probable that the oxygen concentration in pellet interiors is too low to support lignin metabolism; if this is true, then pellet surfaces, representing only a fraction of the total hyphal surface area, are the only part of the mycelium able to metabolise lignin. The reason for the deleterious effect of 100% O_2 with agitation is not known.

In all the investigations described above, similar results were obtained with lignins labelled at side chain-β, γ, U-ring*, or methoxyl carbon, although production of $^{14}CO_2$ from methoxyl carbon was faster than from ring or side-chain carbons, which were approximately equal. Also, recent studies demonstrated that these various results are also true for lignin degradation in a lignocellulosic substrate (Yang, Effland & Kirk, 1980).

From this research it is now known how to obtain good lignin degradation in a chemically defined medium, making more detailed physiological investigations possible. Efforts have since been directed at

* U-ring: uniformly labelled in the ring.

Table 3. *Characteristics of cultures of* Phanerochaete chrysosporium *during 6 days of growth with ammonium tartrate as nitrogen source*[a]

Culture age (days)	Mycelium dry wt (mg)[b]	Glucose (μmol ml^{-1})[b]	DNA (μg per culture)	Extracellular NH_4^+ (μmol ml^{-1})[b]	NH_4^+ permease (μmol min^{-1} g^{-1})[c]	Ligninolytic activity[d] ($^{14}CO_2$, d.p.m. h^{-1} per culture)
0	0	56		2.1 ± 0		
0.25	0	56		2.1 ± 0		
0.5	0	56	3	1.6 ± 0.5		
1	2.9 ± 1.1	50 ± 1	19	0	ND[e]	0
1.25			21			
2	6.2 ± 0.5	44 ± 1	19	0	0.7	<1
3	7.9 ± 0.8		18		0.55	<1
4	8.7 ± 0.9		17		0.56	32 ± 15
5	10.8 ± 1.3		17		0.60	94 ± 36
6	13.2 ± 0.9	25 ± 1	16		0.52	93 ± 21

[a] Each culture contained, in a total of 10 ml of basal medium: 1.2 mM ammonium tartrate, 0.1 M sodium *o*-phthalate buffer (pH 4.5) and 56 mM glucose. They were grown under an atmosphere of ≈100% O_2, at 39 °C, without agitation. Culture acidity remained constant at pH 4.5.
[b] Mean ± standard deviation for three replicate cultures.
[c] Each value was obtained with six combined cultures.
[d] Assay was based on $^{14}CO_2$ evolved during 6 h under 100% O_2 at 39 °C following addition of 5×10^4 d.p.m. of ring-labelled [^{14}C]lignin per culture. Values ± standard deviation for six replicate cultures.
[e] ND, not detected.

understanding the factors regulating the appearance of ligninolytic activity in cultures.

Regulation of ligninolytic activity

Degradation of the lignin polymer would be expected to be regulated and controlled at many stages. However, since little is known about the pathways involved, discussion of regulation important in the degradation can only be in terms of the total process (lignin → CO_2). Nevertheless, culture studies with the [^{14}C]lignin to $^{14}CO_2$ assay have revealed certain important regulatory controls that impinge on the formation and levels of activity of the ligninolytic system.

Table 3 summarises some of the characteristics of cultures during six days under the conditions optimised for lignin degradation as described above (excepting the buffer). Extracellular NH_4^+, the nitrogen source, was depleted from cultures within 24 h after inoculation. Primary growth began between 12 and 24 h after inoculation and was maximal prior to the second day. Mycelial weight increased throughout the 6-day experiment and involved production of an extracellular glucan after day 2 (unpublished data). Increase in total DNA ceased after approximately

30 h. Ammonia permease activity, not detected in 1-day cultures, was maximal in 2-day cultures, indicating nitrogen-starvation. Ligninolytic activity appeared between days 3 and 4 and was maximal after 5 days. These observations with whole cultures provided the background for further study of regulation.

Failure of lignin to induce ligninolytic activity. It was soon discovered that ligninolytic activity is not induced by lignin; cultures develop similar activity whether or not lignin is present during growth. Experiments showed that protein synthesis is required for, and occurs just prior to, the development of ligninolytic activity in cultures, and that development of essential activity(ies) for attack of the lignin polymer does not require new protein synthesis after addition of the polymer (Keyser *et al.*, 1978). Earlier work gave no evidence that exogenous low molecular weight phenolics can induce ligninolytic activity (Kirk, Yang & Keyser, 1978) and the development of activity appears to be independent of the nature of the carbon source used for growth.

Failure of lignin to induce ligninolytic activity is not a total surprise. The complex three-dimensional nature of lignin, its heterogeneous and racemic structure, and lack of repetitive units reduce the chances of its recognition by a specific lignin polymer receptor molecule, or of binding to a specific ligninolytic repressor molecule, which would be predictable if ligninolytic activity were controlled by a classical induction mechanism.

Nitrogen and carbon effects on appearance of ligninolytic activity. Experiments have been conducted to determine the influence of various nutrients on the appearance of ligninolytic activity in cultures. Changes in the concentration (ten-fold range) of the mineral salts and vitamins in the basal medium had no effect on the time of appearance or level of ligninolytic activity (Kirk, Schultz *et al.*, 1978). Similarly, variations in the level of phosphate did not have a significant influence (Kirk, Yang & Keyser, 1978). Only the concentration of nutrient nitrogen was found to have a profound effect upon initiation of activity. In cultures containing initially 2.4 mM nitrogen and excess growth substrate (e.g. glucose), ligninolytic activity first appeared in 3–4 days (Table 3), 2 days after depletion of the extracellular nitrogen. The appearance of activity could be delayed by addition of extra NH_4^+ prior to the onset of degradation (Fig. 5). Moreover, addition of fresh NH_4^+ to ligninolytic 6-day-old cultures resulted in a decrease of approximately 50% in subsequent rates of decomposition to carbon dioxide during the next 16 h. This suppression was transient, and degradation rates returned to control rates after the excess NH_4^+ had been metabolised. These effects are interpreted to mean that an essential component(s) of the lignin-degrading system is repressible by NH_4^+, i.e. the system develops in response to nitrogen starvation (Keyser *et al.*, 1978). *Repression* is used here to denote decreased activities in response to certain nitrogen compounds without implying details of molecular mechanisms.

Research now to be discussed was aimed at elucidating how *P. chrysosporium* processes its nitrogen, and at determining a mechanism(s) whereby nitrogen repression operates with respect to lignin degradation (P. Fenn & T. K. Kirk, unpublished).

Analysis of the free intracellular amino acids showed that *P. chrysosporium* contains approximately 250 nmol of amino acids per 10 ml

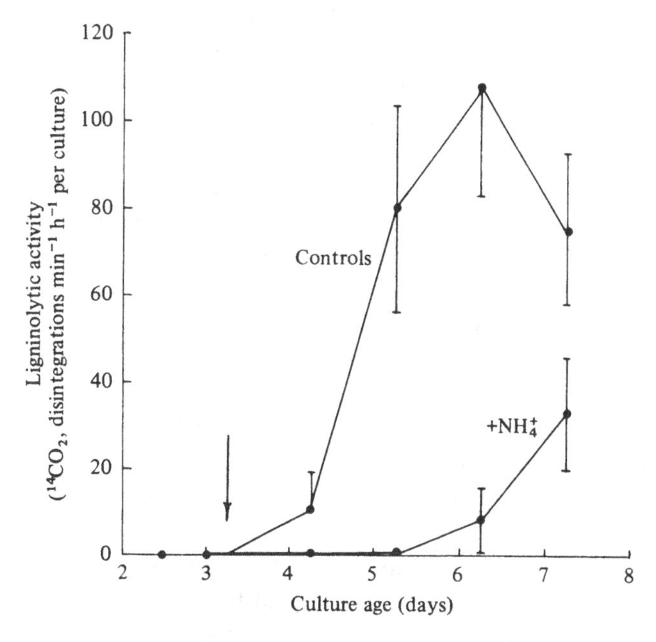

Fig. 5. Effect of NH_4^+ addition on appearance of ligninolytic activity in cultures of *Phanerochaete chrysosporium*. Cultures were grown from spores in a medium containing 1.2 mM ammonium tartrate. Prior to appearance of ligninolytic activity (arrow), an equivalent amount of ammonium tartrate was added (total culture nitrogen was doubled). Control cultures received water only. (Adapted from Keyser *et al.*, 1978.)

culture (initially containing 21 μmol nitrogen) after 1 day's growth. Glutamate-glutamine, arginine, glycine, aspartate-asparagine and alanine accounted for >90% of the total, with arginine accounting for the greatest amount, 35%, of the total amino acid nitrogen. This suggests that during primary growth, the fungus stores some nitrogen as arginine. Perhaps *Phanerochaete* is similar to *Neurospora*, which preferentially stores nitrogen as arginine and glutamine in a non-metabolic compartment (Weiss, 1973).

At the time of appearance of ligninolytic activity (3–4 days) the total amino acid pool had increased to approximately 580 nmol per culture, with the free pool levels of most amino acids increasing two-fold or more. From 5 to 8 days the total pool dropped to 300 nmol per culture. Protein determinations showed a minimum at 2 days. Elevated amino acid pools in *Neurospora* were shown to occur in response to nitrogen starvation or to inhibition of protein synthesis (Mora, Espin, Willus & Mora, 1978). The appearance of ligninolytic activity concomitantly with increased amounts of intracellular amino acids, which when added to cultures repress ligninolytic activity (see below), suggests that the onset of activity does not respond to intracellular nitrogen, or that these amino acids are compartmentalised so that they do not suppress the onset of activity.

Additions of various nitrogen sources to cultures actively degrading lignin showed that repression of the rate of degradation was measurable after 6–9 h. The degree of repression became maximal before 24 h, but the level of repression was dependent on the nitrogen source added (Table 4). Growth studies with various sources showed no relationship between the rate of growth on these nitrogen compounds and their differential effects on repression of activity. Glutamate and glutamine were not utilised as rapidly as NH_4^+ (glucose as carbon source), but were much more effective as repressors. Histidine, a strong repressor, was used only very slowly as a nitrogen source; in fact, it is as poor a source of nitrogen for growth as NO_3^- which is a poor repressor of ligninolytic activity. Such differential effects of nitrogen compounds on nitrogen repressible systems are not uncommon and suggest that there may be several mechanisms to account for nitrogen-based regulation in microorganisms (van de Poll, 1973; Hynes, 1974; Dantizig, Zurowski, Ball & Nason, 1978).

Evidence from work on nitrogen regulation in the bacterium *Klebsiella aerogenes* and with *Aspergillus nidulans* showed that nitrogen repression and catabolite repression by carbon (glucose) do not operate as

Table 4. *Effects of various nitrogen sources on ligninolytic activity of* Phanerochaete chrysosporium

Nitrogen sources	Rate of $^{14}CO_2$ production[a] $n = 6$ (% of control)
None (control)	100
Nitrate	77
L-proline	64
L-arginine	58
NH_4Cl	47
L-alanine	27
L-histidine	24
L-glutamine	23
L-glutamic acid	17

[a] Rate of $^{14}CO_2$ production from ring-[^{14}C]lignin, 6–48 h after addition of the nitrogen sources to 6-day-old ligninolytic cultures. Nitrogen sources were added at concentrations equivalent to 2.8 mM nitrogen (total culture nitrogen was doubled).

independent regulatory mechanisms. Prival & Magasanik (1971) showed that in *Klebsiella* glucose repression of several enzymes involved in nitrogen metabolism only occurred if a surplus of nitrogen was available to the cells. If cells were starved of nitrogen, carbon catabolite repression was relieved. Similarly, proline oxidase in *Aspergillus* was repressed by NH_4^+ when glucose was a carbon source, but not with the non-repressing carbon source, glycerol (Arst & Bailey, 1977). In *Phanerochaete chrysosporium*, fruit-body formation was repressed by glucose at physiological levels of nitrogen, but not at growth-limiting levels (Gold & Cheng, 1979). Because ligninolytic activity in *P. chrysosporium* develops in response to nitrogen starvation in glucose cultures and can be repressed by added nitrogen, it was of interest to determine whether these are true responses to nitrogen or represent changes in the levels of glucose catabolite repression.

In cultures grown on a variety of carbon sources, ligninolytic activity first appeared between 3 and 4 days of culture, and in all cases the activity could be repressed by additions of NH_4^+ or glutamate to the ligninolytic cultures (Table 5). With all carbon sources, glutamate was

Table 5. *Effects of growth substrate (carbon source) on the initial appearance and nitrogen repression of ligninolytic activity*[a]

Carbon source[b]	Onset of ligninolytic activity (days)	Ligninolytic activity, % of controls[c] $n = 6$	
		NH_4^+	L-glutamate
Glucose	3–4	52	25
Xylose	4	70	33
Glycerol	4	60	36
Ethanol	4	50	30
Succinate	3–4	58	—[d]
Malate	4	53	—[d]

[a] Experimental conditions were as previously described (Kirk, Schultz *et al.*, 1978).
[b] Cultures initially contained the various carbon sources at a concentration equivalent to 336 mM carbon.
[c] Each culture contained 5 × 10[4] d.p.m. of ring-labelled [^{14}C]lignin from time of inoculation. $^{14}CO_2$ was collected once each day. When ligninolytic activity was established in the cultures, NH_4Cl or L-glutamate at 2.8 mM, or water (controls) was added. The rate of $^{14}CO_2$ evolution was then monitored after 24 h and 48 h.
[d] Not measured.

more repressive than NH_4^+. Since NH_4^+ and glutamate were repressive in cultures with ethanol or glycerol as growth substrate, which are not known to be repressive carbon sources, it seems unlikely that ligninolytic activity is carbon-catabolite repressed. The repression seen with NH_4^+ and some amino acids, is apparently, therefore, the result of a nitrogen-based regulatory mechanism (or mechanisms).

The catabolism of low molecular weight phenolics, presumed to be possible products of lignin depolymerisation, by fungi has recently been reviewed by Cain (1980). Various levels of regulation of catabolism have been identified, including induction, co-ordinate induction, and carbon catabolite repression. In most of these studies, however, the phenol was used as the sole or predominant carbon and energy source. In these cases all phases of metabolic control and regulation of primary growth and metabolism of carbon would be expected to come into play. The relevance of these controls to degradation of the lignin polymer is not clear. However, degradation of low molecular weight aromatics such as vanillic acid, produced from some structural units in lignin during fungal attack, might of course be regulated in classical ways.

In the final stages of degradation, because lignin-derived carbon undoubtedly enters the glycolytic pathway and/or tricarboxylic acid cycle to be oxidised to carbon dioxide, controls that operate there, e.g. respiratory controls, become important. Because these controls also affect the metabolism of the growth substrate supplied with the lignin, changes in regulation of the metabolism of the growth substrate (e.g. as its concentration changes or when different carbon sources are used) are seen as apparent changes in rates of lignin degradation when complete metabolism to carbon dioxide is being studied. This could lead to an inability to separate the mechanisms more directly involved in regulating the final steps in degradation of lignin, or its metabolites, from those mechanisms controlling metabolism of the growth substrate.

Lignin degradation as a secondary metabolic event

Bu'Lock (1975) defined as secondary 'all these processes which are intensified in "limited" or sub-optimal growth . . . It includes corresponding patterns of enzyme production, *a priori* for enzymes affecting secondary processes . . . For any given system it embraces a whole complex of processes . . . All these processes are governed by an overall control mechanism which is linked to the degree (and character) of growth-limitation'.

Thus all secondary metabolic processes have in common the synthesis of new enzyme systems in response to some limitation on primary growth. Most of the newly made secondary metabolic enzyme systems can be considered anabolic, functioning to synthesise a diversity of products, most of them with uncertain rôles in the lives of the organisms (Weinberg, 1970; Demain, 1974). However, often the anabolic events and syntheses of new enzymes require scavenging of carbon and nitrogen for which new degradative enzymes (e.g. proteases) are synthesised at the onset of secondary metabolism (Weinberg, 1970).

Secondary metabolism is triggered by various limitations in different microbes. Quite frequently, as in penicillin synthesis by *Penicillium* species, it is carbon source limitation, but limitation of phosphate, as in the synthesis of ergot alkaloid by *Claviceps* species (Taber & Vining, 1963) and cephalosporins by *Streptomyces clavuligerus* (Aharonowitz & Demain, 1977), or limitation of nitrogen, as in the synthesis of gibberellins and bikaverins by *Gibberella fujikuroi* (Bu'Lock, Detroy, Hostalek & Munum-Al-Shakarchi, 1974), can cause the metabolic shift. As discussed above, ligninolytic activity is triggered by depletion of extracellular nitrogen.

Characterisation of lignin degradation as a secondary metabolic event seems straightforward. Ligninolytic activity appears, after primary growth is over, in response to nitrogen depletion, and activity can be regulated by nitrogen supply. The conclusion that lignin degradation is secondary-metabolic is further strengthened by the discovery that the appearance of ligninolytic activity is paralleled by synthesis of a typical secondary metabolite, veratryl (3,4-dimethoxybenzyl) alcohol (Lundquist & Kirk, 1978; Keyser *et al.*, 1978). Like ligninolytic activity, veratryl alcohol synthesis is repressed by addition of glutamate, and to a lesser extent by NH_4^+ (Fig. 6). A direct relationship between ligninolytic activity and veratryl alcohol is not apparent; addition of the alcohol to young, non-ligninolytic, cultures does not induce early appearance of, or enhance the subsequent levels of, ligninolytic activity.

Curiously, the structure of veratryl alcohol is very similar to the guaiacyl structural units of lignin, but veratryl alcohol is synthesised by

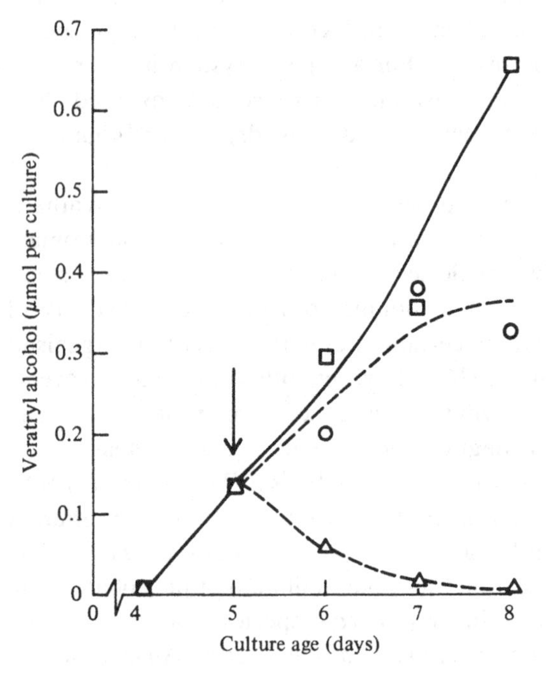

Fig. 6. Effect of ammonium and glutamate addition on veratryl alcohol synthesis in cultures of *Phanerochaete chrysosporium*. Nitrogen additions (arrow) were as described in Fig. 5, and culture conditions were as described in Lundquist & Kirk (1978). Veratryl alcohol in culture extracts was determined quantitatively by gas chromatography. Control, water added (□); NH_4^+ added (○); L-glutamate added (△).

cultures irrespective of the presence of lignin. In glucose-grown cultures, all carbon in veratryl alcohol is derived from glucose (Lundquist & Kirk, 1978). However, in cultures containing either [ring-U-^{14}C]- or [side-chain-β-γ^{14}C]-lignin in addition to glucose, the veratryl alcohol becomes labelled; specific activities and yields of the alcohol showed that not all the carbon is derived from lignin (unpublished data). The label from the [ring-^{14}C]-lignin could conceivably be derived via incorporation of intact guaiacyl moieties, but the labelled positions (β, γ) in the side chains do not correspond to any in the alcohol. The fact that ^{14}C from the β- and γ-positions in lignin was incorporated into veratryl alcohol is good evidence that the carbon of lignin, as expected, enters the central metabolic system of the fungus.

A second metabolic event is also correlated temporally with the appearance of the ligninolytic system: phenol-oxidising enzyme (PO) activity. On the basis of studies with PO-negative mutants, these enzymes have been reported to be necessary for lignin degradation (Ander & Eriksson, 1978). Shimada (1980) showed that the enzymes oxidise, indirectly, α-carbinol groups to α-carbonyl groups in phenolic units in lignin, thus 'destroying' asymmetric centres at C-α, and secondarily at C-β through enolisation. Whether this is important in lignin degradation remains to be determined, and at present the rôle of PO activity is not known. We found (unpublished) that addition of hydrogen peroxide and peroxidase to 1-day-old cultures of *P. chrysosporium* did not influence the time when cultures became competent to degrade lignin (3 to 4 days). The effects of NH_4^+, glutamate, etc., on PO activity have not yet been evaluated.

In any case, evidence indicates that lignin degradation, like veratryl alcohol synthesis and probably appearance of phenol oxidase activity, is a secondary metabolic event. Undoubtedly these are only three of many physiological events that comprise nitrogen-mediated secondary metabolism in *P. chrysosporium*. The total physiological shift to secondary metabolism, with all its manifestations, is initiated and regulated by nitrogen nutrition. A specific effect of nitrogen nutrition on development of ligninolytic activity is unlikely.

Concluding remarks

We assume that the findings presented here and elsewhere for *Phanerochaete chrysosporium* (with the possible exception of veratryl alcohol synthesis) are representative of white-rot fungi in general. As pointed out earlier, comparative studies between *P. chrysosporium* and

Coriolus versicolor revealed no differences (Kirk, Connors & Zeikus, 1976; Kirk, Schultz *et al.*, 1978). The suppressive effect of nutrient nitrogen on lignin degradation, in particular, was noted as being similar. Consequently, the following discussion is assumed to be relevant to white-rot fungi in general, even though only two species have been examined.

Lignin metabolism, a *carbon* metabolic process, is regulated by nitrogen nutrition. When first discovered, this seemed a most curious phenomenon indeed, but further work, as described above, showed that the effect of nitrogen is not on lignin degradation *per se*, but rather on the establishment and maintenance of a secondary metabolic state, which has several manifestations.

Why is lignin degradation a secondary metabolic event under nitrogen regulation? We might speculate that what are now key degradative activities (presumably oxidative), which affect the lignin polymer, evolved with (and still might have) some other rôle as a part of nitrogen-mediated secondary metabolism. Because these activities clearly give certain fungi a competitive ecological advantage (access to cellulose and hemicellulose), evolution has favoured maintenance of the capacity to degrade lignin.

It may then be asked why, if ability to degrade lignin has given certain fungi access to abundant supplies of cellulose and hemicelluloses, has lignin degradation remained a secondary metabolic event? Again we can only speculate. We suggest that the explanation is related to the very low nitrogen content of woody substrates (Merrill & Cowling, 1966). When a white-rot fungus invades a woody substrate, primary growth is probably only a very transient stage, involving establishment of the hyphae. Non-structural components of the wood serve as a substrate for primary growth. Nitrogen rapidly becomes limiting and secondary metabolism, including lignin degradation, begins. Removal of lignin exposes cellulose and hemicelluloses, and progressive decay of all wood components ensues. The ability to degrade lignin would seem to be no advantage to the organism during its evanescent primary metabolic stage.

Whatever the evolutionary basis, the fact is that a secondary metabolic phenomenon, which is basidiomycetous, degradative and regulated indirectly by nitrogen, has assumed a tremendously important ecological rôle in the maintenance of the earth's carbon cycle.

Acknowledgements. This research was supported by National Science

Foundation Grants PCM 76-11144 and AER-75-22789-AO2; by the College of Agricultural and Life Sciences, University of Wisconsin, Madison, and by the Forest Products Laboratory, Forest Service, US Department of Agriculture. One author (P.F.) was supported through the Department of Bacteriology, University of Wisconsin, Madison.

References

Adler, E. (1977). Lignin chemistry – past, present and future. *Wood Science and Technology*, **11**, 169–218.

Aharonowitz, Y. & Demain, A. L. (1977). Influence of inorganic phosphate and organic buffers on cephalosporin production by *Streptomyces clavuligerus*. *Archives of Microbiology*, **115**, 169–73.

Ander, P. & Eriksson, K.-E. (1978). Lignin degradation and utilization by microorganisms. *Progress in Industrial Microbiology*, **14**, 1–58.

Arst, H. N., Jr. & Bailey, C. R. (1977). The regulation of carbon metabolism in *Aspergillus nidulans*. In *Genetics and Physiology of Aspergillus*, ed. J. E. Smith & J. A. Pateman, pp. 131–46. New York and London: Academic Press.

Bassham, J. A. (1975). The substrate: general considerations. In *Cellulose as a Chemical and Energy Resource*, ed. C. R. Wilke, pp. 9–19. Biotechnology and Bioengineering Symposium No. 5. New York: John Wiley & Sons.

Björkman, A. (1956). Studies on finely divided wood. 1. Extraction of lignin with neutral solvents. *Svensk Papperstidning*, **59**, 477–85.

Bu'Lock, J. D. (1975). Secondary metabolism in fungi and its relationship to growth and development. In *The Filamentous Fungi* ed. J. E. Smith & D. R. Berry, vol. 1, pp. 33–58. London: Edward Arnold; New York: John Wiley & Sons.

Bu'Lock, J. D., Detroy, R. W., Hošťálek, A. & Munim-Al-Shakarchi, A. (1974). Regulation of secondary biosynthesis in *Gibberella fujikuroi*. *Transactions of the British Mycological Society*, **62**, 377–89.

Burdsall, H. H. & Eslyn, W. E. (1974). A new *Phanerochaete* with a *chrysosporium* imperfect state. *Mycotaxon*, **1**, 123–33.

Cain, R. B. (1980). The uptake and catabolism of lignin-related aromatic compounds and their regulation in microorganisms. In *Lignin Biodegradation: Microbiology, Chemistry and Potential Applications*, ed. T. K. Kirk, T. Higuchi & H.-m. Chang, vol. 1, pp. 21–60. Boca Raton, Florida: CRC Press.

Chang, H.-m., Chen, C.-L. & Kirk, T. K. (1980). The chemistry of lignin degradation by white-rot fungi. In *Lignin Biodegradation: Microbiology, Chemistry and Potential Applications*, ed. T. K. Kirk, T. Higuchi & H.-m. Chang, vol. 1, pp. 215–30. Boca Raton, Florida: CRC Press.

Cowling, E. B. (1961). Comparative biochemistry of the decay of sweetgum sapwood by white-rot and brown-rot fungi. *USDA Technical Bulletin*, No. 1258, 1–79.

Crawford, R. L., Robinson, L. E. & Cheh, A. M. (1980). ^{14}C-labeled lignins as substrates for the study of lignin biodegradation and transformation. In *Lignin Biodegradation: Microbiology, Chemistry and Potential Applications*, ed. T. K. Kirk, T. Higuchi & H.-m. Chang, vol. 1, pp. 61–76. Boca Raton, Florida: CRC Press.

Dantizig, A. H., Zurowski, W. K., Ball, T. M. & Nason, A. (1978). Induction and repression of nitrate reductase in *Neurospora crassa*. *Journal of Bacteriology*, **133**, 671–9.

Demain, A. L. (1974). How do antibiotic-producing micro-organisms avoid suicide? *Annals of the New York Academy of Sciences*, **235**, 601–12.

Eslyn, W. E., Kirk, T. K. & Effland, M. J. (1975). Changes in the chemical composition of wood caused by six soft-rot fungi. *Phytopathology*, **65**, 473–6.

Fenn, P. & Kirk, T. K. (1979). Ligninolytic system of *Phanerochaete chrysosporium*: inhibition by *o*-phthalate. *Archives of Microbiology*, **123**, 307–9.

Freudenberg, K. (1968). The constitution and biosynthesis of lignin. In *Constitution and Biosynthesis of Lignin* ed. K. Freudenberg & A. C. Neish, pp. 47–122. New York: Springer-Verlag.

Gold, M. H. & Cheng, T. M. (1978). Induction of colonial growth and replica plating of the white-rot basidiomycete *Phanerochaete chrysosporium. Applied and Environmental Microbiology*, **35**, 1223–5.

Gold, M. H. & Cheng, T. M. (1979). Conditions for fruit body formation in the white-rot basidiomycete *Phanerochaete chrysosporium. Archives of Microbiology*, **121**, 37–41.

Gold, M. H., Cheng, T. M., Krisnangkura, K., Mayfield, M. B. & Smith, L.M. (1980). Genetic and biochemical studies on *Phanerochaete chrysosporium* and their relation to lignin degradation. In *Lignin Biodegradation: Microbiology, Chemistry and Potential Applications*, ed. T. K. Kirk, T. Higuchi & H.-m. Chang, vol. 2, pp. 65–71. Boca Raton, Florida: CRC Press.

Haider, K. & Trojanowski, J. (1975). Decomposition of specifically ^{14}C-labeled phenols and dehydropolymers of coniferyl alcohol as models of lignin degradation by soft and white-rot fungi. *Archives of Microbiology*, **105**, 33–41.

Haider, K. & Trojanowski, J. (1980). A comparison of the degradation of ^{14}C-labeled DHP and corn stalk lignins by micro- and macrofungi and by bacteria. In *Lignin Biodegradation: Microbiology, Chemistry and Potential Applications*, ed. T. K. Kirk, T. Higuchi & H.-m. Chang, vol. 1, pp. 111–34. Boca Raton, Florida: CRC Press.

Hata, K. (1966). Investigations on lignins and lignification. XXXIII. Studies on lignins isolated from spruce wood decayed by *Poria subacida* B 11. *Holzforschung*, **20**, 142–7.

Higuchi, T. (1971). Formation and biological degradation of lignins. *Advances in Enzymology*, **34**, 207–83.

Higuchi, T. (1980). Microbial degradation of dilignols as lignin models. In *Lignin Biodegradation: Microbiology, Chemistry and Potential Applications*, ed. T. K. Kirk, T. Higuchi & H.-m. Chang, vol. 1, pp. 171–93. Boca Raton, Florida: CRC Press.

Higuchi, T., Kawamura, I. & Kawamura, H. (1955). Properties of the lignin in decayed wood. *Journal of the Japan Forestry Society*, **37**, 298–302.

Hynes, M. J. (1974). Effects of ammonium, L-glutamate, and L-glutamine on nitrogen catabolism in *Aspergillus nidulans. Journal of Bacteriology*, **120**, 1116–23.

Keyser, P., Kirk, T. K. & Zeikus, J. G. (1978). Ligninolytic enzyme system of *Phanerochaete chrysosporium*: synthesized in the absence of lignin in response to nitrogen starvation. *Journal of Bacteriology*, **135**, 790–7.

Kirk, T. K. (1975). Effects of a brown-rot fungus, *Lenzites trabea*, on lignin in spruce wood. *Holzforschung*, **29**, 99–107.

Kirk, T. K. & Adler, E. (1970). Methoxyl-deficient structural elements in lignin of sweetgum decayed by a brown-rot fungus. *Acta Chemica Scandinavica*, **24**, 3379–90.

Kirk, T. K. & Chang, H.-m. (1974). Decomposition of lignin by white-rot fungi. I. Isolation of heavily degraded lignins from decayed spruce. *Holzforschung*, **28**, 217–22.

Kirk, T. K. & Chang, H.-m. (1975). Decomposition of lignin by white-rot fungi. II. Characterization of heavily degraded lignins from decayed spruce. *Holzforschung*, **29**, 56–64.

Kirk, T. K., Connors, W. J., Bleam, R. D., Hackett, W. F. & Zeikus, J. G. (1975). Preparation and microbial decomposition of synthetic [^{14}C]lignins. *Proceedings of the National Academy of Sciences, USA*, **72,** 2515–9.
Kirk, T. K., Connors, W. J. & Zeikus, J. G. (1976). Requirement for a growth substrate during lignin decomposition by two wood-rotting fungi. *Applied and Environmental Microbiology*, **32,** 192–4.
Kirk, T. K., Connors, W. J. & Zeikus, J. G. (1977). Advances in understanding the microbiological degradation of lignin. In *Recent Advances in Phytochemistry*, ed. F. A. Loewus & V. C. Runeckles, vol. 11, pp. 369–94. New York: Plenum Press.
Kirk, T. K., Higuchi, T. & Chang, H.-m. (eds.) (1980). *Lignin Biodegradation: Microbiology, Chemistry and Potential Applications*, vols. 1 & 2. Boca Raton, Florida: CRC Press.
Kirk, T. K., Schultz, E., Connors, W. J., Lorenz, L. F. & Zeikus, J. G. (1978). Influence of culture parameters on lignin metabolism by *Phanerochaete chrysosporium. Archives of Microbiology*, **117,** 277–85.
Kirk, T. K., Yang, H. H. & Keyser, P. (1978). The chemistry and physiology of the fungal degradation of lignin. In *Developments in Industrial Microbiology*, ed. L. Underkofler, vol. 19, pp. 51–61. Washington, DC: American Institute of Biological Sciences.
Lindeberg, G. (1944). Über die Physiologie lignin-abbauender Bodenhymenomyzeten. *Symbolae Botanicae Upsalienses*, **8,** 1–183.
Lundquist, K. & Kirk, T. K. (1978). De-novo synthesis and decomposition of veratryl alcohol by a lignin-degrading basidiomycete. *Phytochemistry*, **17,** 1676.
Lundström, H. (1973). Studies of the wood-decaying capacity of the soft rot fungi *Allescheria terrestris, Phialophora (Margarinomyces) luteo-viridis* and *Phialophora richardsiae*. Royal College of Forestry, Department of Forest Products, Research Notes, No. R87. Stockholm.
Merewether, J. W. T., Samsuzzaman, L. A. M. & Calder, I. C. (1972). Studies on a lignin–carbohydrate complex. II. Characterization of the water-soluble lignin–carbohydrate complex. *Holzforschung*, **26,** 180–5.
Merrill, W. & Cowling, E. B. (1966). Role of nitrogen in wood deterioration: amounts and distribution of nitrogen in tree stems. *Canadian Journal of Botany*, **44,** 1555–80.
Mora, Y., Espin, G., Willus, K. & Mora, J. (1978). Nitrogen accumulation in mycelium of *Neurospora crassa. Journal of General Microbiology*, **104,** 241–50.
Prival, M. & Magasanik, B. (1971). Resistance to catabolite repression of histidase and proline oxidase during nitrogen limited growth of *Klebsiella aerogenes. Journal of Biological Chemistry*, **246,** 6288–96.
Sarkanen, K. V. & Ludwig, C. H. (eds.) (1971). *Lignins: Occurrence, Formation, Structure, Chemical and Macromolecular Properties, and Utilization*. New York: Wiley-Interscience.
Shimada, M. (1980). Stereobiochemical approach to lignin biodegradation: possible significance of nonstereospecific oxidation catalyzed by laccase for lignin decomposition by white-rot fungi. In *Lignin Biodegradation: Microbiology, Chemistry and Potential Applications*, ed. T. K. Kirk, T. Higuchi & H.-m. Chang, vol. 1, pp. 195–213. Boca Raton, Florida: CRC Press.
Taber, W. A. & Vining, L. C. (1963). Physiology of alkaloid production by *Claviceps purpurea*: correlation with changes in mycelial polyol, carbohydrate, lipid and phosphorus-containing compounds. *Canadian Journal of Microbiology*, **9,** 1–14.
van de Poll, K. W. (1973). Ammonium repression in a mutant of *Saccharomyces carlsbergensis* lacking NADP dependent glutamate dehydrogenase activity. *FEBS Letters*, **32,** 265–6.

Weinberg, E. D. (1970). Biosynthesis of secondary metabolites: roles of trace metals. *Advances in Microbial Physiology*, **4**, 1–44.
Weiss, R. L. (1973). Intracellular localization of ornithine and arginine pools in *Neurospora*. *Journal of Biological Chemistry*, **248**, 5409–13.
Wolter, K., Harkin, J. M. & Kirk, T. K. (1974). Guaiacyl lignin associated with vessels in aspen callus cultures. *Physiologia Plantarum*, **31**, 140–3.
Yamasaki, T., Hata, K. & Higuchi, T. (1978). Isolation and characterization of syringyl component rich lignin. *Holzforschung*, **32**, 44–7.
Yang, H. H., Effland, M. J. & Kirk, T. K. (1980). Factors influencing fungal degradation of lignin in a representative lignocellulosic, thermomechanical pulp. *Biotechnology and Bioengineering*, **22**, 65–77.

5 The movement of *Serpula lacrimans* from substrate to substrate over nutritionally inert surfaces

D. H. JENNINGS
Botany Department, The University, Liverpool L69 3BX, England

Introduction

Serpula lacrimans invades timber either by basidiospores which are aerially dispersed or by mycelium which has grown from another piece of timber. In the former case invasion is by a propagule about 5–10 μm in diameter (see Kramer: Chapter 2); in the latter case the hyphae reaching the new piece of timber may be associated further back with mycelium which is differentiated into histologically complex linear structures (Falck, 1912) which for the moment are best referred to as strands. Garrett (1970) has postulated that the aggregation of hyphae into strands or rhizomorphs is an evolutionary development in response to the need for an inoculum potential adequate to overcome resistance of roots of woody perennials to infection. One may doubt whether such plant organs can be equated with building timber; nevertheless it is a striking fact that, when considering the rôle which Garrett has postulated for strands, a small germling is as capable of colonising timber as are hyphae which have the support of translocating mycelium in the form of strands. Thus, although this chapter is concerned with the ability of *S. lacrimans* to grow over inert surfaces, such as building material, from infected timber to another piece of timber, I hope that this seemingly man-made situation can provide useful insights for those studies which are concerned with the spread of basidiomycetes through the soil.

S. lacrimans is a brown-rot fungus. Its economic importance in the decay of timber is discussed by Dickinson (Chapter 10). As far as we know it can only utilise hemicellulose and cellulose in timber as a carbon source. We can do no more than guess at the source of nitrogen. If hyphae, which are about to invade a new piece of timber, are to produce the appropriate enzyme(s) for breaking down the cellulose, then sufficient nitrogen must be translocated (probably as amino acids or amides) to the hyphal apices so that those enzymes can be synthesised. This will require release of nitrogen compounds from the timber in which the mycelium is growing and in turn a large quantity of cellulose will have to be degraded for those nitrogen compounds to be released. If competition from other microbes is to be minimised the soluble products of the hydrolysis of cellulose need to be rapidly absorbed by the hyphae of *S. lacrimans.* I hope to show that much of what we know about growth of the mycelium from one food source to another over nutritionally inert surfaces becomes more readily understandable when we take into account the need for the fungus to accommodate to the high C:N ratio of timber.

Translocation, and its relation to the water potential of the environment

The best point of departure is to establish how the metabolites produced at the food source eventually reach the growing mycelial front. Jennings, Thornton, Galpin & Coggins (1974) have suggested that known rates of growth of mycelium over inert surfaces can be maintained only if solutes are translocated along hyphae by some mechanism which produces a bulk flow of material. This mechanism might be conceived as a process whereby the compounds are packaged in some way and moved via the mediation of a contractile mechanism. It seems fairly certain that nuclei migrate along hyphae by this means and there is some evidence that translocation of phosphorus occurs in this way along the hyphae of a mycorrhizal member of the Endogonaceae to the host root (Tinker, 1975).

While we cannot say for certain that translocation in *S. lacrimans* does not occur in this way, it seems more likely that the bulk flow is brought about by hydrostatic pressure generated within the hyphae at the food base. The process is akin to the generation of a hydrostatic pressure within the xylem vessels of higher plant roots or to mass flow (if it occurs) in phloem. Essentially, solute accumulates in the hyphae at the food base and this leads to the production of an osmotic potential which is much less than that of the external environment. This leads to water flow across the hyphal membrane and this volume flow (Jv) is dissipated in a vectorial manner – that is , along the hyphae to the mycelial front growing away from the infected wood. The volume flow involves not only solvent but solutes which will provide new cytoplasm at the growing hyphal tips.

The strongest evidence in favour of this mechanism is the presence of droplets at the hyphal tips of mycelium growing away from a food base to an inert non-absorbent surface. One presumes that it is the coalescence of these droplets which leads to those drops of liquid which are readily visible on mycelium in a humid environment. We do not know what solutes are present in the droplets, but from analysis of drops collected from mycelium growing on an agar plate the concentration cannot be more than 0.01 g cm^{-3}. If we assume that the solute has a low molecular weight, say sodium chloride, the concentration would be 0.13 mol l^{-1}. If the solute has a higher molecular weight, the concentration would be correspondingly lower.

It seems unlikely that solute is secreted from the tips such that

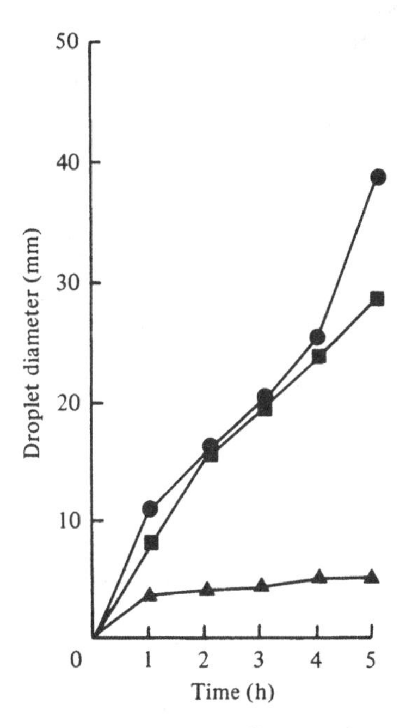

Fig. 1. The effect of the addition of 30 μl water (■), 0.1 (●) or 1.0 (▲) molal potassium chloride to the mycelium of *Serpula lacrimans* at 20 °C (on a 22 mm diameter, 6 mm thick, plug of 5% w/v malt extract agar) on the increase in diameter of droplets produced on those hyphal tips which have grown off the agar onto the surface of a plastic Petri dish. The results are the mean values for ten droplets each from a replicate Petri dish.

droplets are formed either by water following movement of solute across the hyphal tip membrane or by distillation from the surrounding environment. The simplest hypothesis is that the droplets are a visible manifestation of an internal hydrostatic pressure generated within the hyphae as a result of absorption of solutes at the food base.

Droplet formation at the hyphal tip is dependent upon its connection to the hyphal system, since severance results in a decrease in droplet size. The same occurs when the metabolic inhibitor sodium azide is applied to mycelium at the food base (Jennings, 1976). The rate of increase of droplet size can be changed by altering the osmotic potential of the medium at the food base (Fig. 1). High molalities (1.0) of sucrose or potassium chloride reduce the rate, while somewhat lower molalities (0.1) can increase the rate. These observations are consistent with the hypothesis that droplets are the consequence of a hydrostatic pressure produced in the hyphae. High molalities in the medium result in a reduced rate of water uptake with a consequent reduction of turgor within the hyphae. Those molalities which enhance the rate of droplet increase presumably allow sufficient solute absorption to both decrease the osmotic potential of the cytoplasm and reduce the concentration of

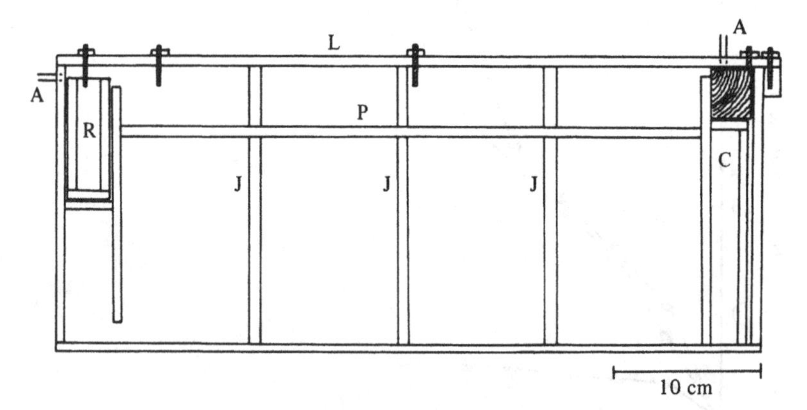

Fig. 2. A vertical longitudinal section through a Perspex cabinet used for determining the position of the front of mycelium of *Serpula lacrimans* growing over nutritionally inert surfaces. A, adaptor for gas inlet or outlet; C, chair to support infected wood; P, plate to support plaster; R, reservoir for the appropriate solution required to alter the humidity gradient along the chamber. Observations are made through the lid (L) which is screwed on to the body of the cabinet. The position of the plate can be altered by jacks which fit into grooves (J) in the side walls. Plasticine is used to seal the gaps between the chair, plate, reservoir and side walls. (See Coggins & Jennings, 1975.)

solute in the medium. In this way a water potential gradient across the hyphal membrane may be generated which is greater than that which existed before the osmoticum was added to the medium. One must presume also that the absorption of solute by the hyphae, whether it be sucrose or potassium chloride, requires metabolic energy, hence the inhibitory effect of sodium azide. There is good evidence from other fungi that this presumption is correct (Jennings, 1976). Finally, if water is to be forced preferentially out of the hyphal tip, the membrane there must have a higher hydraulic conductivity (Lp) than elsewhere.

The hydrostatic pressure generated will of course be responsible not only for droplet formation but also for the generation of turgor for growth. However, the generation of turgor will be dependent upon the water potential of the environment, since if it is too low turgor at the tip will be dissipated. R. W. Clarke in my laboratory has investigated this by studying growth of *S. lacrimans* over plaster of varying water content. The experiments were carried out in Perspex chambers (Fig. 2), such that the growth on the plaster could be readily observed and the osmotic potential surrounding the food base could be changed. Fig. 3 shows the 48 h growth increments for mycelium growing from wood blocks over what, as later results will show, can be termed relatively dry plaster (approximately 0.5% w/w water content). As the mycelium spreads away from the wood over the plaster, the growth rate decreases. There is a changing pattern of growth as is evidenced by the changing profile of the mycelial front. When the water content of the plaster was increased, the pattern of growth was more regular and the mycelial front tended to be parallel to the edge of the infected blocks. I shall return later to this matter of the irregular pattern of growth.

Addition of 1.0 molal sucrose to the wood block resulted in a reduction in the 48 h growth increment, while addition of water had the reverse effect (Fig. 4). This has been observed consistently in a number of experiments. Thus the response of growth to changes in osmotic potential at the food base is analogous to what has been described about the effect of similar changes on droplet size. Thus the water potential of the food source is an important determinant not only of the flow of solution to the hyphal tip but probably also for the maintenance of turgor for growth. The water potential of the medium over which the mycelium is growing will also be important (Table 1). Clarke also showed that when the water content of the plaster is raised from approximately 0.5 to 1.0% and above, the growth rate is increased from 1.25 mm day^{-1} to between 6.9–8.1 mm day^{-1}. One presumes that, for those plaster samples to which water has been added, the water potential is such (sufficiently close to zero) that little water moves out of the hyphae. This supposition is confirmed by the growth rate over the non-absorptive Perspex. In these considerations, we can ignore evaporation of water into the surrounding air because of its relatively small volume (5 mm between the lid of the chamber and the surface over which the mycelium was growing). Thus, if the water potential is

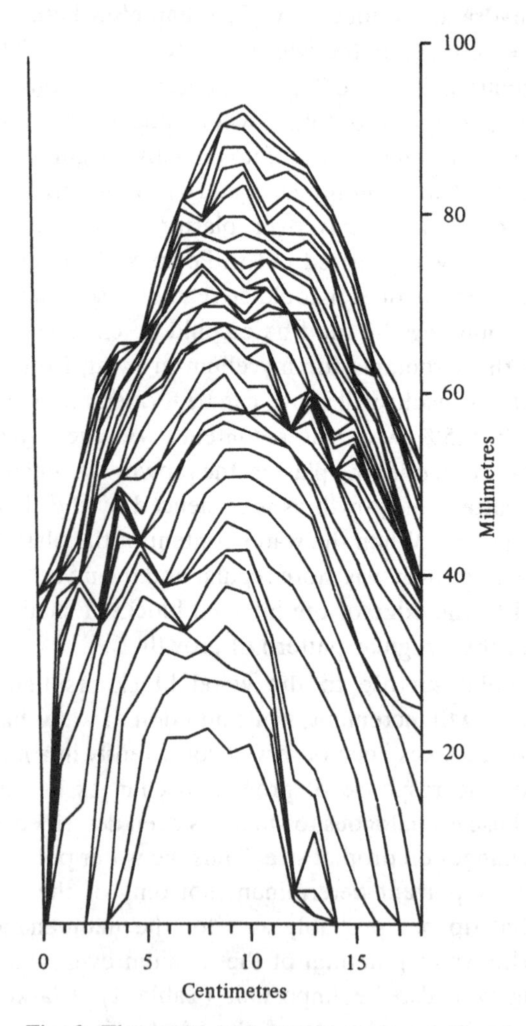

Fig. 3. The position at 48 h intervals of the mycelial front of *Serpula lacrimans* growing at 20 °C in a cabinet shown in Fig. 2, over aged lime-sand plaster to which no water had been added.

Table 1. *Mean growth rates*[a] *of mycelium of* Serpula lacrimans *over non-nutrient surfaces at 20 °C*

	Growth rate (mm day^{-1})	Period of growth measurements (days)
Lime-sand plaster		
'Dry'	1.25	70
1% water content	7.9	20
2% water content	6.9	18
10% water content	8.1	30
Perspex	6.9	16

[a] Single experiments.

less than that of the hyphae, water will move into the medium, tending to dissipate turgor. On the other hand as water moves into the medium, given that it is of limited volume or the rate of movement outwards is slow, the water potential of the medium will rise. Clarke has shown that the water potential of lime-sand plaster can increase from about −900 bar to about −30 bar as the result of growth over the surface by mycelium of *S. lacrimans*.

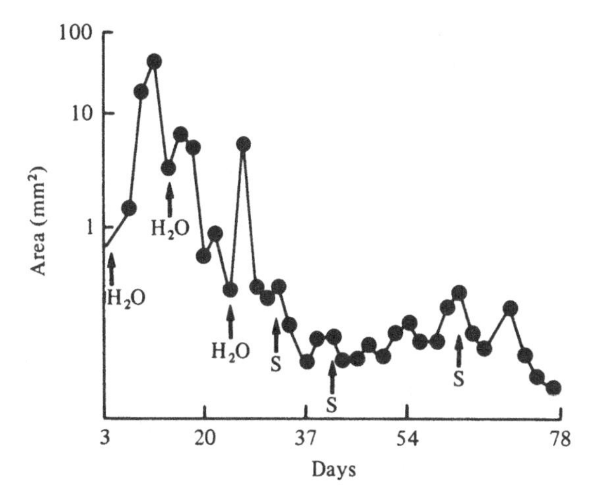

Fig. 4. The effect of additions of 20 cm^3 water or 1.0 molal sucrose (S) solution to a wood block infected with *Serpula lacrimans* on the 48 h increment of growth (area of mycelium produced) made (in a cabinet shown in Fig. 2) by the front of the mycelium extending from the block across aged lime-sand plaster, to which no water had been added. The bulk of any water or solution already associated with the block was removed before a new addition was made.

Metabolic production of water

I will delay for a moment discussing the significance of this rise in the water potential of the environment because I wish first to consider the idea of Miller (1932) that water for mycelial growth of *S. lacrimans* might come from the metabolic breakdown of cellulose. Essentially, we are considering the reaction:

$$(CH_2O)_n \longrightarrow [CO_2 + H_2O]_n,$$

and how this relates to growth over plaster.

Miller showed that 55.6% of the cellulose is converted to water if decay by *S. lacrimans* is complete. Thus, if wood consists of around 58% cellulose, then 0.32 cm^3 of water will be produced per gram of wood decayed. In the experiments carried out by Clarke the wood blocks weighed 18 g which means 17.4 cm^3 of water would be produced. Now the rate of weight loss of softwood during decomposition by *S. lacrimans* is something like 0.47% per day (Anon., 1972) which would give 94.5% decay of the blocks in 580 days. During this time if all the water remained in the fungus it would grow at 1.7 cm day^{-1}. In Clarke's experiment growth was only 0.125 cm day^{-1}, which would lead to the mycelial front being only 72.5 cm from the wood blocks after 580 days. The reduced growth rate can only mean that there is a flow of water from the hyphae into the plaster. Since we can calculate the volume of the mycelium, the difference between that and the total volume of water produced gives the volume of water entering the plaster, namely 16.2 cm^3.

In the period of the experiment the moisture content of the plaster rose from 0.5% to 1.35% w/w. If this rise is produced by 16.2 g of water, the weight of dry plaster which gets infiltrated by water is 2 kg and, if it has a density of 2.5, its volume will be 800 cm^3. From this we can derive a depth to which the plaster achieves 1.35% water content, and this depth is 4.95 cm. In the above calculations evaporation into air has been ignored, because its volume above the mycelium in the experimental chamber is relatively insignificant. There is no doubt from these calculations that sufficient water is produced by the metabolism of cellulose both to generate the requisite turgor and to allow for loss into the environment.

How does this hypothesis compare with reality? Firstly, one must draw attention to the fact that hydrolysis of cellulose requires water. Secondly, the compound which seems to be translocated, namely arabitol, and which can reach a concentration of 25% of the dry weight, is unlikely to release significant amounts of water until it is catabolised at the hyphal front. Indeed, as the mycelium spreads away from the colonised wood over a nutritionally inert surface, the arabitol moving to the front will be increasingly converted into polysaccharide in the form of wall material as a consequence of the development of strands. The situation is shown diagrammatically in Fig. 5. This means that if the bulk of the water for growth is to come from cellulose then its rate of breakdown at the food source must be sufficient to provide not only water for growth but also the arabitol to be translocated to the hyphal front and to the sites of wall accretion. The reader will realise that Miller's measurements relate to water production by mycelium on, or close to, the timber which is infected such that strand formation may not have been very significant. Further, since the experiment was continued until the cellulose in the wood had been broken down completely some water production might have occurred as a result of the autolysis of hyphae because the amount of carbohydrate became insufficient to maintain growth.

These comments suggest that Miller's results may not be so relevant to the growth of mycelium of *S. lacrimans* away from colonised timber as was once envisaged. But there is a more serious criticism. It is implicit in Miller's ideas – certainly in the way that they have been interpreted by others – that *S. lacrimans* can attack relatively dry timber and produce water by the breakdown of cellulose, making the environment more favourable for further growth. This means that the fungus must be

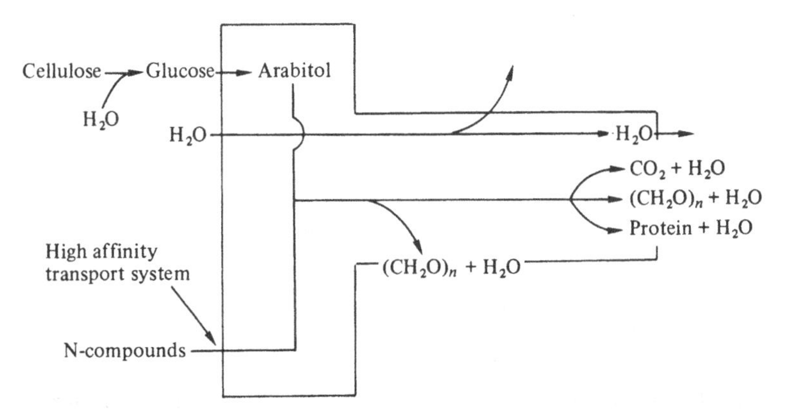

Fig. 5. A diagram to show some of the major metabolic and transport processes postulated as occurring in the mycelium of *Serpula lacrimans*, growing from wood over a nutritionally inert surface.

xerophilic to make the initial attack. All the evidence is to the contrary, and it is clear that *S. lacrimans* cannot withstand water potentials in the external medium much below −60 bar (Fig. 6). The results for growth over agar of different water potentials are in keeping with the results of Brown, Fahim & Hutchinson (1968) who showed that *S. lacrimans* will not grow over glass at relative humidities below 90% at 20 °C. One can only conclude that fairly moist wood is needed for infection to occur. Since this is so, it follows that although some water which may move along those hyphae growing away from the infected timber may be metabolically produced, a significant proportion has to come from that already in the environment. This view is supported by the experiments described in the previous section. Moreover, Weigl & Ziegler (1960) found that *S. lacrimans* produced no more water than other wood-decay fungi.

Strands and the nitrogen economy of the mycelium

We do not know the exact pathway taken by the translocation stream in mycelium which has not yet or only just begun to differentiate

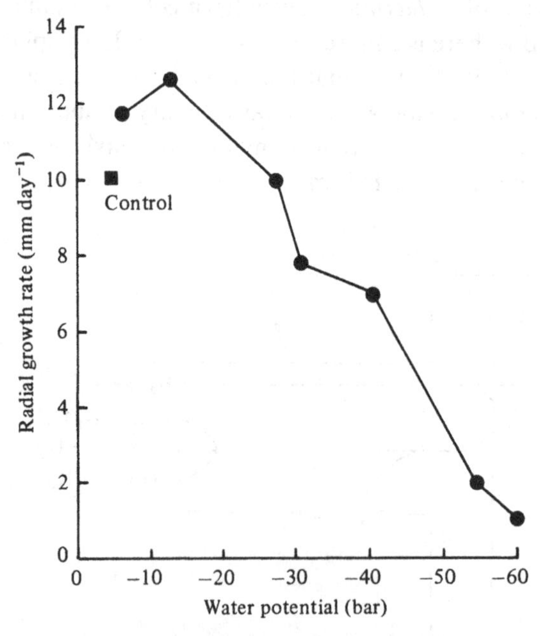

Fig. 6. The effect of varying water potential on the radial growth rate at 20 °C of *Serpula lacrimans*, growing out from a 25 mm diameter, 5 mm thick, wood-block inoculum over 0.5% (w/v) malt agar, sucrose being used to vary the water potential.

into characteristic cell-types. The so-called main hyphae round which narrower or *tendril* hyphae twist (Falck, 1912; Butler, 1958) seem likely to be the channels for movement. When strands become differentiated as distinct entities, *vessel* hyphae and the so-called *hose* hyphae (Falck, 1912) are the most likely candidates for the translocation channels. They appear to be free of cytoplasm and without septa – although beam thickening is present (Falck, 1912; G. Sharples, unpublished data). This indicates that they would be low-resistance pathways for the mass flow of solution.

Production of strands depends upon the nitrogen content of the medium (Watkinson, 1975). The number increases both as the nitrogen concentration decreases and the C:N ratio increases. This relationship between strand formation and nitrogen nutrition is worth exploring further. We know that as a strand matures the walls of many hyphae increase in thickness. Also very thick-walled fibrous hyphae (Falck, 1912) lie increasingly in the outer regions of the strand until they form a distinct rind in older strands (Falck, 1912; G. Sharples, unpublished data).

This direction of metabolism towards wall synthesis may be a mechanism which removes excess carbohydrate from the translocation stream. I have pointed out already that a considerable amount of cellulose will have to be hydrolysed and the soluble products absorbed if sufficient nitrogen for growth is to be obtained from wood. The high concentration of arabitol which ensues helps to generate the appropriate water potential gradient across the plasmalemma at the food source, such that water is absorbed and the hydrostatic pressure to drive translocation generated. It seems likely that the C/N composition of the translocation stream reflects that in wood. This will result in an excess of carbohydrates in the growing region of the mycelium. This is because the amount of cytoplasm synthesised is limited by the amount of nitrogen reaching the region. It seems a reasonable supposition that excess will be removed as wall material. Increases in wall thickness will also have a beneficial effect on translocation since they will help to prevent the dissipation of the hydrostatic pressure (Jennings *et al.*, 1974).

However one must be cautious in pushing this idea too far at this stage. Falck (1912) describes how, as the mycelium ages, thinner strands disappear, their contents apparently absorbed by the thicker strands which continue to increase in diameter. The developing thicker strands are well spaced, and the intervening mycelium disappears. The excess of carbon over nitrogen will therefore depend on the extent to which nitrogen compounds move from (or leak out of) other parts of the mycelium into the translocation stream (Watkinson, 1971). However, this in itself does not argue against the hypothesis. We must not forget that fungal cell walls can themselves contain significant amounts of nitrogen, particularly in the form of chitin. There is one set of measurements on the composition of fungal cell walls which allows us to put the matter in perspective. Wessels (1965) found that in the vegetative mycelium of the basidiomycete *Schizophyllum commune* chitin comprises not more than 5% of the total polysaccharide of the cell walls. The remainder is glucan. The value of 5% gives a C:N ratio for the wall polysaccharide of 160:1. This ratio is a good deal lower than that of wood which is in the range 350–500:1 (Cowling & Merrill, 1966). Nevertheless the ratio for *S. commune* vegetative wall polysaccharide is sufficiently high to indicate that chemical analysis of the walls of strands of *S. lacrimans* is urgently required.

The flux of translocation in *S. lacrimans* determined from growth measurements has a value of 1.27×10^4 pmol cm^{-2} s^{-1} (Jennings *et al.*, 1974). That is a minimum value because it takes no account of incorporation of carbohydrate into wall material and metabolism to carbon dioxide. Even so the flux is two or three orders of magnitude greater than the likely flux of glucose into the mycelium across the plasmalemma subsequent to cellulose hydrolysis. It follows that the loading area must be of sufficient magnitude to bring to the start of the translocation pathway sufficient arabitol to allow translocation to proceed at the measured rate.

The question that now arises is how a sufficient concentration of nitrogen compound(s) is maintained to supply the translocation stream such that a constant concentration moves to the hyphal front. I think we might anticipate that those transport systems which are responsible for absorbing nitrogen compounds from the medium will have a very high affinity for those compounds and will be capable of scavenging them as they are released to the medium. Such a system exists in *Neurospora crassa* where System II, responsible for the uptake of basic, acidic and neutral amino acids, has indeed been shown to have a scavenging function, the affinity of the system for the amino acids transported being less than 50 μmol l^{-1} (Pall, 1969). Release of nitrogen compounds by living or autolysing hyphae, which run alongside the translocation pathways, will also help to maintain the concentration of nitrogen compounds within the translocation stream at an appropriate level. This is speculation; we clearly need data for the uptake and translocation of nitrogen compounds by *S. lacrimans*. This will be difficult because of the unavailability of a radioactive isotope for nitrogen. The situation with regard to phosphorus which has a readily available radioactive isotope might be explored with profit. The phosphorus economy of the fungus as it grows over non-nutrient surfaces may well be analogous to that of nitrogen.

Pattern of growth

Earlier in this paper, I indicated that on relatively dry plaster there is an irregular pattern of growth. It is not exactly clear why this occurs. However, it may be related to a phenomenon that can be observed on agar plates where sectors develop which are capable of more rapid growth than the rest of the mycelium (Coggins, 1977). The sectors arise from very small areas of mycelium, both at the edge and from within the colony (Fig. 7). Because the sectors arise on a small front, we have termed the phenomenon 'point growth'. The faster growing mycelium maintains its growth rate when transferred to fresh medium, and also differs from the parent mycelium by being more closely adpressed to the agar surface and losing its original colour more rapidly under staling conditions. There is some evidence that 'point growth' occurs more frequently under adverse conditions, e.g. on highly alkaline plaster. We do not know the underlying mechanism for the phenomenon. It may be related to what one can see at the microscopic level in growing cultures of *S. lacrimans*, namely the tendency of individual hyphae to grow rapidly ahead of the majority. Falck (1912) described how the surface of any substrate colonised by the fungus appears to have a mycelium in which the hyphae are running parallel but on closer examination can be seen to be composed of bundles which originate irregularly from single points on the surface. The phenomenon may also be analogous to the situation for rhizomorph-producing fungi, where the rhizomorph has a much greater growth rate than the undifferentiated colony (Garrett, 1953; Rishbeth, 1968).

I have already referred to the similarity between strands and strand-like organs and rhizomorphs which Garrett (1970) has highlighted with respect to their capacity to provide an adequate inoculum potential. The similarity between the strands of *S. lacrimans* and rhizomorphs may be of a more fundamental kind. The similarity between point growth and rhizomorphs with respect to growth rate could be an aspect of this. Also the apparent organisation of an outer cortex consisting of fibre hyphae

(Falck, 1912; G. Sharples, unpublished data) may be comparable with the organisation in rhizomorphs (Townsend, 1954).

In view of the possibility that there might be a fundamental similarity between strands and rhizomorphs, it is pertinent to point out that there are also important differences. Falck (1912) uses the term *syrotia* for the strands of *S. lacrimans*, distinguishing them from those stranding systems that are looser aggregations of hyphae. It may be timely to consider the re-use of this term by Falck. It seems pertinent to do so now that I have suggested that the component parts of the mature organ function in an integrated manner. The strands of *S. lacrimans* differ significantly from those of *Psalliota hortensis* (= *Agaricus bisporus*), which although they continue to increase in thickness through the accretion of further individual hyphae in the region where the strands originated, nevertheless remain as simple and undifferentiated aggregations of hyphae (Garrett, 1954).

The growth of mycelium of *S. lacrimans* and the concept of inoculum potential

Finally, it is appropriate to consider how the term inoculum potential applies to *S. lacrimans* in the light of the information that I have presented. Garrett (1970) defines inoculum potential for a saprophyte as 'the energy of growth of a fungus available for colonisation of a substrate at the surface of substrate to be colonised'. At face value it is not clear how the energy for growth is quantified, especially when one remembers that colonisation of wood can be achieved as effectively by a

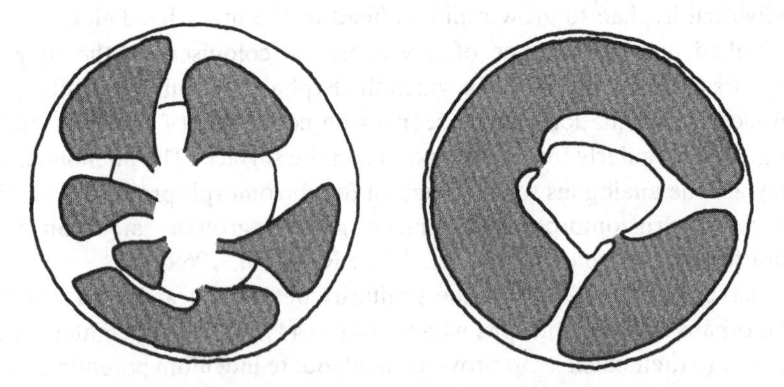

Fig. 7. Drawings from Petri dish cultures of *Serpula lacrimans* growing on glucose (0.5% w/v)–tryptone (0.3% w/v) Oxoid agar, showing the sectoring characteristics of what is termed point growth (Coggins, 1977).

spore as by mycelium. From one point of view, it could be wrong to make this comparison, because Garrett's definition explicitly relates to a competition, and none of the data presented here concern competitive interactions. Nevertheless, one idea is worth pursuing: that the energy within an inoculum potential should be partitioned not only into that which allows growth on the new substrate but also into that which is required to transport sufficient biomass (whether it be spore or mycelium) to infect a new food source competitively. If we take this viewpoint we can see that the energy required to produce and transport spores to a new food base may be of similar magnitude to that required to bring about the advance of mycelium from one food source to another. In the first instance energy has to go into the production of the fruit body and its complex of tissues and the spores themselves. Translocation will be involved since nutrients will have to move from the food source to growing cells some distance away. As with translocation of nutrients to hyphal tips in vegetative mycelium, there must be some mechanism which ensures that the translocation stream taking nutrients to the developing spores has the appropriate C:N ratio to synthesise the protoplasm and wall of the infective propagules.

One can crystallise the comparison by reference to the division made by Gregory (1966) of spores into xenospores and memnospores. The former are often small and are dispersed from their place of origin (see Kramer: Chapter 2), whereas the latter tend to be large and remain in their place of origin, being released by lysis. The spores released from the fruit bodies of *S. lacrimans* are clearly xenospores. But might we not think of the hyphal tips which invade fresh timber as being equivalent to memnospores? Certainly the invading mycelium can be considered as a large infective propagule. It is virtually stationary when compared with the distance that can be traversed in a matter of minutes by a xenospore, and the invading hyphae become separated to a degree from the parent mycelium through partial lysis of the cytoplasm in the strands. It would be wrong to extend further the analogy between a true memnospore and the invading hyphal tips, but it seems to me that the analogy is sufficiently strong to allow us to clarify our ideas about what is meant by the term inoculum potential.

Relevance of these studies to those on rhizomorphs

I have pointed out already that there are similarities between strands and rhizomorphs. There is no doubt that rhizomorphs are capable of translocating nutrients over long distances. Findlay (1951) records the presence of two rhizomorphs of *Armillaria mellea* – one about 30 m long and the other about 90 m – in a water tunnel, 65 m below the ground surface. Since they were growing on a nutritionally inert surface, nutrients must have been moving along them from some food base to the growing tip. Yet we do not know the mechanism responsible for the movement. It is time that physiological studies were initiated. They would be assisted by the considerable amount of information now available concerning the morphology and anatomy of rhizomorphs (Motta, 1969; Schmid & Liese, 1970, Botton & Dexheimer, 1977) and there is the very elegant investigation of Smith & Griffin (1971) of diffusion of oxygen in the rhizomorphs of *Armillariella elegans*, showing the power of physiological studies in helping to understand the ecology of these basidiomycete structures. Their ability to translocate and the mechanism whereby translocation takes place ought to be accessible to some of the experimental procedures described in this chapter.

When considering rhizomorphs we need to remember that they are found growing through soil. So, unlike those situations that I have been considering for the strands of *S. lacrimans*, nutrients are available to a rhizomorph in the intervening spaces between one infected piece of higher plant material and another. In view of this, it seems likely that a rhizomorph is not only an organ for translocation but also for nutrient absorption from the soil. It should be an easy matter to investigate the efficiency of rhizomorphs as nutrient absorbing organs; the information obtained could be very illuminating and help to increase our understanding of the ecology of basidiomycetes in the soil.

Acknowledgements. I am most grateful to Dr Ursula Hornung for her very helpful comments on the manuscript and to Dr George Sharples of the Biology Department, Liverpool Polytechnic, for allowing me to see his unpublished findings.

Since this Chapter was written, we have found that trehalose is the carbohydrate which is translocated, being converted to arabitol at the mycelial front. This new information does not alter the arguments presented here.

References

Anon. (1972). Laboratory tests of natural decay resistance of timber. Timberlab Paper No. 50, Building Research Establishment, Princes Risborough Laboratory, Princes Risborough, UK.

Botton, B. & Dexheimer, J. (1977). Ultrastructure des rhizomorphes du *Sphaerostilbe repens* B. et Br. *Zeitschrift für Pflanzenphysiologie*, **85**, 429–43.

Brown, J., Fahim, M. N. & Hutchinson, S. A. (1968). Some effects of atmospheric humidity on the growth of *Serpula lacrimans*. *Transactions of the British Mycological Society*, **51**, 507–10.

Butler, G. M. (1958). The development and behaviour of mycelial strands in *Merulius lacrimans* (Wulf.) Fr. II. Hyphal behaviour during strand formation. *Annals of Botany*, **22**, 219–36.

Coggins, C. R. (1977). Aspects of growth of *Serpula lacrimans* the dry rot fungus. PhD thesis, University of Liverpool.

Coggins, C. R. & Jennings, D. H. (1975). A cabinet designed for fungal growth studies and an example of its use in investigating the sensitivity of *Serpula lacrimans* to zinc oxychloride. *International Biodeterioration Bulletin*, **11**, 64–6.

Cowling, E. B. & Merrill, W. (1966). Nitrogen in wood and its role in wood deterioration. *Canadian Journal of Botany*, **44**, 1539–54.

Falck, R. (1912). Die Merulius-Faule des Bauholzes. *Hausschwammforschungen*, **6**, 1–405.

Findlay, W. P. K. (1951). The development of *Armillaria mellea* rhizomorphs in a water tunnel. *Transactions of the British Mycological Society*, **34**, 146.

Garrett, S. D. (1953). Rhizomorph behaviour in *Armillaria mellea* (Vahl) Quel. I. Factors controlling rhizomorph initiation by *A. mellea* in pure culture. *Annals of Botany*, **17**, 63–79.

Garrett, S. D. (1954). The function of the mycelial strands in substrate colonisation by the cultivated mushroom *Psalliota hortensis*. *Transactions of the British Mycological Society*, **37**, 51–7.

Garrett, S. D. (1970). *Pathogenic root-infecting fungi.* Cambridge University Press.

Gregory, P. H. (1966). The fungus spore: what it is and what it does. In *The Fungus Spore*, ed. M. F. Madelin, pp. 1–13. (18th Symposium of the Colston Research Society, March–April 1966). London: Butterworth.

Jennings, D. H. (1976). Transport and translocation in filamentous fungi. In *The Filamentous Fungi*, ed. J. E. Smith & D. R. Berry, vol. 2, pp. 32–64. London: Edward Arnold.

Jennings, D. H., Thornton, J. D., Galpin, M. F. J. & Coggins, C. R. (1974). Translocation in fungi. In *Transport at the Cellular Level* ed. M. A. Sleigh & D. H. Jennings, pp. 139–56. Symposia of the Society for Experimental Biology 28. Cambridge University Press.

Miller, V. V. (1932). *Points in the biology and diagnosis of house fungi.* Leningrad: State Forrestal Technical Publishing Office.

Motta, J. J. (1969). Cytology and morphogenesis in the rhizomorph of *Armillaria mellea*. *American Journal of Botany*, **56**, 610–19.

Pall, M. L. (1969). Amino acid transport in *Neurospora crassa*. 1. Properties of two amino acid transport systems. *Biochimica et Biophysica Acta*, **173**, 113–27.

Rishbeth, J. (1968). The growth rate of *Armillaria mellea*. *Transactions of the British Mycological Society*, **51**, 575–86.

Schmid, R. & Liese, W. (1970). Finstruktur der Rhizomorphen von *Armillaria mellea*. *Phytopathologische Zeitschrift*, **68**, 221–31.

Smith, A. M. & Griffin, D. M. (1971). Oxygen and the ecology of *Armillariella elegans* Heim. *Australian Journal of Biological Sciences*, **24**, 231–62.

Tinker, P. B. H. (1975). Effects of vesicular-arbuscular mycorrhizas on higher plants. In *Symbiosis* ed. D. H. Jennings & D. L. Lee, pp. 325–49. Symposia of the Society for Experimental Biology 28. Cambridge University Press.

Townsend, B. B. (1954). Morphology and development of fungal rhizomorphs. *Transactions of the British Mycological Society*, **37**, 222–33.

Watkinson, S. C. (1971). The mechanism of mycelial strand induction in *Serpula lacrimans*: a possible effect of nutrient distribution. *New Phytologist*, **70,** 1079–88.
Watkinson, S. C. (1975). The relation between nitrogen nutrition and formation of mycelial strands in *Serpula lacrimans*. *Transactions of the British Mycological Society*, **64,** 195–200.
Weigl, J. & Ziegler, H. (1960). Wasser-haushalt und Stoffleitung bei *Merulius lacrimans* (Wulf) Fr. *Archiv für Mikrobiologie*, **37,** 124–33.
Wessels, J. G. H. (1965). Morphogenesis and biochemical processes in *Schizophyllum commune* Fr. *Wentia*, **13,** 1–113.

6 Population structure in wood-decomposing basidiomycetes

A. D. M. RAYNER* and N. K. TODD†

*School of Biological Sciences, University of Bath, Claverton Down, Bath, England

†Department of Biological Sciences, University of Exeter, Washington Singer Laboratories, Perry Road, Exeter, England

Introduction – the need for population studies

As yet, relatively few reasonably detailed population studies have been undertaken with basidiomycetes, nor indeed with many other fungi. This is unfortunate because such studies have considerable potential for the mycologist. They can help to give direction to laboratory-based investigations by placing them in their proper, natural context, and they can also be important in their own right. If we consider the first point, many physiological and genetical studies in the laboratory have relied on use of just one or a few isolates. Whilst amply justified in certain detailed work dependent on previously well-characterised strains, this practice has obvious drawbacks. Either the significance of some phenomena may be given incorrect emphasis, or important features characteristic of the natural population may be overlooked (see, for example, the discussion of heterokaryosis, parasexuality and heterokaryon incompatibility in Chapter 7).

In broader terms, population studies can in themselves be of considerable value by providing information about a wide range of important genetical, evolutionary and ecological issues. These include variation, opportunities for recombination, isolating mechanisms, the nature of selection pressures and patterns of propagation, dispersal and colonisation.

Here we will attempt to show how basidiomycete populations in decaying wood can be defined, and their structure and composition ascertained. In Chapter 7 we explore ways in which an ecogenetical approach to population studies can provide insight of various important aspects of basidiomycete ecology.

Delimitation of individuals

Problems of definition – what is an individual?

A 'population' is here defined as an assemblage of *individuals* of the *same* species considered together because of their proximity in space and time. By contrast, assemblages of *different* species are probably better regarded as 'communities' (Rayner & Todd, 1979; Hedger & Basuki: Chapter 15; Mercer: Chapter 8). Exactly what is meant by 'proximity' varies according to the approach adopted: thus we can recognise populations ranging in size from very local ones in individual stumps or logs to those of world-wide distribution. An alternative approach might be to consider the 'breeding' population – that group of individuals which are potentially capable of interbreeding.

Whatever the definition of a population, its basic units are *individuals*. For fungi this poses immediate problems since it is generally considered that fungal individuals are difficult or impossible to recognise with certainty. In part these difficulties are attributable to the colonial growth form of filamentous fungi; some would regard each hyphal tip as an individual, others would consider the entire colony as such. Even if the latter view is taken, it is not always easy to define the limits of individual colonies, since by hyphal anastomosis (which occurs readily in higher fungi) colonies that were originally separate may merge and become part of a single unit. Buller (1931) demonstrated the formation of such a 'unit mycelium' with *Coprinus sterquilinus*. He argued that the size of fruit bodies produced by *C. sterquilinus* on dung balls could not be accounted for unless fusion had occurred between colonies originating from numerous separate germinating basidiospores.

Here we shall digress briefly to define certain terms we shall use to describe sexual incompatibility systems in basidiomycetes. *Diaphoromixis* (= multiple allelic heterothallism) is that condition whereby dikaryon formation can be accomplished only following conjugation between homokaryons (primary mycelia derived from single basidiospores) differing at one (unifactorial, bipolar) or two (bifactorial, tetrapolar) unlinked mating-type loci; these latter exist as numerous alleles in the natural population. *Homodiaphoromixis* (= secondary homothallism or amphithallism) is a derived condition whereby single basidiospores can germinate to produce a dikaryon directly because they are heterokaryotic for mating-type. *Homomixis* (= primary homothallism) is that condition whereby single, homokaryotic basidiospore isolates are capable of sexual reproduction.

Returning to Buller's work, *C. sterquilinus* is homomictic and so it is conceivable that basidiospores derived from a single fruit body may be of very similar, or identical, genotype. There is no doubt that genetically identical mycelia of higher fungi will fuse if meeting in or on a common substrate.

The possibility that genetically distinct mycelia of the same basidiomycete species could fuse to form a physiological and ecological, but genetically heterogeneous, unit was suggested by Burnett & Partington (1957) as a result of studies with two polypores, *Coriolus versicolor* and *Piptoporus betulinus*. Individual fruit bodies of *P. betulinus* collected over a period of years from a single birch trunk were obviously derived from different dikaryons, but they contained mating-type factors in common. Similarly adjacent fruit bodies of *C. versicolor* growing on a stump contained different permutations and combinations of the same mating-type factors. Whilst recognising that these results could be viewed in terms of physically separate dikaryotic entities, Burnett & Partington favoured the interpretation that a single genetically heterogeneous mycelium existed in the wood, partly because hyphal anastomosis occurred readily between genetically different isolates in the laboratory. Unfortunately they did not investigate the detailed structure of the mycelium in the wood.

If genetically heterogeneous unit mycelia were to occur in nature, then certainly the terms 'individual' and 'population' would be difficult, or impossible, to apply. However, the existence of such mycelia is questionable (Rayner & Todd, 1979), and there are good grounds for believing that for many, if not all, wood-decaying basidiomycetes, including *C. versicolor* and *P. betulinus*, genetically distinct mycelia behaving both physiologically and ecologically as individuals do occur in nature, and that these are readily recognisable.

Intraspecific antagonism

There are several reports that when isolates of the same species of certain wood-decaying basidiomycetes from different origins are paired in culture, antagonistic (this term is used here in a broad sense) reactions variously described as 'barrages', 'aversion phenomena', 'interaction zones' or 'lines of demarcation' develop along the line of contact between colonies. Most reports describe the formation of clear zones containing relatively few hyphae and flanked on either side by dense aerial mycelium; later the zones themselves or the adjacent margins of the interacting mycelia may become pigmented, usually dark brown or black.

Amongst some of the earlier reports of such 'intraspecific antagonism' are those of Schmitz (1925) and Mounce (1929) for *Fomitopsis pinicola*, Campbell (1938) for *Oudemansiella radicata* and Childs (1937) for *Phaeolus schweinitzii*. Several early workers assumed that antagonism was evidence of genetical dissimilarity between isolates, and began to use the phenomenon in ecological studies. Thus Roth (1952) detected 'multiple infections' of *Phellinus pini* in Douglas fir (*Pseudotsuga menziesii*) poles, by observing antagonism between isolates collected from different positions along the bole. Later Childs (1963) used antagonism to determine whether isolates of *Phellinus weirii* isolated from different trees in a stand of Douglas fir belonged to the same or different 'clones'. His results suggested that each individual clone occupied at least several adjacent trees.

A feature of many of these early studies was confusion over the genetical basis of antagonism, particularly regarding its occurrence between homokaryotic and dikaryotic isolates, and its relation to sexual incompatibility. In some cases antagonistic effects were noted between homokaryons, in others between dikaryons and in yet others in interactions involving both homokaryons and dikaryons. Between homokaryons of some tetrapolar diaphoromictic species such as *Schizophyllum commune*, a barrage is closely related to sexual incompatibility (Raper, 1966), but in other cases the interaction is dependent on genetical differences which may be unrelated to mating type. It seems likely that antagonistic phenomena of apparently similar general appearance but of different origin may often occur in one and the same species and that both sexual and vegetative incompatibility mechanisms may be operating. Only detailed genetical studies with several natural isolates can help us to discriminate between these alternatives.

The first detailed work of this kind was undertaken by Adams & Roth (1967) with *Fomitopsis cajanderi* which causes a brown heart rot in Douglas fir. They obtained four dikaryotic isolates from two ice-damaged trees, and made various pairings between these, and between monokaryons, and synthesised dikaryons derived from fruit bodies obtained in culture. They showed that development of a pigmented demarcation line was characteristic of pairings involving dikaryons, and was based on genetical difference between isolates. The intensity of the interaction was highly variable and decreased in pairings made between closely related strains. Between monokaryons, the expected bipolar diaphoromictic mating pattern was demonstrated, 100% dikaryon formation occurring between those derived from different parent dikaryons. They therefore concluded that formation of 'demarcation lines' in paired cultures was a reliable indicator of genetical differences between dikaryons. Similar findings were made by Barrett & Uscuplic (1971) in an equally detailed study with *Phaeolus schweinitzii*.

An issue left in doubt by these studies concerns the operation and manifestation of intraspecific antagonism in nature. In an early observation, Brodie (1935) noticed development of a 'barrage' between two 'mycelia' on a herbarium specimen of *Corticium calceum* on rotted *Pinus* wood. Much later Adams & Roth (1969) obtained evidence of the operation in nature of antagonism between dikaryotic isolates of *F. cajanderi* by incubating sections of wood, cut from three infected trunks of Douglas fir, for five months in polythene bags. At the end of the incubation period, mycelial mats of *F. cajanderi*, separated by demarcation zones, had formed on the wood surface. By pairing isolates from each mat, Adams & Roth concluded that the spatial patterns observed could be due to competition between genetically different dikaryons invading simultaneously from an infection court. They realised the importance of relating antagonistic phenomena to spatial

---

Fig. 1. (pp. 114, 115) Evidence of intraspecific antagonism. (*a*) Section through part of a *Quercus* log colonised by *Stereum hirsutum*. The sapwood is extensively rotted and appears as a mosaic of pale decay areas separated by narrow dark interaction zones (from Rayner, 1976). (*b*) Intraspecific antagonism between cultures of different dikaryons of *Coriolus versicolor* on 3% malt agar (from Rayner & Todd, 1977). (*c*) Portion of a log colonised by *Radulomyces confluens* showing correspondence in position of a barrage zone between fruit bodies and an interaction zone line in the wood (arrowed). (*d*) Close-up of a section through a *Quercus* log extensively colonised by *S. hirsutum* after incubation. Notice the crust-like areas found between interacting mycelia (from Rayner & Todd, 1979).

Fig. 2. (pp. 116, 117) Polymorphism in *Coriolus versicolor*, and its relation to intraspecific antagonism (from Rayner & Todd, 1978). (*a*) Top of stump bearing clumps of fruit bodies. Within each group the fruit bodies are of uniform appearance, but there is considerable variation between groups. The marked fruit bodies (black arrowheads) are of *Bjerkandera adusta*. (*b*) Wood underlying the fruit bodies in (*a*), showing several decay regions separated by narrow dark zones (black arrowheads). The position of the larger decay regions could readily be correlated with that of different groups of fruit bodies at the surface. (*c*) Top of stump bearing numerous, generally small, highly polymorphic fruit bodies. (*d*) Wood underlying (*c*) with numerous small decay zones. (*e*) Log with a relatively uniform assemblage of fruit bodies. (*f*) Section showing more or less uniform decay, those lines present (black arrowheads) being discontinuous or due to *Stereum hirsutum*.

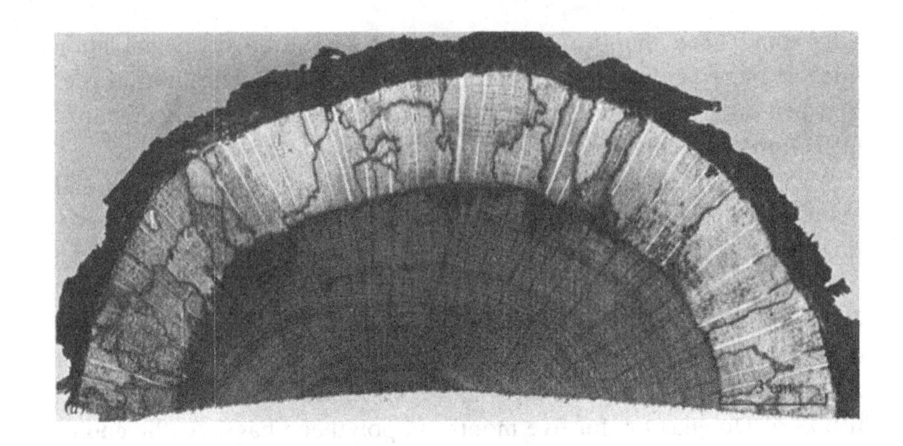

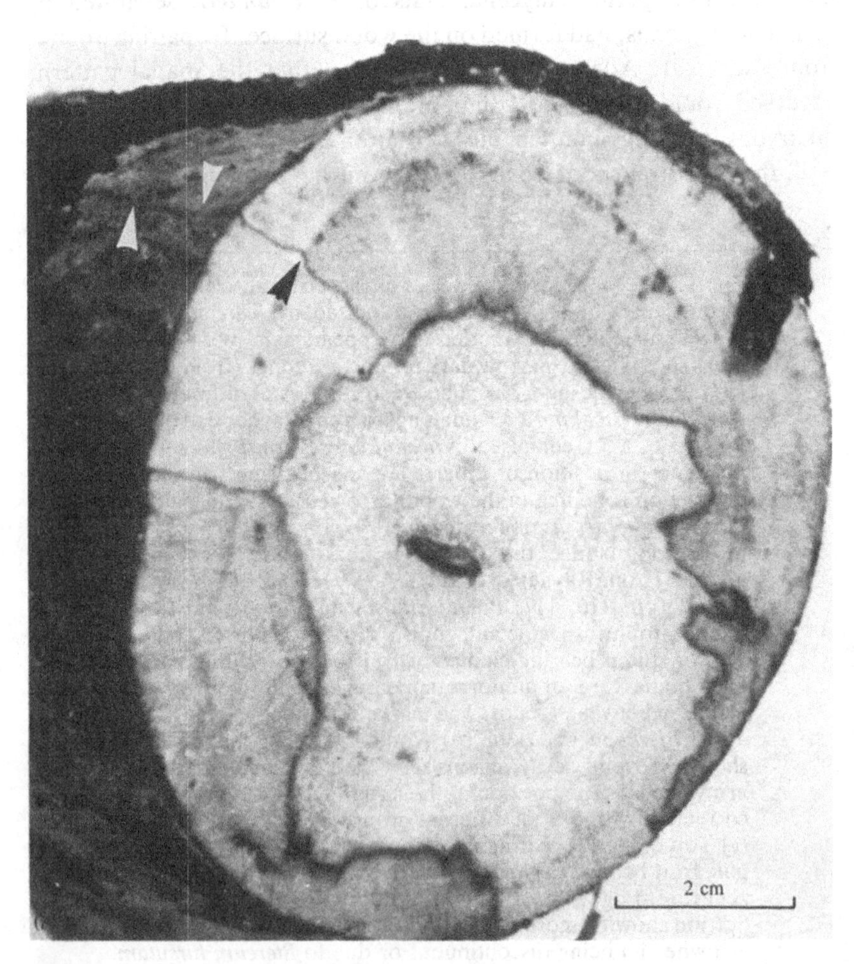

Fig. 1

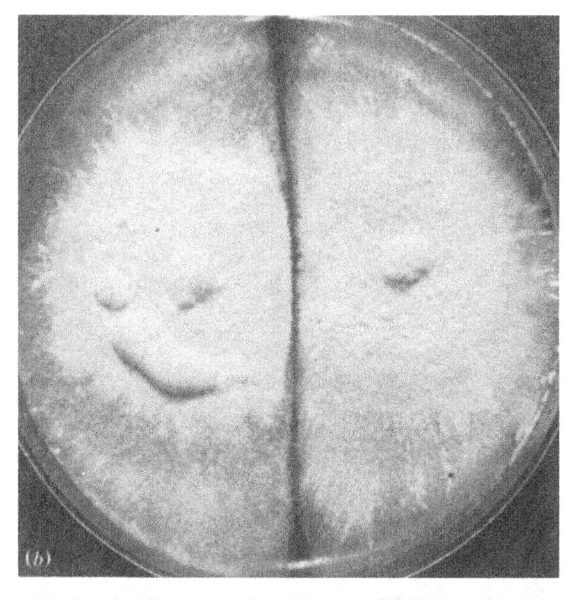
(b)

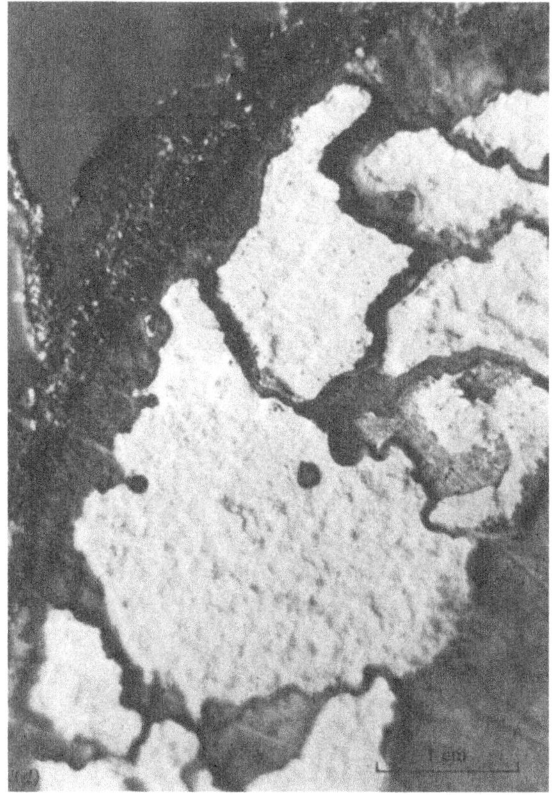
1 cm
(d)

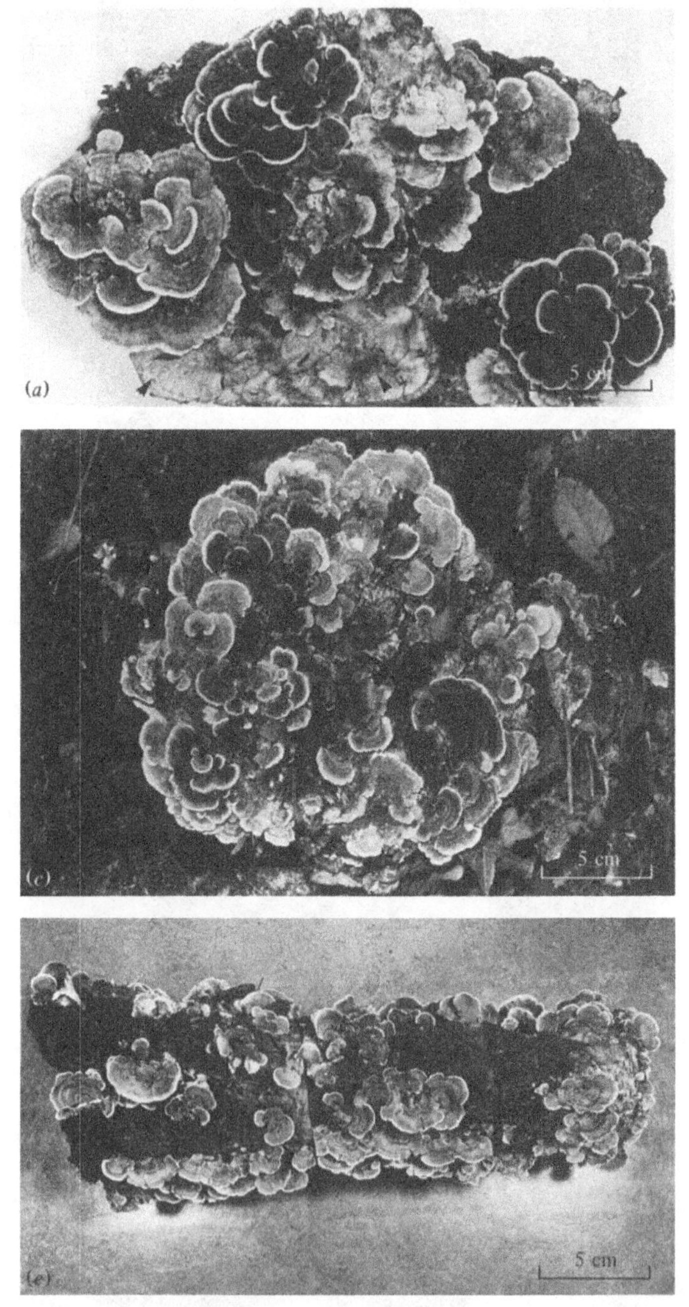

Fig. 2

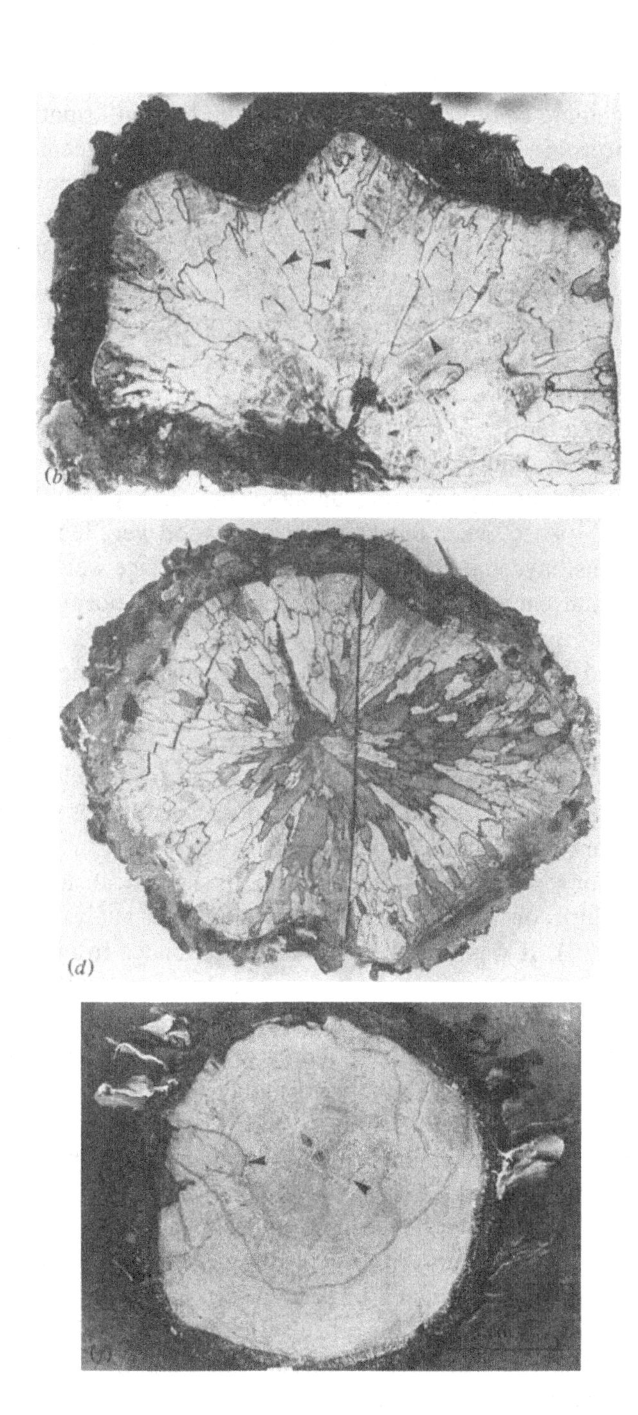

(b)
(d)

distribution of isolates in wood, and with hindsight there is little doubt that their interpretations were correct.

By the beginning of the 1970s a substantial body of information was available concerning intraspecific antagonism. However, it seems fair to say that the existence of the phenomenon was not widely known, nor were its implications fully grasped.

In 1976, one of us (Rayner, 1976) noticed that white-rotted wood occupied by such basidiomycetes as *Coriolus versicolor, Bjerkandera adusta, Phlebia radiata* and *Stereum hirsutum* often contained narrow dark zones of relatively undecayed wood (seen as a network in transverse section) between decay regions (Fig. 1*a,c,e* and Fig. 2*b,d,f*). That the zones were sites of intraspecific antagonism between individual mycelia of the basidiomycetes was indicated by the fact that isolates from adjacent decay regions were antagonistic when paired in agar culture (Fig. 1*b*). Also, when blocks of decayed wood were incubated in moist conditions, mycelial mats developing at the surface were strongly demarcated from each other at sites corresponding in position to the dark zones (Fig. 1*d*).

We followed up these observations by investigating the detailed structure of a population of *C. versicolor* on a *Betula* stump (Rayner & Todd, 1977; Todd & Rayner, 1978). Here it was possible to correlate the position of morphologically distinct groups of fruit bodies at the surface with that of *longitudinally continuous* decay columns separated by narrow, dark, interaction zones in the wood. This enabled us to adopt a sampling procedure whereby both the genetical and spatial inter-relationships of the mycelial population were analysed simultaneously (Fig. 3). It was shown that dikaryotic isolates from separate decay columns were invariably antagonistic in culture, whilst those from the same column merged readily (Table 1). Monokaryons derived from basidiospores of fruit bodies corresponding to different decay columns were interfertile, almost invariably forming dikaryons with clamp connections when paired in culture (this is to be expected as *C. versicolor* is tetrapolar diaphoromictic). Apart from demonstrating conspecificity this allowed dikaryons of various genetical constitutions to be synthesised. Pairings between genetically different synthesised dikaryons were invariably antagonistic (Table 2), although the intensity of interaction declined with increased relatedness, paralleling the results of Adams & Roth (1967) and Barrett & Uscuplic (1971). It therefore seemed likely that each decay column contained a single, genetically uniform, dikaryon whose integrity was maintained by intraspecific antagonism with its neighbours. This was confirmed by genetical experiments which indicated that nuclear transfer into an established dikaryon was not possible.

Virtually identical results have been obtained by analysing populations of the bipolar species *Bjerkandera adusta* (Rayner & Todd, unpublished), and current studies indicate that similar mechanisms operate in *Hypholoma fasciculare, Phlebia radiata, Piptoporus betulinus, Radulomyces confluens* and *Stereum hirsutum*. Further, C. M. Brasier (personal communication) has detected antagonism between genetically different dikaryons obtained from a natural population of *Schizophyllum commune* (for details of this population see Brasier, 1970).

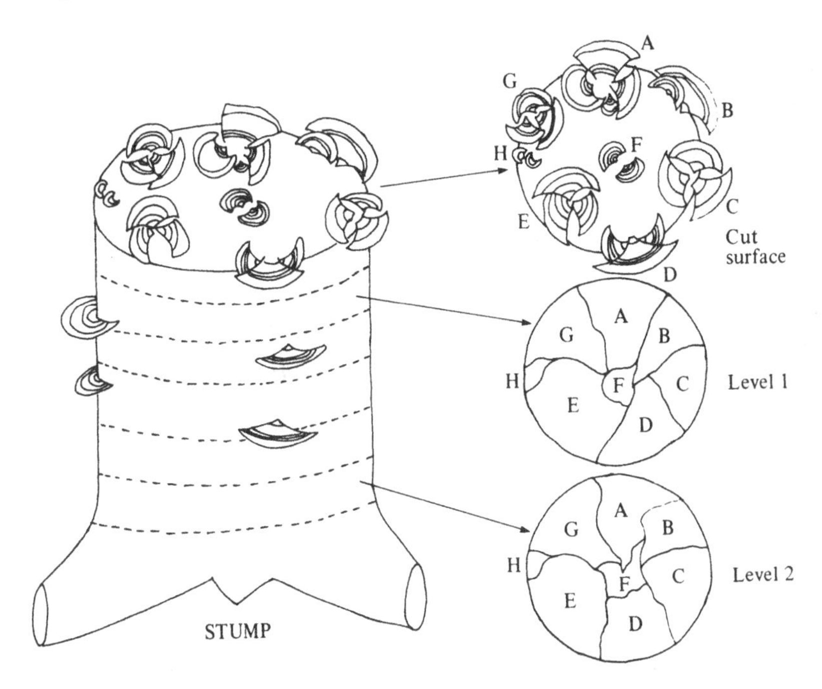

Fig. 3. Diagram illustrating procedure for analysis of a population of *Coriolus versicolor* in a stump. The stump is first cut into transverse slices as indicated by the dotted lines. The position of the different decay columns, delimited by narrow dark zones, on separate slices is noted, and correlated with that of fruit bodies at the surface. The decay columns and their corresponding fruit bodies are assigned codes, and isolates made from them as follows: (i) from fruit-body tissue; (ii) from single basidiospores from fruit bodies used for tissue isolation; (iii) from the decayed wood at different levels. Pairings can then be made between isolates in various combinations (adapted from Todd & Rayner, 1978).

Table 1. *Pairings between isolates of* Coriolus versicolor *from a population in a* Betula *stump.* (After Rayner & Todd, 1977)

(*a*) Pairings between isolates from decay columns

(i) From different decay columns

	Y2	X1	O1	N1	L1	J1	H1	G1	C1	A1
A1	○	△	△	△	○	△	○	○	○	●
C1	△	△	△	△	△	△	△	○	●	
G1	△	△	△	○	○	△	○	●		
H1	△	△	△	△	○	○	●			
J1	△	△	△	△	△	●				
L1	△	△	△	△	●					
N1	○	△	△	●						
O1	△	△	●							
X1	△	●								
Y2	●									

(ii) From the same decay columns

A1 × A2●
C1 × C2●
G1 × G2●
H1 × H2●
J1 × J2 ●
N1 × N2●
O1 × O2●
X1 × X2●

(*b*) Pairings between isolates from fruit-body tissue

	O	N	H	F	C	A
A	△	○	○	△	△	●
C	△	△	○	△	●	
F	△	△	△	●		
H	△	△	●			
N	△	●				
O	●					

(*c*) Pairings between isolates from fruit-body tissue (*left*) with isolates from decay columns (*right*)

A × A1●
C × C1●
H × H1●
O × O1●
N × N1●
A × J1 △
C × O1△
H × J1 △
O × N1△
C × L1 △

Isolate codes: the letters indicate the isolates or the decay column from which the isolates were obtained, and the numbers 1 and 2 refer to different levels within a decay column.

Symbols: ○, antagonism; △, antagonism accompanied by pigment production; ●, complete intermingling of isolates.

The type of interaction between antagonistic isolates varies amongst these different decay fungi. Most frequently a clear zone of variable width (normally < 3 mm) develops between the colonies; subsequently it may become pigmented if the interaction is intense (pigmentation according to Rayner (1970): *C. versicolor* – sepia; *P. radiata* – orange to brick-red; *H. fasciculare* – luteous; *B. adusta* – ochreous; *S. hirsutum* – various shades of yellow). In most cases the interaction

Table 2. *Occurrence of antagonism in pairings on 3% malt agar between synthesised dikaryons of* Coriolus versicolor. (After Todd & Rayner, 1978)

Pairing type	No. of pairings	No. △	No. ○	% △
(A) Dikaryons not sibcomposed				
(*a*) all monokaryotic components from different parents (e.g. A26/O6 × C1/N4)	180	144	36	80.0
(*b*) one pair monokaryotic components in common (e.g. A26/O6 × A26/N31)	79	48	31	60.8
(*c*) one pair monokaryotic components sibrelated (e.g. A26/O6 × A29/J13)	160	104	56	65.0
(*d*) two pairs monokaryotic components sibrelated (e.g. A26/O6 × A23/O4)	13	8	5	61.5
(B) Dikaryons sibcomposed				
(*a*) dikaryons from different parents (e.g. N4/N31 × J19/J6)	50	39	11	78.0
(*b*) dikaryons from same parents (e.g. N4/N31 × N8/N36)	67	3	64	4.5

Isolate codes: the letters indicate the dikaryotic parent from which the monokaryons were derived; the numbers are the reference numbers of each monokaryon. Each dikaryon is given a code (e.g. A26/O6), indicating the monokaryotic components from which it was synthesised by pairing them in culture. Symbols: ○, antagonism; △, antagonism accompanied by pigment production.

zone remains as a definite 'ditch' between colonies, but in *B. adusta* it may become over-arched with aerial mycelium, as reported for *Phaeolus schweinitzii* by Barrett & Uscuplic (1971).

Our investigations of the physiological basis of antagonism are still at an early stage. At least in *C. versicolor* and *B. adusta*, hyphal anastomosis occurs prior to expression of antagonism, but subsequently the cells derived by fusion change in refractive properties and swell to form spindle-shaped structures separated from parent hyphae by septa. Autoradiographic experiments using ^{86}Rb demonstrated that there was no exchange of this ion between paired, antagonistic, isolates of *C. versicolor*, but that there was free exchange between genetically identical colonies, presumably indicating active translocation following hyphal fusion.

Clearly much further work is required, concerning both the nature and occurrence of intraspecific antagonism in natural populations of a wide range of wood-decaying basidiomycetes and other higher fungi, before a full assessment of its importance can be made. From results obtained so far, two features seem particularly significant. Firstly, the quite large number of species, of diverse mating and nuclear behaviour (Boidin, 1971; Kühner, 1977), in which the phenomenon has now been reported suggests that it may be of general rather than exceptional occurrence. Secondly, the fact that monokaryons derived from genetically different, mutually antagonistic, dikaryons are mostly interfertile suggests that the system operates for *vegetative separation of individual mycelia* rather than for reproductive isolation of potential sexual conjugants.

In conclusion we are tempted to suggest that natural populations of most, if not all, sexually reproducing wood-decaying basidiomycetes, and perhaps higher fungi in general, consist of numerous, genetically distinct, individuals (dikaryons in diaphoromictic species), which when in close juxtaposition are maintained separate by intraspecific antagonism. Should this prove true, then population studies of these organisms may acquire new dimensions.

Field identification of individual mycelia within natural populations

The most obvious direct field evidence for the presence of different, mutually antagonistic, individuals is the occurrence of narrow dark interaction zones within wood decayed by a single species. It is important to be able to distinguish these zones from other types of zone line (i.e. narrow dark regions seen as lines in transverse section), and from relic lines left behind by fungi which have been replaced. Generally there seem to be at least four possible causes for zone lines (Rayner & Todd, 1979): (i) interspecific antagonism; (ii) intraspecific antagonism; (iii) laying down of single, closely interwoven sheets of (usually) pigmented mycelium termed pseudosclerotial plates or PSPs (Lopez-Real, 1975) by a single colony, either in the course of its normal growth in wood or in response to environmental stimuli such as fluctuating moisture levels, desiccation and high temperatures; (iv) as a host reaction to fungal invasion of living tissues. This is not an authoritative statement: zone lines remain incompletely understood and their further investigation seems warranted. In identifying zones due to intraspecific antagonism (IA zones for short) the most likely confusion arises over distinguishing between these and zone lines due to PSPs produced by a single colony.

IA zones are generally slightly broader and more diffuse than PSPs and consist of undecayed wood (which nonetheless may be densely filled with mycelium) rather than obvious sheets of mycelium. PSPs, such as those of *Armillaria mellea*, are often easy to dissect from well-rotted wood and generally contain a high proportion of inflated ('bladder') hyphae. IA zones are often much paler than PSPs, most commonly being pale brown. Sometimes, however, IA zones may be darkened due to colonisation by various dematiaceous hyphomycetes including several *Rhinocladiella* spp., the *Catenularia* state of *Chaetosphaeria myriocarpa*, a *Cladosporium* sp., and an *Endophragmiella* sp. (Rayner, 1976). These fungi appear to grow specifically between mutually antagonistic colonies of decay fungi in wood, and their presence is probably diagnostic of interaction zone lines. A further feature of IA zones is that they usually delimit longitudinally extensive columns of decay. In contrast PSPs produced by a single colony are often of much more irregular distribution, and frequently either surround numerous adjacent pockets of decay, each of which is inextensive longitudinally, as with *A. mellea*, or are produced adjacent to exposed wood surfaces. PSPs may, however, be produced sometimes between antagonistic colonies of decay fungi. This may result in two PSPs being formed on either side of an interaction zone, and, if the gap between them is sufficiently wide, could well account for reports of 'double zone lines'. It will be apparent that whilst we think the term 'PSP' usefully describes a single sheet of mycelium, we do not accept the view that it should be applied to all zone lines (Lopez-Real, 1975).

Another opportunity for direct field recognition of individuals within a population arises when species of basidiomycete form closely interwoven mycelial structures widely effused over the wood surface. These may be actual fruit bodies in the case of many resupinate forms, or simply sheets of vegetative mycelium. Barrages are often seen between such structures where different individuals of the same species meet, and can often be shown to correlate very closely in position with narrow dark interaction zones in the underlying wood (Fig. 1*c*).

Fruit bodies of many wood-decaying basidiomycetes show varying degrees of polymorphism which, if genetically based, will reflect differences between individuals. We have investigated this possibility with *Coriolus versicolor*, a species well-known for its polymorphic fruit bodies (Rayner & Todd, 1978). In numerous examples that we examined of wood colonised by *C. versicolor* we found no exception to the generalisation that, where fruit bodies differing in appearance were present at the surface, interaction zones occurred in the wood. The delimited individual decay columns were equivalent in number and position to the various morphological types. In contrast, where uniform collections of fruit bodies occurred, few, if any, interaction zones were found (Figs. 2*a–f*). From this and work with other fungi it appears that genetic differences between individuals in a population of the same species are often externally manifested by polymorphism between their fruit bodies. Therefore an estimate of the number of morphological types can act as a general guide to the number of individuals in a population.

Methods of sampling and analysing natural populations

A potentially exciting aspect of intraspecific antagonism is the opportunity which knowledge of its existence and manifestation provides for identifying the spatial distribution of individual mycelia within natural populations of species of wood-decaying fungi. Lack of such knowledge has significantly restricted the scope of most of the population studies with wood-decaying basidiomycetes so far reported (e.g. Roshal, 1950; Raper, Krongelb & Baxter, 1958; Miles, Takemaru & Kimura, 1966; Brasier, 1970, with *Schizophyllum commune*; Burnett & Partington, 1957 with *Piptoporus betulinus* and *Coriolus versicolor*; Eggertson, 1953 with *Polyporus obtusus* (= *Spongipellis unicolor*); Ullrich & Raper, 1974 with *Sistotrema brinkmannii*; Lindsey & Gilbertson, 1977 with *Poria carnegiea*). Evidence for this can be seen in Ullrich's (1977) discussion of non-random distributions of mating-type alleles amongst multiple isolates of wood-decay fungi from individual substrates. He noted that in their studies with *S. commune* both Roshal (1950) and Raper *et al.* (1958) examined isolates from only one fruit body per substrate, and that as a result mating-type factor repeats occurred with equal frequencies in populations of different size. However, work with the same fungus by Miles *et al.* (1966) included multiple isolations from the same substrate and this resulted in an increased frequency of factor repeats. These and similar factor repeats reported in isolates from individual substrates or from very small areas were interpreted as suggesting either that non-random distribution of spores is restricted to very small areas, or that non-randomness arises from multiple isolations from the same mycelium. Knowledge of the spatial distribution of mycelia would obviously distinguish between these possibilities!

A primary objective in future population studies must therefore be to use sampling procedures which allow for adequate recording of spatial interrelationships between isolates. We like to think that suitable procedures fulfilling this requirement can be based on that illustrated in Fig. 3 which we have used, variously modified, to study population structure in *C. versicolor* and several other basidiomycetes. These procedures were designed for use in very localised populations, as in individual stumps and branches. If larger populations are to be sampled it may be useful to undertake detailed analysis of the mycelial populations within individual substrates at a few well-separated localities, and elsewhere to sample on the basis of one isolate per substrate.

Having adopted suitable sampling procedures, we must then decide on which parameters to study. Here there is little doubt that mating-type factors are particularly suitable genetic markers for population studies with diaphoromictic fungi. Firstly, they can be characterised quickly by simple mating experiments in the laboratory. Further, because they have multiple alleles they can automatically convey far more information about a population than simple biallelic loci. Finally, since only these factors control dikaryon formation and hence production of new individuals, they are primary determinants of population structure. Their distribution may therefore provide insight into the processes occurring during the establishment and maintenance of population structure.

It is apt, therefore, that many population studies so far made with wood-decaying basidiomycetes have concerned the number and distribution of mating-type alleles (Roshal, 1950; Eggertson, 1953; Burnett & Partington, 1957; Raper, *et al.*, 1958; Miles *et al.*, 1966; Ullrich & Raper, 1974; Lindsey & Gilbertson, 1977) and it is to be hoped that such work will be continued. However, fungi do not have only mating-type factors, and for a deeper understanding of their workings in natural populations we must reach to a broader canvas and seek other characters for study. These may include continuously variable attributes such as growth rate and fruiting ability. Simchen & Jinks (1964) first developed suitable biometrical techniques for quantitative investigations of such characters and these were applied to the progeny of small numbers of wild isolates of *S. commune* and *Flammulina velutipes* (Croft & Simchen, 1965; Simchen, 1966). Subsequently, the value of these techniques in a much larger population of *S. commune* was demonstrated by Brasier (1970) and Williams, Verma, Jinks & Brasier (1976).

A remarkably neglected feature of fungi, which is of potential value for future population studies, is polymorphism. This has often been used very successfully in higher organisms, where, amongst other things, it has sometimes had a considerable bearing on problems of taxonomic status and ecological identity – as we may expect it to do with fungi also (cf. Watling: Chapter 1). In the first instance polymorphism studies may be done at the gross morphological level, especially where very obvious polymorphisms occur, as between fruit bodies of *C. versicolor* (Rayner & Todd, 1978). Increasingly, however, more sophisticated chromatographic and electrophoretic procedures may become worthwhile.

Concluding comments

There would seem to be little doubt now that the communal exploitation of substrates by physiologically unitary mycelia envisaged by Buller (1931) and Burnett & Partington (1957) (p. 111) is questionable, at least where genetically different entities are involved. It seems more plausible instead to envisage that wood colonised by a single basidiomycete species may often contain populations of mutally antagonistic individuals between which little or no exchange of organelles, cytoplasm or nutrients can occur. We are just beginning to obtain confirmation of this from various physiological experiments, but some direct field evidence has been obtained with *Coriolus versicolor*. Here we noticed, in a range of woody substrates, that where the spacing between interaction zones was large, and the amount of wood occupied by each individual correspondingly great, numerous sizeable fruit bodies are frequently present (Fig. 2*a,b*). Conversely, when the spacing is small, numerous but very much smaller, and often stunted, fruit bodies occur (Fig. 2*c,d*). (Rayner & Todd, 1977).

It may be fair to suggest that knowledge of the existence of intraspecific competition between individuals in basidiomycete populations completely alters the ecological picture of processes occurring in decaying wood, and it seems very likely that this will soon be extended to other substrates, including litter. So far, *inter*specific competition has had considerable emphasis, particularly as a mechanism directing successional changes or community development (Rayner & Todd, 1979; see also Hedger & Basuki: Chapter 15), but intraspecific competition may soon be regarded as playing at least an equal rôle in decomposition processes.

References

Adams, D. H. & Roth, L. F. (1967). Demarcation lines in paired cultures of *Fomes cajanderi* as a basis for detecting genetically distinct mycelia. *Canadian Journal of Botany*, **45**, 1583–9.

Adams, D. H. & Roth, L. F. (1969). Intraspecific competition among genotypes of *Fomes cajanderi* decaying young growth Douglas-fir. *Forest Science*, **15**, 327–31.

Barrett, D. K. & Uscuplic, M. (1971). The field distribution of interacting strains of *Polyporus schweinitzii* and their origin. *New Phytologist*, **70**, 581–98.

Boidin, J. (1971). Nuclear behaviour in the mycelium and the evolution of the Basidiomycetes. In *Evolution in the Higher Basidiomycetes*, ed. R. H. Petersen, pp. 129–48. Knoxville: University of Tennessee Press.

Brasier, C. M. (1970). Variation in a natural population of *Schizophyllum commune. American Naturalist*, **104**, 191–204.

Brodie, H. J. (1935). The occurrence in nature of mutual aversion between mycelia of hymenomycetous fungi. *Canadian Journal of Research*, **13**, 187–9.

Buller, A. H. R. (1931). *Researches in Fungi*, vol. 4, London: Longmans Green.

Burnett, J. H. & Partington, M. (1957). Spatial distribution of mating-type factors. *Proceedings of the Royal Physical Society of Edinburgh*, **26**, 61–8.

Campbell A. H. (1938). Contribution to the biology of *Collybia radicata* (Relh.) Berk. *Transactions of the British Mycological Society*, **22**, 151–9.

Childs, T. W. (1937). Variability of *Polyporus schweinitzii* in culture. *Phytopathology*, **27**, 29–50.

Childs, T. W. (1963). *Poria weirii* root rot. In *Symposium on root diseases of forest trees Corvallis, Oregon, 1962. Phytopathology*, **53**, 1124–7.

Croft, J. H. & Simchen, G. (1965). Natural variation among monokaryons of *Collybia velutipes. American Naturalist*, **99**, 451–62.

Eggertson, E. (1953). An estimate of the number of alleles at the loci for heterothallism in a local concentration of *Polyporus obtusus. Canadian Journal of Botany*, **31**, 710–59.

Kühner, R. (1977). Variation of nuclear behaviour in the homobasidiomycetes. *Transactions of the British Mycological Society*, **68**, 1–16.

Lindsey, J. Page & Gilbertson, R. L. (1977). Some aspects of bipolar heterothallism and other cultural characters of *Poria carnegiea. Mycologia*, **69**, 761–72.

Lopez-Real, J. M. (1975). Formation of pseudosclerotia ('zone lines') in wood decayed by *Armillaria mellea* and *Stereum hirsutum*. 1. Morphological aspects. *Transactions of the British Mycological Society*, **64**, 465–71.

Miles, P. G., Takemaru, T. & Kimura, K. (1966). Incompatibility factors in a natural population of *Schizophyllum commune*. 1. Analysis of the incompatibility

factors present in fruit bodies collected within a small area. *Botanical Magazine, Tokyo*, **79**, 693–705.
Mounce, I. (1929). Studies in forest pathology. II. The biology of *Fomes pinicola* (Sw.) Cooke. *Bulletin of the Canadian Department of Agriculture*, **111**, 1–77.
Raper, J. R. (1966). *Genetics of Sexuality in Higher Fungi.* New York: Ronald Press.
Raper, J. R., Krongelb, G. S. & Baxter, M. G. (1958). The number and distribution of incompatibility factors in *Schizophyllum. American Naturalist*, **92**, 221–32.
Rayner, A. D. M. (1976). Dematiaceous hyphomycetes and narrow dark zones in decaying wood. *Transactions of the British Mycological Society*, **67**, 546–9.
Rayner, A. D. M. & Todd, N. K. (1977). Intraspecific antagonism in natural populations of wood-decaying basidiomycetes. *Journal of General Microbiology*, **103**, 85–90.
Rayner, A. D. M. & Todd, N. K. (1978). Polymorphism in *Coriolus versicolor* and its relation to interfertility and intraspecific antagonism. *Transactions of the British Mycological Society*, **71**, 99–106.
Rayner, A. D. M. & Todd, N. K. (1979). Population and community structure and dynamics of fungi in decaying wood. *Advances in Botanical Research*, **7**, 333–420.
Rayner, R. W. (1970). *A Mycological Colour Chart.* Kew, Surrey: Commonwealth Mycological Institute and British Mycological Society.
Roshal, J. Y. (1950). Incompatibility factors in a population of *Schizophyllum commune.* PhD thesis, University of Chicago.
Roth, L. F. (1952). Factors affecting the rate of decay of old-growth Douglas-fir by *Fomes pini.* (Abstract). *Phytopathology*, **42**, 518–9.
Schmitz, H. (1925). Studies in wood decay. V. Physiological specialization in *Fomes pinicola* Fr. *American Journal of Botany*, **12**, 163–77.
Simchen, G. (1966). Fruiting and growth rate among dikaryotic progeny of single wild isolates of *Schizophyllum commune. Genetics*, **53**, 1151–65.
Simchen, G. & Jinks, J. L. (1964). The determination of dikaryotic growth rate in the Basidiomycete *Schizophyllum commune*: a biometrical analysis. *Heredity, London*, **19**, 629–49.
Todd, N. K. & Rayner, A. D. M. (1978). Genetic structure of a natural population of *Coriolus versicolor* (L. ex Fr.) Quél. *Genetical Research*, **32**, 55–65.
Ullrich, R. C. (1977). Natural distribution of incompatibility factors in basidiomycetous fungi. *Mycologia*, **69**, 714–9.
Ullrich, R. C. & Raper, J. R. (1974). Number and distribution of bipolar incompatibility factors in *Sistotrema brinkmannii. American Naturalist*, **108**, 507–18.
Williams, S., Verma, M. M., Jinks, J. L. & Brasier, C. M. (1976). Variation in a natural population of *Schizophyllum commune. Heredity, London*, **37**, 365–75.

Note added in proof: For recent references pertinent to this Chapter and Chapter 7 see list on p. 355.

7 Ecological genetics of basidiomycete populations in decaying wood

A. D. M. RAYNER* and N. K. TODD†

*School of Biological Sciences, University of Bath, Claverton Down, Bath, England

†Department of Biological Sciences, University of Exeter, Washington Singer Laboratories, Perry Road, Exeter, England

Introduction

Population studies provide a meeting ground for geneticists and ecologists where each can benefit from the other's approach. For the fungal geneticist, work with a variety of isolates obtained from natural populations can often yield unexpected and interesting results. It may even be that, on occasions, an entirely new perspective may be provided of traditional concepts which have been developed as a result of work with just one or a few isolates. A good example, albeit not relating specifically to basidiomycetes, concerns the occurrence of heterokaryosis and parasexuality in fungi. These phenomena had often been thought (and still are by some) to be a common and important source of flexibility in natural populations of many ascomycetes and fungi imperfecti. However, Caten & Jinks (1966) challenged this assumption on the basis that most studies had been conducted either by forcing between mutants with complementary nutritional requirements, i.e. auxotrophs, or by using mutant strains derived from the same isolate and differing only at a few genetic markers. For example, the mutant strains of *Aspergillus nidulans* used by Pontecorvo and his associates (Pontecorvo *et al.*, 1953) were derived from a single isolate, designated A69 by Yuill (1939, 1950) and NRRL 194 by Thom & Raper (1945). When the natural population of this species was sampled (Grindle, 1963*a,b*) it was discovered that many of the isolates did not readily produce heterokaryons when paired, i.e. they were heterokaryon-incompatible. Heterokaryon incompatibility (sometimes also referred to as vegetative incompatibility) has since been demonstrated in many other aspergilli (Caten, 1971) as well as in other ascomycetous fungi such as the destructive pathogen of American chestnut (*Castanea dentata*), *Endo-*

thia parasitica (Anagnostakis, 1977), and shown to be due to genetical differences between isolates. The position with *A. nidulans* has recently been reviewed by Croft & Jinks (1977) who stated that in the wild this species is divided into a number of subpopulations, each of which is a clonally related group of strains in which evolution may be proceeding independently, little gene exchange occurring between groups.

This work on heterokaryon incompatibility in ascomycetous fungi closely parallels certain recent developments with basidiomycetes (Rayner & Todd, 1979), and highlights the value of ecological considerations to fungal geneticists. However, in this chapter we shall be more concerned with the ways in which a genetically based approach to population studies can be of use to the basidiomycete ecologist. We hope to show that this approach can give insight into such important determinants of the ecology of basidiomycetes as their means of perennation, propagation and dispersal, and patterns of establishment in a substrate. Whilst our discussion will focus primarily on basidiomycetes causing wood decay, many of the arguments will be equally appropriate for litter decomposers. Our discussion is divided into four closely interconnected themes concerning outbreeding and inbreeding mechanisms, genetical variability, modes of propagation and survival, and patterns of establishment.

For descriptions of such terms as 'population' and 'individual', the reader is referred to Chapter 6.

Functioning of systems for outbreeding and inbreeding in natural populations of basidiomycetes

In homobasidiomycetes a variety of mechanisms can determine the types of nuclei which associate and ultimately fuse prior to meiosis. The best known are the mating systems, of which according to Burnett (1975) three types, *diaphoromixis, homodiaphoromixis* and *homomixis*, are prevalent (for definitions see Rayner & Todd: Chapter 6). Burnett's nomenclature is preferred here to the more traditional 'homothallism' and 'heterothallism' because it distinguishes more adequately between the types of behaviour. In diaphoromictic species an essential feature is that the mating-type loci (designated A factors in bipolar species, A and B factors in tetrapolar ones) can occur as numerous different alleles within the natural population. This contrasts with *dimixis* found in other fungi where only two mating-type alleles occur, and allows an *outbreeding bias* (defined by Burnett (1965) as the ratio of potential non-sister to sister matings) approaching 2.0 (bipolar) or 4.0 (tetrapolar) to be developed within the population.

Diaphoromixis effectively strongly favours outbreeding and occurs in most homobasidiomycetes (*c*. 90% according to Raper, 1966). In contrast, the other patterns, homodiaphoromixis and homomixis, may result in a greater or lesser degree of inbreeding. In homodiaphoromictic species, whilst a proportion (usually small) of basidiospores may germinate to produce homokaryons with bipolar or tetrapolar incompatibility, others form dikaryons directly, being heterokaryotic for mating-type. These fungi often have two-spored basidia, each spore receiving two compatible nuclei. As the name implies, homodiaphoromixis is therefore clearly derived from diaphoromixis, but this is not the case with homomictic species in which single, homokaryotic, basidiospore isolates are capable of sexual reproduction.

Inbreeding, in addition, sometimes may be favoured by a mechanism of sexual incompatibility that is due to genetical *differences* between strains, i.e. heterogenic. Such incompatibility may be truly a reproductive isolating mechanism, resulting in inter-sterile groups within species, and ultimately perhaps in new species. It should not be confused (as it often is) with *vegetative* incompatibility resulting in delimitation of individuals, but not precluding sexual conjugation, and which may also be heterogenic. Inter-sterile groups are known in several wood-decaying basidiomycetes including *Fomitopsis pinicola* (Mounce & Macrae, 1938), *Hirneola auricula-judae* (Barnett, 1937), *Serpula himantioides* (Harmsen, 1960) and *Armillaria mellea* (Ullrich & Anderson, 1978; Korhonen, 1978). Whilst it is assumed that the basis for these inter-sterile groups is heterogenic, in no case has the system been worked out in such detail as in Esser's classical studies with the ascomycete *Podospora anserina* (e.g. Esser, 1965).

A primary objective of population studies should be to obtain information concerning the functioning of these various systems for outbreeding and inbreeding in nature. By determining the types of individual which can exchange genetic information these systems will be instrumental in directing patterns of establishment, attainment and maintenance of equilibrium, and changes in population structure. By the same token, their manner and efficiency of operation will be affected by such factors as distances and patterns of basidiospore dispersal, spore germinability and the relative occurrence of asexual as opposed to sexually based means of propagation within local populations. Information about these other factors may therefore be provided indirectly by knowledge of the functioning of the incompatibility systems, and vice versa.

If we consider diaphoromixis first, its efficiency as an outbreeding mechanism can be influenced by both genetical and ecological factors. The main genetical considerations are the numbers of mating-type alleles available in populations, and the possibility of recombination occurring within the mating-type loci. Both of these can be determined readily, if tediously, by simple laboratory pairings between wild isolates. Regarding the first point, Burnett (1965) has suggested that once 20–30 mating-type alleles are available in the population, the outbreeding bias asymptotically approaches its maximum value of 2 or 4, noting in passing that the much larger numbers of alleles found in many species must effectively be redundant in this respect. The question of possible recombination within the mating-type loci arises from the fact that in tetrapolar species the A and B factors have been shown to be subdivided into α and β subloci. Recombination between these generates different factors. If linkage between the subloci is weak, this may significantly increase the inbreeding potential of a population by allowing a larger number of compatible sister matings. Such linkage may be genetically controlled, e.g. in *Schizophyllum commune*, but here the dominant alleles for low recombination are the most frequent in natural populations, so that the inbreeding potential is about 25% (see Burnett, 1975).

Such concepts as outbreeding and inbreeding potentials and outbreeding bias are inevitably abstract if based on theoretical considerations only, and it is important to recognise that they may be considerably influenced by ecological factors. It is therefore vital to understand the workings of very local populations adequately before attempting to apply such concepts to the natural situation. Thus restrictions on wind-dissemination of basidiospores may be far more important in determining the levels of inbreeding and outbreeding than the nature of the mating system. We can visualise this by imagining, for example, a tree stump suddenly exposed to the elements. How heterogeneous will be the 'spore rain' descending onto its surface? Is it likely that an even mixture from different dikaryons arrives, or are spores from a nearby fruit body of a single dikaryon likely to predominate? Observations of aerobiologists such as Kramer (Chapter 2) on basidiospore dispersal and the fates of spore clouds may help here.

Whilst with diaphoromictic species we need to discover how effectively the immense genetical potential for outbreeding and variation is exploited naturally, with systems favouring (or at least not restricting) inbreeding we wish to know the extent to which genetical exchange occurs *in spite of* the breeding system. Here knowledge of variation in local populations may help to provide an answer.

Homomictic species provide particularly interesting possibilities for study. Theoretically a fruit body developing from a homokaryotic haploid mycelium must inevitably produce homozygous (genetically identical) progeny. Unless there is opportunity for some form of genetical exchange prior to meiosis, the tendency would therefore be towards genetically uniform populations or subpopulations. These possibilities can readily be checked by comparisons of the progeny of single fruit bodies, or assessment of variation within natural populations. Little work of this kind seems to have been done. Burnett (1975) suggested that in certain homomictic ascomycetes such as *Sordaria fimicola* and *Aspergillus nidulans* a 'variety of situations and rather imprecise regulating systems can result in heterozygosity'. However, the evidence was based on closely related strains, capable of heterokaryosis.

Amongst wood-decaying basidiomycetes, members of the genus *Stereum* have generally been held to be 'homothallic'; most of them exhibit 'holocoenocytic' behaviour, such that the mycelium is plurinucleate and has verticillate (whorled) clamp connections (Boidin, 1971). In preliminary studies with *Stereum hirsutum* (Rayner & Todd, 1979 and unpublished) we found that this species frequently forms populations containing individual mutually antagonistic mycelia in nature (Chapter 6, Fig. 1*a,d*). We were anxious to trace the source of the genetical heterogeneity which we presumed to be the basis of the antagonism. Early studies showed that variable amounts of antagonism occurred when single-spore isolates from the same fruit body were paired in culture. When several isolates from the same fruit body were spaced regularly on a plate, a repeatable mosaic pattern of interacting and non-interacting colonies developed, some interaction zones apparently developing *after* fusion between certain isolates (Rayner & Todd, 1979). Very recently, further studies (Coates, Rayner & Todd, 1981) with a range of isolates have provided a possible explanation for these findings. These have indicated that *S. hirsutum* generally behaves as a heteromictic species with a modified bipolar diaphoromictic mating pattern. Monosporous primary mycelia give rise to a morphologically distinctive secondary mycelial phase in compatible matings between isolates from the same or different fruit bodies.

Thanatephorus cucumeris (= *Rhizoctonia solani*), which is also plurinucleate, is, so far as we know, the only other allegedly homomictic basidiomycete to have been subjected in any detail to the type of analysis we have described. Here single-basidiospore isolates from any one homokaryon are uniform morphologically. However, field isolates generally produce basidiospores giving rise to cultures of variable morphology, and so are probably heterokaryotic (Bolkan & Butler, 1974). Genetical control of heterokaryosis appears to be rather complex. Anastomosis between isolates can only occur if they belong to the same anastomosis group (AG). So far four AGs have been identified (Parmeter, Sherwood & Platt, 1969), within which heterokaryons can only be formed between isolates with different so-called H-factors (Anderson, Stretton, Groth & Flentje, 1972). Parmeter *et al.* (1969) noted an antagonistic or lethal reaction following anastomosis between all their wild heterokaryotic isolates within an AG group. Anderson *et al.* (1972) noted this reaction also in certain homokaryotic pairings. The mechanism controlling this lethal reaction is apparently unrelated to the H-factor system controlling heterokaryosis (Anderson *et al.*, 1972). At one time it was believed that only closely related compatible homokaryons would form heterokaryons (Flentje, Stretton & McKenzie, 1970), but Bolkan & Butler (1974) have since obtained evidence that certain heterokaryotic field isolates could interact with each other to produce new heterokaryons. The latter grew out as tufts of aerial mycelium between parent colonies confronted in agar culture.

More work is obviously required to clarify this rather confused situation, but the results so far obtained clearly indicate the potential of this type of approach for understanding the functioning and structure of natural populations of homomictic fungi.

Genetical variability in local populations

This subject is intimately connected with the operation of breeding mechanisms, and as such we have inevitably touched on it previously. Suffice it here to remind ourselves that knowledge of genetical variability is vital to understanding the functioning of fungi and the action of selection pressures in natural populations, and to emphasise the fact that, so far, little real attempt has been made to study such variation systematically. An indication of the value of variation studies is seen in the work mentioned earlier on *Aspergillus nidulans* reviewed by Croft & Jinks (1977). Here it was found that within each heterokaryon incompatibility group there was very little variation in such characteristics as growth rate and penicillin titre, although there was wide variation between groups. It was for this reason that each group was considered to be a 'clonally related group of strains', sexual reproduction, especially between groups, being of little or no significance in natural populations.

So far the only reasonably detailed study of variation within a natural population of a wood-decaying basidiomycete is that of Brasier (1970). Here 77 dikaryotic isolates were obtained from a large but local population of *Schizophyllum commune.* It was shown that considerable variation in growth rates, fruiting characteristics and cultural morphology was present. Evidence was obtained that growth rate was subject to stabilising selection (i.e. with extreme variations tending to be eliminated), whilst directional selection, with a stabilising component, was indicated for early fruiting and small numbers of fruit bodies.

We have suggested that for a fungus such as *Coriolus versicolor*, polymorphism may be a useful indicator of genetic variability. We have not yet undertaken any systematic surveys, but it would seem that the amount of variation occurring amongst fruit bodies varies in different localities, especially concerning the occurrence of widely differing forms on the same substrate. So far, we think that the situation (illustrated in Chapter 6, Fig. 2*c,d*) where large numbers of very different individuals occur on the same stump may be rather unusual. Such patterns may be explained by a combination of high genetical variability of local spore sources and favourable conditions for establishment at the time of spore germination (Rayner & Todd, 1978).

Modes of propagation and survival

Basically, two modes of propagation and survival are available to basidiomycetes: sexual and asexual. The sexual mode is characterised by production and liberation of basidiospores. The asexual mode can involve production of asexual spores (oidia, conidia, chlamydospores) which may be dispersed or remain *in situ.* Alternatively sclerotia may be formed or the mycelium itself may function as an agent of spread, or at least perennation; production of mycelial cords or rhizomorphs by many wood-decaying fungi appears to be an important means of spread across nutritionally inert substrates (Jennings: Chapter 5).

It is obviously important in understanding the ecology of wood-decaying basidiomycetes to be able to distinguish between these possibilities, and here population studies again offer a useful approach. We suggest that, in general, predominance of the sexual mode will result in numerous, genetically distinct, individuals within the population, whilst the asexual mode will not. If we remind ourselves of the work with *Aspergillus nidulans*, a noteworthy feature, apart from the lack of variation within each heterokaryon incompatibility group, was that isolates from the same group could be found in geographically widely separated locations in the British Isles (Croft & Jinks, 1977). This is probably a consequence of the dominance of the asexual mode in the life cycle, and associated long-range dispersal.

The application of population studies to these questions, and incidentally to identification of the spatial delimitation of individual mycelia in nature, thus presents exciting, but largely unexploited prospects. Two approaches have so far been used. The first has involved investigation of the distribution of mating-type factors in closely associated individual fruit bodies. Thus Burnett & Partington (1957) found that in *Coprinus comatus, Flammulina velutipes* and *Hypholoma fasciculare* the same factors were recovered from all fruit bodies collected from a specific location. Similar findings were later obtained by Burnett & Evans (1966) for fruit bodies within individual fairy rings of *Marasmius oreades*, and were interpreted as evidence for single extensive dikaryons. For *M. oreades* these must have been both very extensive and highly stable – their genetic integrity being maintained for up to 100 years or more. As we have seen in the previous chapter (p. 111), very different observations and interpretations were made for *Coriolus versicolor* and *Piptoporus betulinus*.

A much simpler approach has been to use intraspecific antagonism (see previous chapter), and, provided the genetical basis for antagonistic effects is properly understood, this is highly attractive. The basis of this approach is simply that whilst antagonism indicates genetical dissimilarity between isolates, ready intermingling of colonies indicates their genetical identicality. We have noticed that for fungi colonising predominantly aerially-exposed wood surfaces, such as *Stereum hirsutum, Coriolus versicolor* and *Bjerkandera adusta*, numerous intraspecific interaction zone lines are often present in the wood, whilst for those colonising as mycelium from soil or litter (e.g., as mycelial cords) such zones are less common, e.g. *Hypholoma fasciculare, Phallus impudicus, Phanerochaete velutina* and *Tricholomopsis platyphylla* (Carruthers & Rayner, 1979; Rayner & Todd, 1979).

Childs (1963) was probably the first to use antagonism to define the mode of spread of a decay fungus (*Phellinus weirii*) in a stand of Douglas fir (*Pseudotsuga menziesii*). He found groups of trees, sometimes more than 50 m wide, isolates from within a group of which all fused readily, but between which exhibited antagonism in the form of a dark demarcation line when paired in culture. Each group was therefore described as containing a 'clone' originating presumably from vegetative mycelium of a single colony. More recently, Adams (1974) provided evidence, based on dark demarcation zone formation in culture, for the existence of three 'clones' of *Armillaria mellea* amongst 243 isolates of the fungus collected from a field site. However, Rishbeth (1978) did not observe any obvious interaction between isolates from different infection foci of *A. mellea* in East Anglia, and Ullrich & Anderson (1978) have reported that formation of a dark interaction zone occurs only in pairings between *sexually inter-sterile* groups of *A. mellea*. This confused situation may be clarified by some important recent work by Korhonen (1978). He distinguished among at least five inter-sterile 'biological species' within what is regarded as the *A. mellea* complex, including *A. mellea* sensu stricto, *A. bulbosa* (= *Armillariella bulbosa*) and three more, designated A, B and C, present in Finland. This compared with the six inter-sterile groups identified by Ullrich & Anderson (1978) in North America. According to Korhonen, whilst a dark demarcation zone is characteristic of pairings between inter-sterile groups, pairings between genetically different isolates of the same sterility group often result in a zone of demarcation (= antagonism in our terminology) lacking pigmentation. Using this feature, and analysis of mating-type factors, Korhonen was able to identify six clones of *A. mellea*, the largest being 120–150 m in diameter, at a heavily infected locality in a mixed *Picea, Betula* and *Pinus* forest near Helsinki. These and similar studies could shed new light on the biology of this important pathogen, which is unusual amongst basidiomycetes in having an extended diploid phase in its life cycle (Korhonen & Hintikka, 1974).

Following Childs' work, Barrett (1967) and Barrett & Uscuplic (1971) have used antagonism in a study of the aetiology of infection by *Phaeolus schweinitzii* in British forests. It was found that isolates from different, closely adjacent trees were invariably antagonistic (contrasting with Childs' observations with *Phellinus weirii*) although those obtained from within each tree merged readily in culture. The inference was that, whilst each tree contained a single mycelium, vegetative mycelial spread of a single colony between trees could not explain the infection patterns. It was therefore thought that basidiospores in the soil might be the infective agents. This would be a surprising and important finding, since a root-infecting pathogen, especially of a woody host, might be expected to require greater inoculum potential for invasion than could be conferred by basidiospores (cf. Garrett, 1970).

Patterns of establishment in woody substrates

The numbers and spatial distribution of antagonistic individual mycelia within a piece of wood must inevitably reflect patterns of establishment and colonisation from the surface. These in turn can

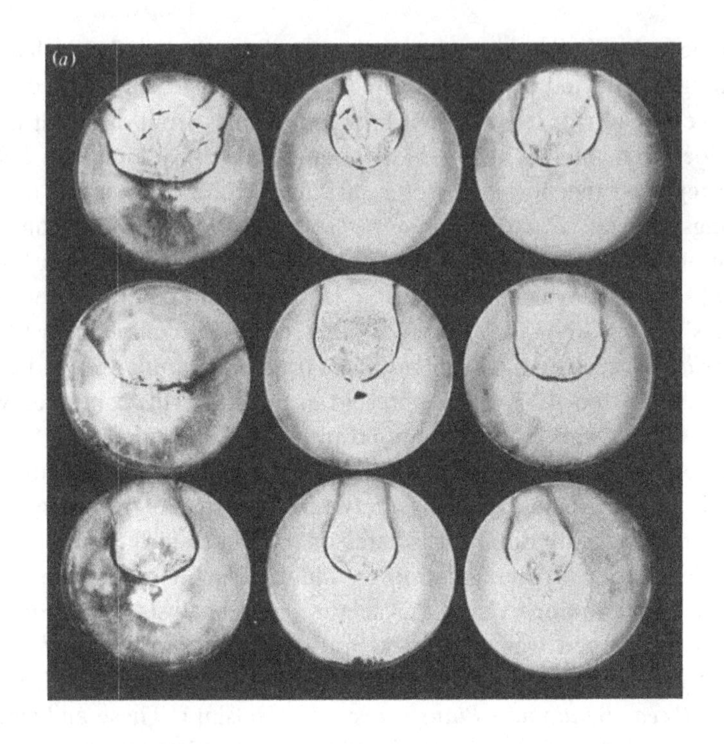
(a)

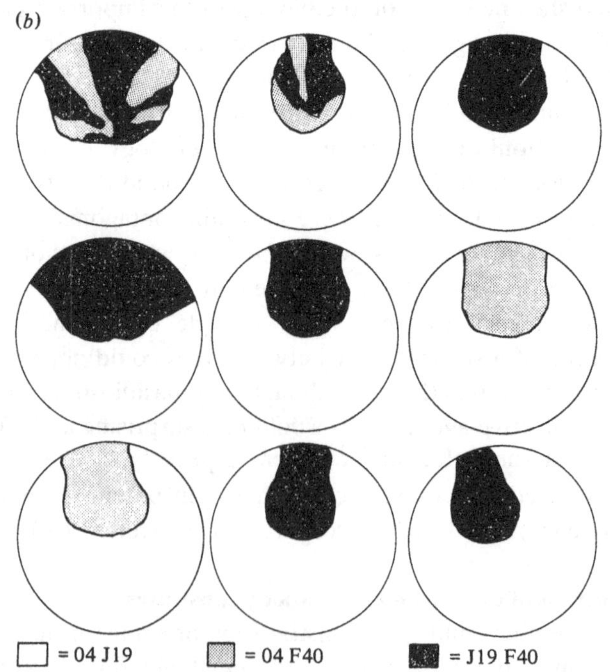
(b)
= 04 J19
= 04 F40
= J19 F40

depend on the type of inoculum – whether spores or mycelium (see previous section), modes of growth of mycelia in or on wood, patterns of dikaryotisation and genetical exchange, and competitive and environmental effects.

Regarding the pattern of growth of mycelia in or on wood, it has been noticed (e.g. Rayner, 1977, 1979; Carruthers & Rayner, 1979) that many decomposer basidiomycetes may be grouped into one or other of two ecological categories. In one group rapid, subcortical, mycelial spread is possible, precluding growth of more than one or a few colonies, and allowing extensive occupation of the peripheral regions of woody substrates with few or no interaction zone lines developing, e.g. *Hypholoma fasciculare.* In other species, e.g. *Coriolus versicolor* and *Stereum hirsutum*, such subcortical growth does not occur so readily, and these usually occupy relatively narrow columns of decay, of which there may be many (Chapter 6, Figs. 1 and 2).

In studies of polymorphism in *C. versicolor* (Rayner & Todd, 1979) we noted that the presence of numerous individuals in a stump (Chapter 6, Fig. 2*c,d*) might reflect favourable conditions at the time of spore germination, whilst occurrence of one or a few dikaryons in uniformly decayed substrates might indicate either limited local spore sources at the time of colonisation, or unfavourable conditions. We found it difficult to understand how rather large substrates (Chapter 6, Fig. 2*e,f*) could apparently sometimes be colonised by a single dikaryon. Since a dikaryon is formed by fusion of two compatible monokaryons it is difficult to see why, if one such pairing is achieved, others should not occur at the same time. We suggested the possibility of competitive effects or preferential germination of basidiospores where monokaryons are already established. Barrett & Uscuplic (1971) found it equally difficult to explain the occurrence of only one mycelium of *Phaeolus schweinitzii* per tree. They suggested that whilst several interacting mycelia might colonise the roots, the one which first reached the stem base might be in a position to monopolise development in the main stem.

An extremely interesting observation reported by Barrett & Uscuplic,

---

Fig. 1. (*a*) Results of nine identical di-mon matings between a dikaryon (monokaryotic components from different parents, designated O4 and J19) and a monokaryon (designated F40, i.e. from a different parent from O4 and J19) of *Coriolus versicolor* on 3% malt agar. Several pairings show track-formation (arrowed). (*b*) Interpretation of patterns of dikaryotisation. (From Rayner & Todd, 1979.)

which might have a bearing on possible competitive effects and establishment at the wood surface, concerned the behaviour of complex polyspore isolates of *P. schweinitzii*. These isolates were obtained by allowing germinated spores of a seeded plate to intermingle for up to 20 days before subculturing. They exhibited abnormal, weak, growth (described as 'appressed') similar to that obtained by isolating from interaction zones between antagonistic colonies, but quite unlike 'normal' isolates from wood or fruit-body tissue. They were not antagonistic when paired in culture. It seems not impossible that many such mycelia might be formed during initial stages of colonisation, and that any 'normal' dikaryotic unions would be at a selective advantage against them.

These considerations raise the question of dikaryon formation at the wood surface. Does this generally occur by chance fusions between compatible homokaryons, or are more complex situations possible? One obvious possibility is the occurrence of dikaryon–monokaryon (di–mon) matings. Although we have not yet been able to confirm this in our own studies, several reports (reviewed by Ullrich, 1977) suggest a greater frequency of mating-type factor repeats amongst multiple isolates from individual substrates than in the population as a whole. One explanation is the occurrence of di–mon matings, which would generate different dikaryons with nuclei in common. Certainly the results of Burnett & Partington (1957) for *Coriolus versicolor* and *Piptoporus betulinus* are amenable to this interpretation (discussed in Chapter 6, p. 111).

We have investigated the possibility of di–mon mating in *C. versicolor* (Todd & Rayner, 1978; Rayner & Todd, 1979; R. Aylmore, N. K. Todd & A. D. M. Rayner, unpublished), and confirmed that it occurs readily in culture. A particularly interesting observation was made in pairings where the monokaryon contained a nucleus differing from both nuclei in the dikaryon. Here a zone of antagonism develops along the line of contact between the colonies, and further such zones ('tracks') develop between dikaryotised sectors in the monokaryon. These sectors are of alternating genetic constitution, each containing one or other of the nuclei present in the dikaryotic parent. Various patterns obtained in a single such combination and their interpretation are shown in Fig. 1. Recently we have made similar observations of track formation in several other fungi, including *Bjerkandera adusta, Coprinus cinereus* and *Piptoporus betulinus* (N. K. Todd, A. D. M. Rayner and others, unpublished data). It is a fascinating conjecture that such processes could occur during establishment of natural populations. One implication, if they do, is that dikaryons may generate their own competitors by contributing nuclei to adjacent monokaryons!

It will be evident, incidentally, from these observations that routine use of polyspore isolates in studies of wood-decaying basidiomycetes is to be regarded with suspicion, since, being genetically heterogeneous, they are likely to behave inconsistently.

References

Adams, D. H. (1974). Identification of clones of *Armillaria mellea* in young-growth Ponderosa pine. *Northwest Science*, **48,** 21–8.

Anagnostakis, S. L. (1977). Vegetative incompatibility in *Endothia parasitica. Experimental Mycology*, **1,** 306–16.

Anderson, N. A., Stretton, H. M., Groth, J. V. & Flentje, N. T. (1972). Genetics of heterokaryosis in *Thanatephorus cucumeris. Phytopathology*, **62,** 1057–65.

Barnett, H. L. (1937). Studies on the sexuality of the Heterobasidiae. *Mycologia*, **29,** 626–49.

Barrett, D. K. (1967). The biology of *Polyporus schweinitzii* Fr. in British forests. Unpublished Diploma thesis, Commonwealth Forestry Institute, Oxford.

Barrett, D. K. & Uscuplic, M. (1971). The field distribution of interacting strains of *Polyporus schweinitzii* and their origin. *New Phytologist*, **70,** 581–98.

Boidin, J. (1971). Nuclear behaviour in the mycelium and the evolution of the Basidiomycetes. In *Evolution in the Higher Basidiomycetes*, ed. R. H. Petersen, pp. 129–48. Knoxville: University of Tennessee Press.

Bolkan, H. A. & Butler, E. E. (1974). Studies on heterokaryosis and virulence of *Rhizoctonia solani. Phytopathology*, **64,** 513–22.

Brasier, C. M. (1970). Variation in a natural population of *Schizophyllum commune. American Naturalist*, **104,** 191–204.

Burnett, J. H. (1965). The natural history of recombination systems. In *Incompatibility in Fungi*, ed. K. Esser & J. R. Raper, pp. 98–113. Berlin: Springer-Verlag.

Burnett, J. H. (1975). *Mycogenetics*. London, New York, Sydney & Toronto: John Wiley & Sons.

Burnett, J. H. & Evans, E. J. (1966). Genetical homogeneity and the stability of the mating-type factors of 'fairy rings' of *Marasmius oreades. Nature, London*, **210,** 1368–9.

Burnett, J. H. & Partington, M. (1957). Spatial distribution of mating-type factors. *Proceedings of the Royal Physical Society of Edinburgh*, **26,** 61–8.

Carruthers, S. M. & Rayner, A. D. M. (1979). Fungal communities in decaying hardwood branches. *Transactions of the British Mycological Society*, **72,** 283–9.

Caten, C. E. (1971). Heterokaryon incompatibility in imperfect species of *Aspergillus. Heredity, London*, **26,** 299–312.

Caten, C. E. & Jinks, J. L. (1966). Heterokaryosis: its significance in wild homothallic Ascomycetes and Fungi Imperfecti. *Transactions of the British Mycological Society*, **49,** 81–93.

Childs, T. W. (1963). *Poria weirii* root rot. In *Symposium on root diseases of forest trees, Corvallis, Oregon, 1962. Phytopathology*, **53,** 1124–7.

Coates, D., Rayner, A. D. M. & Todd, N. K. (1981). Mating behaviour, mycelial antagonism and the establishment of individuals in *Stereum hirsutum. Transactions of the British Mycological Society*, **76,** 41–51.

Croft, J. H. & Jinks, J. L. (1977). Aspects of the population genetics of *Aspergillus nidulans*. In *Genetics and Physiology of Aspergillus*, ed. J. E. Smith & J. A. Pateman, pp. 339–60. London, New York & San Francisco: Academic Press.
Esser, K. (1965). Heterogenic incompatibility. In *Incompatibility in Fungi*, ed. K. Esser & J. R. Raper, pp. 6–13. Berlin: Springer-Verlag.
Flentje, N. T., Stretton, H. M. & McKenzie, A. R. (1970). Mechanism of variation in *Rhizoctonia solani*. In *Rhizoctonia solani: biology and pathology*, ed. J. R. Parmeter, pp. 52–65. Berkeley: University of California Press.
Garrett, S. D. (1970). *Pathogenic root-infecting fungi*. Cambridge University Press.
Grindle, M. (1963*a*). Heterokaryon compatibility of unrelated strains in the *Aspergillus nidulans* group. *Heredity, London*, **18**, 191–204.
Grindle, M. (1963*b*). Heterokaryon compatibility of closely related wild isolates of *Aspergillus nidulans*. *Heredity, London*, **18**, 397–405.
Harmsen, L. (1960). Taxonomic and cultural studies on brown-spored species of the genus *Merulius*. *Friesia*, **6**, 233–77.
Korhonen, K. (1978). Interfertility and clonal size in the *Armillariella mellea* complex. *Karstenia*, **18**, 31–42.
Korhonen, K. & Hintikka, V. (1974). Cytological evidence for somatic diploidization in dikaryotic cells of *Armillariella mellea*. *Archives of Microbiology*, **95**, 187–92.
Mounce, I. & Macrae, R. (1938). Infertility phenomena in *Fomes pinicola*. *Canadian Journal of Research, C*, **16**, 364–76.
Parmeter, J. R., Sherwood, R. T. & Platt, W. D. (1969). Anastomosis grouping among isolates of *Thanatephorus cucumeris*. *Phytopathology*, **59**, 1270–8.
Pontecorvo, G., Roper, J. A., Hemmons, L. M., Macdonald, K. D. & Bufton, A. W. J. (1953). The genetics of *Aspergillus nidulans*. *Advances in Genetics*, **5**, 141–238.
Raper, J. R. (1966). *Genetics of Sexuality in Higher Fungi*. New York: Ronald Press.
Rayner, A. D. M. (1977). Fungal colonization of hardwood stumps from natural sources. II. Basidiomycetes. *Transactions of the British Mycological Society*, **69**, 303–12.
Rayner, A. D. M. (1979). Internal spread of fungi inoculated into hardwood stumps. *New Phytologist*, **82**, 505–18.
Rayner, A. D. M. & Todd, N. K. (1978). Polymorphism in *Coriolus versicolor* and its relation to interfertility and intraspecific antagonism. *Transactions of the British Mycological Society*, **71**, 99–106.
Rayner, A. D. M. & Todd, N. K. (1979). Population and community structure and dynamics of fungi in decaying wood. *Advances in Botanical Research*, **7**, 333–420.
Rishbeth, J. (1978). Infection foci of *Armillaria mellea* in first-rotation hardwoods. *Annals of Botany*, **42**, 1131–9.
Thom, C. & Raper, K. B. (1945). *A Manual of the Aspergilli*. Baltimore: Williams & Wilkins.
Todd, N. K. & Rayner, A. D. M. (1978). Genetic structure of a natural population of *Coriolus versicolor* (L. ex Fr.) Quél. *Genetical Research*, **32**, 55–65.
Ullrich, R. C. (1977). Natural distribution of incompatibility factors in basidiomycetous fungi. *Mycologia*, **69**, 714–9.
Ullrich, R. C. & Anderson, J. B. (1978). Sex and diploidy in *Armillaria mellea*. *Experimental Mycology*, **2**, 119–29.
Yuill, E. (1939). Two new *Aspergillus* mutants. *Journal of Botany, London*, **77**, 174–5.
Yuill, E. (1950). The numbers of nuclei in conidia of Aspergilli. *Transactions of the British Mycological Society*, **33**, 324–31.

Note added in proof: For recent references pertinent to this Chapter and Chapter 6 see list on p. 355.

8 Basidiomycete decay of standing trees

P. C. MERCER

Forestry Commission, Forest Research Station, Alice Holt Lodge, Wrecclesham, Farnham, England

Introduction

Just over a hundred years ago Hartig (1878) discovered that the decay of trees was associated with basidiomycetes. He did not think other organisms were involved. This was understandable as the coincidence of decay and the presence of fruit bodies of basidiomycetes is so common, and isolates from decayed wood yielded cultures of basidiomycetes. Although other fungi and bacteria were found, they were usually dismissed as contaminants.

These classical ideas of decay have persisted for too long. Only in the last fifteen years has the full complexity of the interactions between basidiomycetes, ascomycetes, fungi imperfecti, yeasts and actinomycetes been realised (Monk, 1973; Shigo & Hillis, 1973; Kallio, 1974; Basham & Anderson, 1977; Blanchette & Shaw, 1978; Shortle & Cowling, 1978). Other organisms, such as nematodes (Ruehle, 1964), insects (Whitney & Cobb, 1972) and birds (Shigo, 1963), may also be involved. Another important factor which had been omitted until recently is the reaction of the tree. In evolutionary terms some sort of balance must be maintained between the micro-organisms and the host to ensure the survival of both.

The complexity of the interactions and the long time scale over which they operate are probably the reasons why research into decay has tended to concentrate on the final stages of the decay process. In this chapter it is hoped to redress the balance by laying more emphasis on the initial stages of colonisation and on the response of the host.

The decay process

Entry of basidiomycetes into the host

Although data on the parasitic or saprophytic natures of basidiomycetes are difficult to obtain, it appears that most wood-decaying basidiomycetes invade trees by way of wounds or weakened areas of the tree (Wagener & Davidson, 1954). The majority appear to be able to colonise wood which has already been colonised by other micro-organisms (Etheridge, 1973), but a minority, such as *Stereum* species, require freshly created wounds for entry.

In addition, some basidiomycetes are capable of entering the tree without wounds. These fungi enter mainly through the roots, e.g. *Heterobasidion annosum* and *Armillaria mellea*. Some of them take advantage of dead or dying roots to enter the tree and progress to the trunk via live tissue, whereas some of them, e.g. *H. annosum* and *Ganoderma lucidum*, have been shown to enter uninjured roots directly as complete pathogens (Wagener & Cave, 1946; Rishbeth, 1951; Toole, 1966). However, only basidiomycetes invading through the stem and branches will be discussed in detail in this chapter.

Wounds are the commonest entry points for decay organisms. These may be caused either necessarily as in pruning wounds, or by accident as in damage caused during timber extraction. Wounds are also caused by fire, insects, animals or birds. Another common entry point is through the branches and roots which are continually dying off even in a young vigorously growing tree. This process is accelerated after the tree becomes mature; the dead branches and roots are ideal sites for the entry of basidiomycetes (Basham & Anderson, 1977). Many trees also absciss their lower branches naturally in the course of their growth and the wounds remaining may act as infection courts (Shortle & Cowling, 1978).

The host response and the pattern of decay

As soon as the continuity of a tree's surface is broken a number of responses occur. The first is probably an electrical one. R. Mulhearn (personal communication) in the USA has shown peaks of electrical potential in *Populus* when leaves are excised. The next response is chemical (Sucoff, Ratsch & Hook, 1967; Hart & Johnson, 1970) and results in the laying down of physical and chemical barriers around the wound, which in turn results in a limitation of the spread of invading micro-organisms. The term 'reaction zone' (Fig. 3) was used by Shain (1967) to define these barriers. Shigo (1976) and Shigo & Marx (1977) refined Shain's concept and introduced the theory of compartmentalisation to explain the tree's reaction. The enclosed area around the wound is looked upon as a box or compartment composed of four basic wall types (Fig. 1). Wall 1 is formed by the plugging of vessels above and below the point of wounding; wall 2 is formed by existing annual rings; wall 3 is formed by medullary rays, and wall 4 by a barrier zone which is produced between wood present at the time of wounding and wood formed subsequently. Walls 1 to 3 together make up the reaction zone as defined by Shain. The walls appear to be formed by a combination of resistance due to anatomical structures and antifungal chemicals. All walls are not equally strong. The strength increases from walls 1 to 4, the relative strengths of the various walls giving rise to the compartment

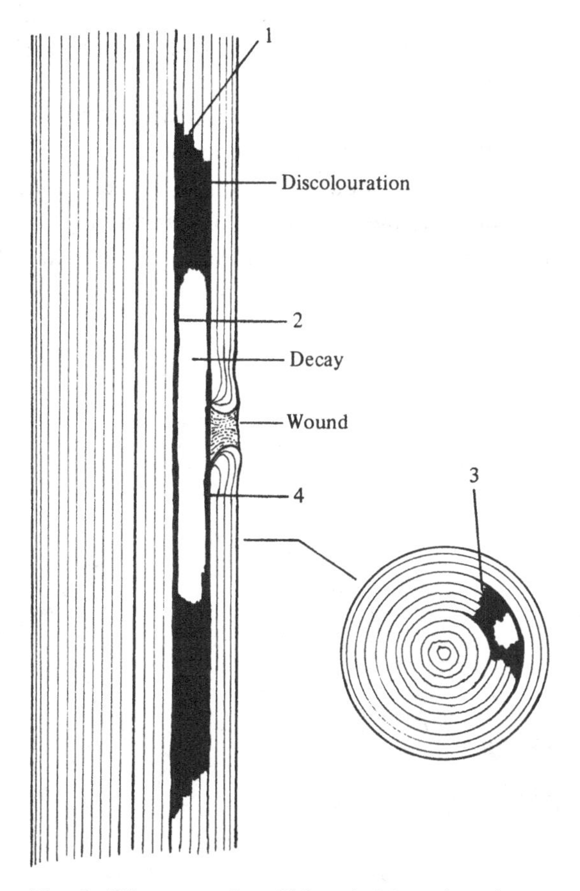

Fig. 1. Diagram, after Shigo & Marx (1977), of a longitudinal and transverse section through a wound, showing the compartment produced by wall types 1 to 4 (see text).

shape which is usually in the form of a wedge, often elongated vertically (Fig. 2). Host reaction is not limited to stems and branches. Shigo (1975) has observed compartmentalisation of *Heterobasidion annosum* in roots of red pine (*Pinus resinosa*). In broadleaved trees, polyphenols form chemical barriers. In many conifer species, copious resin, as well as polyphenols, is produced from special resin canals, which helps to limit the spread of fungi (Gibbs, 1968; Prior, 1975).

The compartmentalisation response varies from tree to tree, even within the same species (Copony & Barnes, 1974; French & Manion, 1975). It does not appear to be closely related to growth rate or tree vigour (Shigo, Shortle & Garrett, 1977), although it is possible that a tree which is under stress may succumb to invasion more readily than one which is not. Shigo *et al.* suggested that the compartmentalisation response might be inherited and that therefore it should be possible to breed for decay resistance.

Compartmentalisation of wounds will tend to restrict the advance of fungi, but compartment walls, with the exception of wall 4, are not usually finite. If one wall is breached, a new one is formed at the next natural anatomical boundary, medullary ray or annual ring, or more plugs may be formed. Shigo & Marx (1977) indicated that the rate of

Fig. 2. Wound in *Quercus* which had been compartmentalised and callused over.

enlargement of the compartment becomes slower with time. This has been partly confirmed by the author (unpublished) in a study of compartment enlargement in *Fagus* stubs. The distance of the reaction zone (Fig. 3) from the cut end of the stub was noted 10, 12 and 19 months after pruning, but significant differences could not be found, suggesting that fairly rapid extension had occurred before the first sampling. The diameter of the wound and the length of the stub had much greater effects than time (58% and 11% of the experimental variation respectively compared with only 0.7% for time). Greater retreat of the reaction zone into the tree results from larger diameters and longer stubs. Why the rate of enlargement of the compartment should slow down is not altogether clear. It is possible that factors such as reduced oxygen concentration at greater depth affect fungal growth. Basidiomycetes entering through roots may be specially adapted to growth at low oxygen concentrations (Hintikka: Chapter 13).

D. R. Houston (personal communication) has also suggested that, at least in stub wounds, the position of the reaction zone may be largely determined by factors already present in the tree before wounding, which are independent of the presence of micro-organisms. He postulated that the reaction zone will tend to form around the position of the

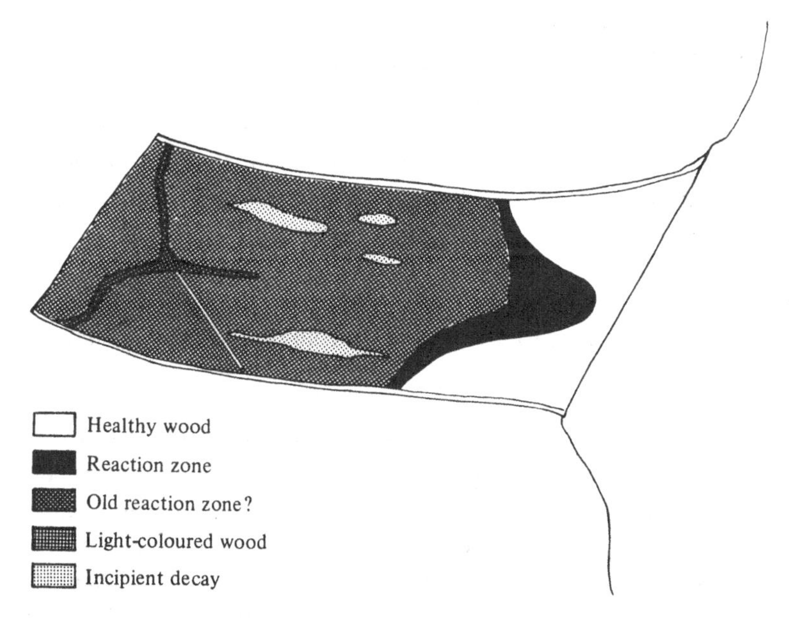

Fig. 3. Diagram of a longitudinal section through a *Fagus* stub 20 months after pruning.

potential callus collar. This would explain the lack of correlation between time and the position of the reaction zone.

When the stubs from the above experiment were examined visually it was possible to see wood in different stages of alteration. Fig. 3 shows a longitudinal section of a typical beech stub 20 months after branch removal. There is a dark greenish-brown band between the dead and live wood – the reaction zone. Above this is an area of wood that is still darker than the healthy wood but considerably lighter than the reaction zone. Some even lighter patches of incipient decay occur within this area. Sometimes, as in this stub, there is a faint dark band close to the surface, possibly indicating the position of the reaction zone at an earlier stage.

The colour differences in the wood are also reflected in changes in electrical resistance as measured by a pulsed current resistance meter – the 'Shigometer' (Skutt, Shigo & Lessard, 1972). A probe with two pins was used to map resistance measurements on the surface of five stub sections (Fig. 3), 20 months after wounding. On average, the lowest resistance occurred in the reaction zone (18 KΩ), confirming findings of other workers (Malia & Tattar, 1978; Sylvia & Tattar, 1978). Electrical resistance is dependent on moisture and cation concentrations. Shain (1971) has shown increases in K, Ca and Mn in reaction zones of Norway spruce (*Picea abies*). Similarly, Blanchard, Smith, Shigo & Safford (1978) found increases of K, Mg and Mn in reaction zones of red maple (*Acer rubrum*). Light-coloured wood had a higher resistance (146 KΩ) than healthy wood (53 KΩ). It was originally thought that the higher figures were due to drying out of the wood, but all readings were re-taken after the surfaces of the stub sections had been soaked in water. Cations released by hyphae from parenchyma cells in the reaction zone must somehow be locked up again, possibly in fungal mycelium (Frankland: Chapter 14). They do not appear to be lost, as areas of incipient decay found in the light-coloured wood again have low resistance (35 KΩ).

Interactions of basidiomycetes with other micro-organisms

The establishment and spread of basidiomycetes are dependent on several factors. If entry is through a wound, the season of wounding may influence the tree's response, alter the inoculum potential of the basidiomycete, and affect competition with other micro-organisms. Micro-climatic factors around the wound will govern the temperature and humidity of the wood close to the surface. The 3–4 cm of a wound closest to the surface are frequently less decayed than further in, presumably because of drying out of the wood. Physical factors are therefore of considerable importance, but interactions with other microorganisms in the wood have received increasing attention recently (e.g. Shigo, 1967; Shortle & Cowling, 1978).

Occasionally a tree may be attacked by a basidiomycete acting alone. Wikström (1976) observed *Phellinus tremulae* causing heart rot of *Populus tremula* without the involvement of other organisms. However, this seems to be very much the exception. Workers frequently report difficulty in re-isolating basidiomycetes after inoculation into trees. Shigo (1968), for example, was unable to recover inoculated basidiomycetes from more than 20% of six species of broadleaved trees after five years. Bacteria and non-basidiomycetes were however isolated from almost every tree. A piece of decayed wood is rarely occupied by a basidiomycete alone. It is frequently divided up into a number of regions, which may contain either different species of fungus or different strains of the same one (Rayner & Todd, 1977, and Chapters 6 and 7). The pockets are frequently separated by dark, single or double, zone lines (Fig. 4). Rayner (1976) also discovered that a range of

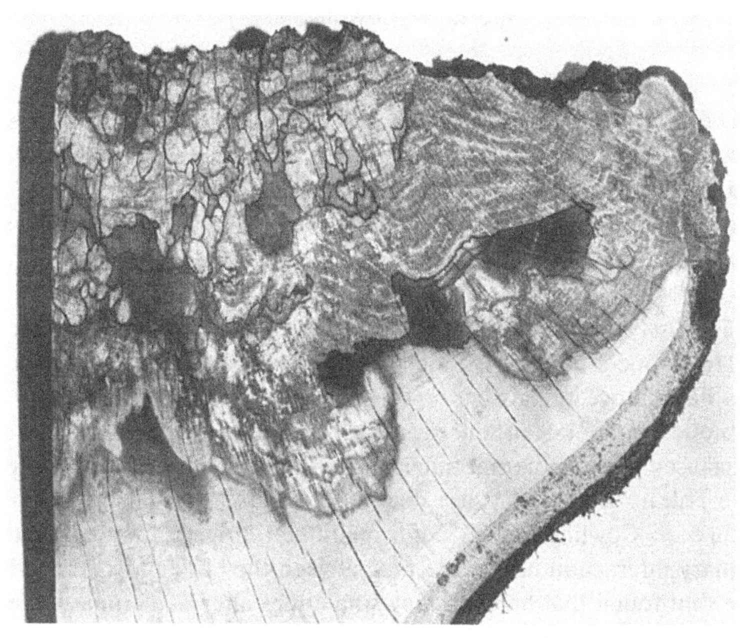

Fig. 4. Zone lines in *Fagus* caused mostly by the ascomycete *Ustulina deusta*.

dematiaceous Hyphomycetes is associated with these zone lines. Bacteria are frequently found throughout decaying wood. There is also a difference between microbial populations in wood that is newly colonised and wood in the last stages of breakdown. The make-up of a microbial population at any point is usually explained by theories of succession (Etheridge, 1961; Ueyama, 1965; Shigo, 1967; Shigo & Sharon, 1968; Rayner, 1977*a*,*b*; Shortle & Cowling, 1978). However, the manner of these successions is a matter for debate. Two possible mechanisms of succession are summarised below.

(i) Injury → Chemical discolouration → Bacteria; pioneer non-decay fungi → Basidiomycetes → Decay
(Shigo, 1967)

(ii) Injury → Basidiomycetes; pioneer non-decay fungi → Discolouration → Phenol-tolerant fungi → Reduction in discolouration → Basidiomycetes → Decay
(Shortle & Cowling, 1978)

The major difference between the two theories concerns the early part of the process. Shigo (1967) suggests that pioneer non-decay fungi and bacteria interact with parenchyma cells, either as part of the discolouration process, which produces the reaction zone, or following chemical discolouration; basidiomycetes follow later. Shortle & Cowling (1978) suggest that basidiomycetes and pioneer non-decay fungi invade first; the basidiomycetes trigger off the discolouration process, and the pioneer non-decay fungi grow in the sapwood but do not persist. The basidiomycetes are strongly suppressed, but do persist until the phenolic compounds produced in the reaction zone are metabolised by specialised phenol-tolerant fungi, such as *Phialophora* species. They are then able to grow and cause decay. In a more recent paper, Shortle, Menge & Cowling (1978) confirmed their original finding that the primary interaction usually occurs between the basidiomycetes and the tree, but found that bacteria may sometimes alter this primary interaction.

The author (unpublished) has observed a succession in pruning

Table 1. *Succession of micro-organisms from pruning wounds of* Fagus. *Average number of species isolated per wood chip* (× 10); *n* = 30

Organism	Time after wounding[a]							
	4 months		8 months		12 months		18 months	
Bacteria	5.7	8.0	5.3	9.0	6.3	6.0	7.7	4.3
Yeasts	3.0	1.3	3.0	0.0	1.7	0.3	0.3	0.0
Fungi imperfecti + ascomycetes + Mucorales	17.3	1.3	24.3	2.0	21.0	9.0	13.3	7.3
Basidiomycetes	4.0	1.3	1.0	0.0	0.0	0.0	1.0	0.7

[a] First figure in each pair is for untreated wounds and the second figure is for wounds treated with copper naphthenate and latex.

Mature *Fagus* trees were pruned in autumn 1976 to give flush wounds. Some wounds were left uncovered and some were treated with copper naphthenate (Green Cuprinol) and covered with a latex paint (Lac Balsam). At intervals, chips of wood (five replicates) were removed aseptically from three positions on the surface of the wound and from the same three positions at a depth of 2 cm from the surface. Individual wounds were sampled only once. The chips were plated out on 2% malt agar, 2% malt agar containing 0.02% of the antibiotic Crystamycin, or 2% malt agar containing antibiotic and 0.002% of carbendazim, and incubated at 23 °C.

wounds of *Fagus* which accords best with the second theory. The numbers of isolates of different types are shown in Table 1. Basidiomycetes are seen to be most common at the beginning of sampling. Numbers of fungi imperfecti, ascomycetes and Mucorales appear to rise to a peak and then fall off. Numbers of yeasts decline, but numbers of bacteria remain relatively constant. The data presented here come from chips of *Fagus* wood taken up to 2 cm from the surface of a wound made by cutting off a branch flush with the trunk. If stubs, such as shown in Fig. 3, were removed and incubated in polythene bags for 2–3 days at room temperature there was a profuse mycelial growth in the reaction zone. This was frequently a basidiomycete, either *Chondrostereum purpureum* or *Bjerkandera adusta*. Light-coloured wood also contains basidiomycetes, but growth does not appear to be as active as in the reaction zone. Fungi imperfecti such as *Phialophora, Penicillium* and *Phoma* species may also be present, as well as bacteria. A series of thin sections were also taken from the region of the reaction zone and plated out. Generally, isolates confirmed the presence of basidiomycetes, but occasionally *Trichoderma viride* was present and basidiomycetes were then absent. The succession can be summarised as below.

Injury → Specialised basidiomycetes;
bacteria; (pioneer non-decay fungi) → Discolouration →

Non-decay fungi;
bacteria; basidiomycetes → Reduction in → Basidiomycetes; → Decay
discolouration bacteria

This succession differs from that of Shortle & Cowling (1978) in that the basidiomycetes seem to be generally very active, and the non-decay fungi less active, in the reaction zone, whereas Shortle & Cowling found the converse. It may be that the succession is dependent on the fungi that colonise initially. *C. purpureum* and *B. adusta* are perhaps particularly well adapted for growth in the reaction zone. Rayner (1977*b*) has noted *C. purpureum* as an early coloniser of *Fagus* and *Betula* stumps, but did not find it so frequently at later stages. On the other hand, a non-decay fungus, such as *T. viride*, may invade first, in which case *C. purpureum* and *B. adusta* are unable to occupy their niche in the reaction zone.

It is probable that more than one form of succession can occur in different species of tree and even within the same piece of wood. Succession in living trees will not necessarily be the same as successions in timber, which have generally been examined much more fully (Dickinson: Chapter 10; Levy: Chapter 9).

Type of decay caused by basidiomycetes

The chemical effects of decay in wood by basidiomycetes have been well studied and documented. Good accounts are given by Cartwright & Findlay (1958), Cowling (1958), Carey (1975), Dickinson: Chapter 10, and Levy: Chapter 9. There are basically two sorts of decay: brown rot in which only the cellulose is attacked, and white rot in which both cellulose and lignin are attacked (Kirk & Fenn: Chapter 4; Montgomery: Chapter 3). However, it should again be borne in mind that basidiomycetes are not usually found in decaying wood in pure culture. Blanchette & Shaw (1978) have shown significant increases in decay (measured by weight loss) in coniferous wood when bacteria and yeasts were added to the basidiomycetes.

Control of decay in wounds

Chemical control

Man has tried for hundreds of years to prevent decay in living trees by basidiomycetes, mostly by the addition of sealing materials to wounds which have been created naturally or by pruning. Many different materials have been used. Mercer (1979*a*) when reviewing the literature of the last sixty years found that ninety-eight different materials had been tried. The aim generally appeared to be to emulate the sealing action of the tree's own natural covering of callus. When callus completely covers a wound microbial activity beneath the callus is greatly reduced, possibly because of lowered oxygen concentration (Fig. 2). Jensen (1969) suggested that decay ceased in *Quercus* species after callusing over of infection courts because of reduced oxygen concentration and increased carbon dioxide concentration, although Thacker & Good (1952) were unable to show any relationships between decay and aeration in sugar maple (*Acer saccharum*). Many of the proprietary products used to treat pruning wounds fail to provide protection from decay probably because the seal is insufficiently good to keep oxygen concentrations low and stop microbial activity. Not only are they usually ineffective as sealants but the moist conditions behind the covering may well stimulate decay rather than prevent it.

The addition of a fungicide to dressings has also been tried, e.g. Marshall (1942), but most materials which were effective fungicides were also phytotoxic to the tree (Mercer, 1979*b*). Many of these materials were first used in timber preservation (Dickinson: Chapter 10). Phytotoxic materials interfere both with the tree's ability to form callus and with the compartmentalisation response, thereby limiting the tree's ability to defend itself. In the experiment on *Fagus* stubs quoted above, the addition of copper naphthenate as a fungicide caused considerably greater enlargement of the compartment than in the controls, and accounted for over 50% of the experimental variation. This would not matter if the affected wood were still fungitoxic, but, although numbers of isolates from wood treated with copper naphthenate were lower than in untreated wood (Table 1), the wood was by no means sterile. In another experiment wounds bore fruit bodies of *Chondrostereum purpureum* growing on wood still clearly impregnated with copper naphthenate, after only one year. Shigo & Wilson (1977) found no beneficial effect after five years' use of a variety of non-fungitoxic dressings on red maple and American elm (*Ulmus americana*). On the other hand Shortle & Shigo (1978) showed that wood was less discoloured after one year when wounds on red maple were covered by plastic sheet, and Houston (1971) using a variety of chemicals showed a reduction in discolouration and decay in red maple and yellow birch (*Betula alleghaniensis*) after three years. The most successful chemical treatments were those, such as glue, which appeared to restrict the entrance of air to the wound. It is possible that if a really good sealing compound could be found decay would be reduced. The old idea of trying to prevent decay by preventing ingress of fungi may well be ineffective, as it has been shown that fungal propagules may already be present in the wood before wounding (Shortle & Cowling, 1978).

Biological control

Another possibility for the prevention of decay is to improve the competitive ability of non-decay fungi at the expense of the basidiomycetes. Workers have studied the antagonism of bacteria and actinomycetes to basidiomycetes (Greaves, 1970; Lapetite, 1970; Ricard, 1970) and of non-decay fungi (Stilwell, 1966; Gibbs, 1967; Hulme & Shields, 1972; Kallio & Salonen, 1972; Morquer & Touvet, 1974; Aufsess, 1976) and of other basidiomycetes (Rishbeth, 1952; Greig, 1976). Practical biological control has been achieved by the use of *Phlebia gigantea* to control *Heterobasidion annosum* in conifer plantations (Greig, 1976). Hunter (1977) has also shown control of silver leaf symptoms caused by *Chondrostereum purpureum* in pear (*Pyrus communis*) trees by *Trichoderma viride*, and Pottle, Shigo & Blanchard (1977) obtained control of basidiomycetes in red maple for up to 21 months using *T. harzianum*.

The author (unpublished) has obtained control of wood-decaying basidiomycetes common in *Fagus* by *T. viride, T. koningii, Fusarium lateritium, Cryptosporiopsis fasciculata* and a *Bacillus* species in laboratory tests on beech wood strips. Some of the results are shown in Table 2, and the techniques are shown in Fig. 5. Tests (*a*) and (*b*) were used to test the fungi imperfecti, and tests (*c*) and (*d*) to test the *Bacillus* species. Results varied with the test: those on nutrient-rich agar tended to favour the antagonists and those on nutrient-free glass beads tended to favour the basidiomycete. *T. viride* was the most effective antagonist, strongly inhibiting nearly all the decay fungi. *Bjerkandera adusta* was the least inhibited of the decay fungi, being able to overcome even *T. viride* in test (*b*). It is obvious in view of the variation between tests that extrapolation of laboratory results to the field should be viewed with caution. It has been shown that antagonism between organisms is

Table 2. *Responses of basidiomycetes to antagonism by fungi imperfecti and a* Bacillus *species*

Basidiomycete	Test[a]	*Trichoderma viride*	*Trichoderma koningii*	*Cryptosporiopsis fasciculata*	*Fusarium lateritium*	*Bacillus* sp.
Bjerkandera	a/c	3.0	3.0	0.0	0.0	2.0
adusta	b/d	0.5	0.0	0.0	0.0	0.5
Chondrostereum	a/c	3.0	3.0	0.7	0.0	0.5
purpureum	b/d	3.0	2.0	0.5	1.5	0.5
Coniophora	a/c	3.0	3.0	0.0	0.0	2.0
puteana	b/d	3.0	3.0	2.8	0.3	3.0
Stereum	a/c	3.0	3.0	2.4	0.0	3.0
hirsutum	b/d	3.0	3.0	1.5	1.8	3.0

[a] Tests for fungi imperfecti and *Bacillus* sp. response (see Fig. 5).

All measurements were taken at 23 °C and reduced to a 0–3 scale: 0, no inhibition; 3, strong inhibition.

dependent on a number of factors, e.g. pH (Gibbs, 1967; Sierota, 1976) and temperature (Tronsmo & Dennis, 1978). It should be possible to alter the balance of competition between the antagonist and basidiomycete in favour of the antagonist by increasing the inoculum potential of the antagonist, altering the pH or changing the nutrient status (e.g. Rishbeth, 1959).

The possibility of controlling decay by improving the compartmentalisation response by breeding has already been discussed. Another possibility is to improve the compartmentalisation response in individual wounds by stimulation with non-decay organisms. Blanchette & Sharon (1975) found that they could do this in wounds of *Betula alleghaniensis* by using suspensions of *Agrobacterium tumefaciens*.

Conclusions

Many basidiomycetes have evolved as wood-decay organisms, either singly or in associations. Levy (1973) has defined wood from the point of view of the micro-organisms colonising it as 'a series of conveniently orientated holes surrounded by food'. However, most trees have evolved responses which minimise the amount of their tissue that can be used as a fungal energy source. This, together with competition from other micro-organisms, ensures that wood-decaying basidiomycetes do not have things all their own way.

This paper has dealt with the means whereby invasion by basidiomycetes of a tree may be limited, but it should not be assumed that a rotten core in an adult tree is always detrimental in the natural environment, although it may be undesirable in commercial forestry.

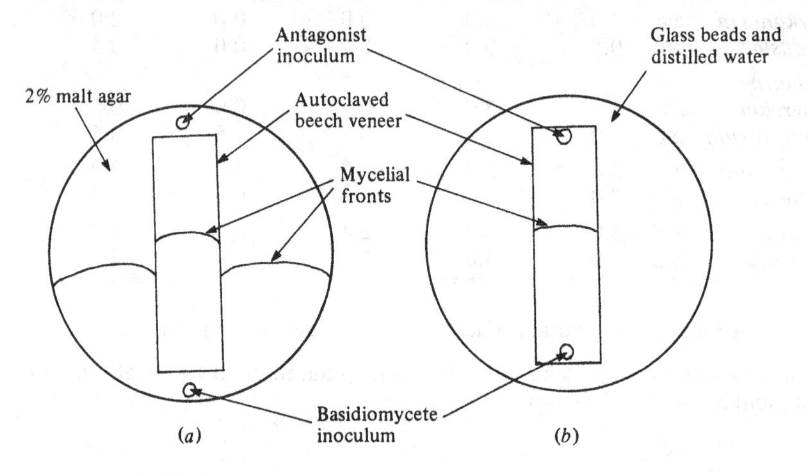

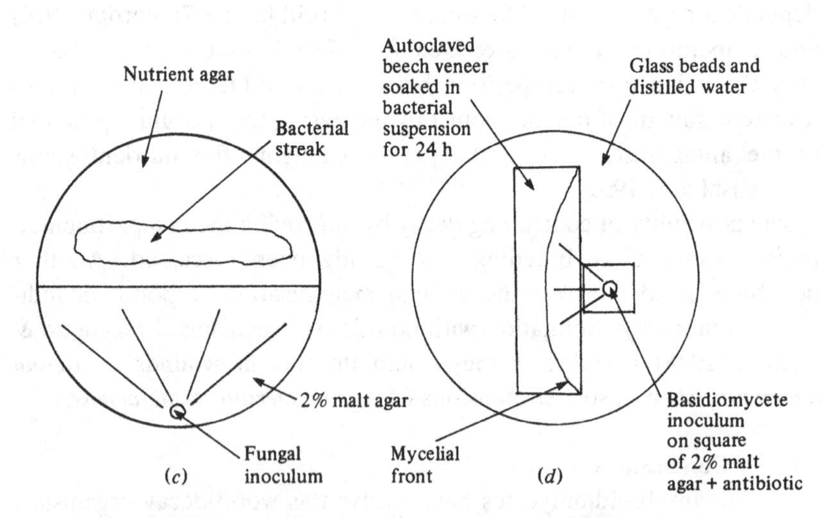

Fig. 5. Diagram of methods used to test for control of basidiomycetes by other fungi and by bacteria. (*a*, *b*), tests for fungi; (*c*, *d*), tests for bacteria. All plates were incubated at 23 °C and all veneers were cut longitudinally. Glass beads were about 300 μm in diameter and were soaked in water. Inocula of the basidiomycete and antagonist were applied simultaneously.

There may be situations where some decay is advantageous to the tree. Janzen (1976) said that 'a rotten hollow core is often an adaptive trait, selected as a mechanism of nitrogen and mineral trapping'. For both trees and wood-decaying basidiomycetes to survive it is essential that a balance be struck between decay and growth of wood. It is probably for this reason that most basidiomycetes are largely saprophytic, breaking down tissue which is already dead, and only becoming parasitic when there is a very high inoculum potential.

Acknowledgement. I would like to thank Dr John Gibbs for his help in preparing the text of this chapter.

References

Aufsess, H. von (1976). The effect of different micro-organisms on the progress of wood decomposition by stem-rotting fungi. *Beihefte zu Material und Organismen*, No. 3, 239–51.

Basham, J. T. & Anderson, H. W. (1977). Defect development in second growth Sugar maple in Ontario. I. Microfloral infection relationships associated with dead branches. *Canadian Journal of Botany*, **55**, 934–76.

Blanchard, R. O., Smith, D., Shigo, A. L. & Safford, L. O. (1978). Effect of soil-applied potassium on cation distribution around wounds in Red maple. *Canadian Journal of Forest Research*, **8**, 228–31.

Blanchette, R. A. & Sharon, E. M. (1975). *Agrobacterium tumefaciens*, a promoter of wound healing in *Betula alleghaniensis*. *Canadian Journal of Forest Research*, **5**, 722–30.

Blanchette, R. A. & Shaw, C. G. (1978). Associations among bacteria, yeasts and basidiomycetes during wood decay. *Phytopathology*, **68**, 631–7.

Carey, J. K. (1975). Notes on the isolation and characterisation of wood-inhabiting fungi. *Building Research Establishment Current Paper*, 93/75. Department of the Environment, UK.

Cartwright, K. St G. & Findlay, W. P. K. (1958). *Decay of timber and its prevention*. London: Her Majesty's Stationery Office.

Copony, J. A. & Barnes, B. U. (1974). Clonal variation in the incidence of *Hypoxylon* canker on trembling aspen. *Canadian Journal of Botany*, **52**, 1475–81.

Cowling, E. B. (1958). A review of the literature on the enzymatic degradation of cellulose and wood. *Report of the United States Forest Products Laboratory, Madison*, No. 2116.

Etheridge, D. E. (1961). Factors affecting branch infection in aspen. *Canadian Journal of Botany*, **39**, 799–816.

Etheridge, D. E. (1973). Wound parasites causing tree decay in British Columbia. *Forest Pest Leaflet of the Pacific Forest Research Centre*, No. 62. Victoria, British Columbia.

French, J. R. & Manion, P. D. (1975). Variability of host and pathogen in *Hypoxylon* canker of aspen. *Canadian Journal of Botany*, **53**, 2740–4.

Gibbs, J. N. (1967). A study of the epiphytic growth habit of *Fomes annosus*. *Annals of Botany*, **31**, 755–74.

Gibbs, J. N. (1968). Resin and the resistance of conifers to *Fomes annosus*. *Annals of Botany*, **32**, 649–65.
Greaves, H. (1970). Effect of selected bacteria and actinomycetes on the decay capacity of some wood-rotting fungi. *Material und Organismen*, **5**, 265–79.
Greig, B. J. W. (1976). Inoculation of pine stumps with *Peniophora gigantea* by chainsaw felling. *European Journal of Forest Pathology*, **5**, 286–90.
Hart, J. H. & Johnson, K. C. (1970). Production of decay-resistant sapwood in response to injury. *Wood Science and Technology*, **4**, 267–72.
Hartig, R. (1878). *Die Zersetzungerscheinungen des Holzes der Nadelholzbaume und der Eiche in forstlicher, botanischer und chemischer Richtung*. Berlin: Springer-Verlag.
Houston, D. R. (1971). Discoloration and decay in Red maple and Yellow birch – reduction through wound treatment. *Forest Science*, **17**, 402–6.
Hulme, M. A. & Shields, J. K. (1972). Effect of a primary fungal infection upon secondary colonisation of birch bolts. *Material und Organismen*, **7**, 177–88.
Hunter, T. (December, 1977). Biological control as a method to be used against silver leaf. *Horticulture Industry*, p. 841.
Janzen, D. H. (1976). Why tropical trees have rotten cores. *Biotropica*, **8**, 110.
Jensen, K. E. (1969). Oxygen and carbon dioxide concentrations in sound and decaying Red oak trees. *Forest Science*, **15**, 246–51.
Kallio, T. (1974). Bacteria isolated from injuries to growing spruce trees *Picea abies* (L.) Karst. *Acta Forestalia Fennica*, No. 137.
Kallio, T. & Salonen, A. (1972). The effect of *Gliocladium deliquescens* Sopp. on the decaying capacity of some decay fungi. *Annales Agriculturae Fenniae*, **11**, 320–2.
Lapetite, M. D. (1970). Antagonistic action of bacteria against wood-destroying fungi studied on wood. *Material und Organismen*, **5**, 229–38.
Levy, J. F. (1973). Colonisation of wood by fungi. *British Wood Preserving Association News Sheet*, No. 130. London.
Malia, M. E. & Tattar, T. A. (1978). Electrical resistance, physical characteristics and cation concentrations in xylem of sugar maple infected with *Verticillium dahliae*. *Canadian Journal of Forest Research*, **8**, 322–7.
Marshall, R. P. (1942). Care of damaged shade trees. *Farmers' Bulletin*, No. 1896. Washington, DC: United States Department of Agriculture.
Mercer, P. C. (1979*a*). Attitudes to pruning wounds. *Arboricultural Journal*, **3**, 457–65.
Mercer, P. C. (1979*b*). Phytotoxicity and fungitoxicity tests for tree wound paints. *Annals of Applied Biology*, **91**, 199–202.
Monk, V. (1973). Microbial colonisation of pruning wounds with particular reference to oak *Quercus robur*, PhD Thesis, University of Surrey.
Morquer, R. & Touvet, A. (1974). A comparison of the action of antagonistic fungi on various hymenomycetes parasitic on coniferous trees. *Comptes Rendus Hebdomadaires des Séances de l'Académie des Sciences, Série D*, **278**, 709–13.
Pottle, H. W., Shigo, A. L. & Blanchard, R. O. (1977). Biological control of wood hymenomycetes by *Trichoderma harzianum*. *Plant Disease Reporter*, **61**, 687–90.
Prior, C. (1975). Resin production and susceptibility to *Heterobasidion annosum* in Corsican pine. *Annals of Botany*, **39**, 1103–9.
Rayner, A. D. M. (1976). Dematiaceous hyphomycetes and narrow dark zones in decaying wood. *Transactions of the British Mycological Society*, **67**, 546–9.
Rayner, A. D. M. (1977*a*). Fungal colonisation of hardwood stumps from natural sources. I. Non-basidiomycetes. *Transactions of the British Mycological Society*, **69**, 291–302.

Rayner, A. D. M. (1977*b*). Fungal colonisation of hardwood stumps from natural sources. II. Basidiomycetes. *Transactions of the British Mycological Society*, **69,** 303–12.

Rayner, A. D. M. & Todd, N. K. (1977). Intraspecific antagonism in natural populations of wood-decaying basidiomycetes. *Journal of General Microbiology*, **103,** 85–90.

Ricard, J. L. (1970). Biological control of *Fomes annosus* in Norway spruce (*Picea abies*) with immunizing commensals. *Studia Forestalia Suecica*, No. 84. Uppsala: Swedish University of Agricultural Sciences.

Rishbeth, J. (1951). Observations on the biology of *Fomes annosus* with particular reference to East Anglian pine plantations. III. Natural and experimental infection of pines and some factors affecting severity of the disease. *Annals of Botany*, **58,** 221–46.

Rishbeth, J. (1952). Control of *Fomes annosus* Fr. *Forestry*, **25,** 41–50.

Rishbeth, J. (1959). Stump protection against *Fomes annosus*. II. Treatment with substances other than creosote. *Annals of Applied Biology*, **47,** 529–41.

Ruehle, J. L. (1964). Nematodes, the overlooked enemies of tree roots. *Proceedings of the International Shade Tree Conference*, **40,** 60–7.

Shain, L. (1967). Resistance of sapwood in stems of Loblolly pine to infection by *Fomes annosus*. *Phytopathology*, **57,** 1034–45.

Shain, L. (1971). The response of sapwood of Norway spruce to infection by *Fomes annosus*. *Phytopathology*, **61,** 301–7.

Shigo, A. L. (1963). Ring shake associated with sapsucker injury. *United States Forest Service Research Paper, North eastern*, No. 8. Upper Darby, Pa.

Shigo, A. L. (1967). The early stages of discoloration and decay in living hardwoods in North Eastern United States: a consideration of wound-initiated discoloration and heartwood. *Proceedings of 14th International Union of Forest Research Organisations Congress*, **9,** Section 41, 117–33. Vienna: UFRO Secretariat.

Shigo, A. L. (1968). Discoloration and decay in hardwoods following inoculations with hymenomycetes. *Phytopathology*, **58,** 1493–8.

Shigo, A. L. (1975). Compartmentalization of decay associated with *Fomes annosus* in trunks of *Pinus resinosa*. *Phytopathology*, **65,** 1038–9.

Shigo, A. L. (1976). Compartmentalization of discolored and decayed wood in trees. *Beihefte zu Material und Organismen*, No. 3, 221–6.

Shigo, A. L. & Hillis, W. E. (1973). Heartwood, discolored wood and micro-organisms in living trees. *Annual Review of Phytopathology*, **11,** 197–222.

Shigo, A. L. & Marx, H. G. (1977). Compartmentalization of decay in trees. *Agriculture Information Bulletin of United States Department of Agriculture Forest Service*, No. 405. Washington, DC.

Shigo, A. L. & Sharon, E. M. (1968). Discoloration and decay in hardwoods following inoculations with hymenomycetes. *Phytopathology*, **58,** 1493–8.

Shigo, A. L., Shortle, W. C. & Garrett, P. W. (1977). Genetic control suggested in compartmentalization of discolored wood associated with tree wounds. *Forest Science*, **23,** 179–82.

Shigo, A. L. & Wilson, C. L. (1977). Wound-dressings on Red maple and American elm; effectiveness after five years. *Journal of Arboriculture*, **3,** 81–7.

Shortle, W. C. & Cowling, E. B. (1978). Development of discoloration, decay and micro-organisms following wounding of Sweetgum and Yellow poplar trees. *Phytopathology*, **68,** 609–16.

Shortle, W. C., Menge, J. A. & Cowling, E. B. (1978). Interaction of bacteria, decay fungi, and live sapwood in discoloration and decay of trees. *European Journal of Forest Pathology*, **8,** 293–300.

Shortle, W. C. & Shigo, A. L. (1978). Effect of plastic wrap on wound closure and internal compartmentalisation of discolored and decayed wood in Red maple. *Plant Disease Reporter*, **62**, 992–1002.

Sierota, Z. A. (1976). Influence of acidity on the growth of *Trichoderma viride* Pers. ex Fr. and on the inhibiting effect of its filtrates against *Fomes annosus* (Fr.) Cke. in artificial cultures. *European Journal of Forest Pathology*, **6**, 302–11.

Skutt, H. R., Shigo, A. L. & Lessard, R. A. (1972). Detection of discolored and decayed wood using a pulsed electric current. *Canadian Journal of Forest Research*, **2**, 54–6.

Stilwell, M. A. (1966). A growth inhibitor produced by *Cryptosporiopsis* sp., an imperfect fungus isolated from Yellow birch, *Betula alleghaniensis* Britt. *Canadian Journal of Botany*, **44**, 259–67.

Sucoff, E., Ratsch, H. & Hook, D. D. (1967). Early development of wound-initiated discoloration in *Populus tremuloides* Michx. *Canadian Journal of Botany*, **45**, 649–56.

Sylvia, D. M. & Tattar, T. A. (1978). Electrical resistance properties of tree tissues in cankers incited by *Endothia parasitica* and *Nectria galligena. Canadian Journal of Forest Research*, **8**, 162–7.

Thacker, D. G. & Good, H. M. (1952). The composition of air in trunks of sugar maple. *Canadian Journal of Botany*, **30**, 475–85.

Toole, E. R. (1966). Root rot caused by *Polyporus lucidus. Plant Disease Reporter*, **50**, 945–6.

Tronsmo, A. & Dennis, C. (1978). Effect of temperature on antagonistic properties of *Trichoderma* species. *Transactions of the British Mycological Society*, **71**, 469–74.

Ueyama, A. (1965). Studies on the succession of higher fungi on felled beech logs (*Fagus crenata*) in Japan. *Beihefte zu Material und Organismen*, No. 1, 326–32.

Wagener, W. W. & Cave, M. S. (1946). Pine killing by the root fungus *Fomes annosus* in California. *Journal of Forestry*, **44**, 47–54.

Wagener, W. W. & Davidson, R. W. (1954). Heart rots in living trees. *The Botanical Review*, **20**, 61–134.

Whitney , H. S. & Cobb, F. W. (1972). Non-staining fungi associated with bark beetles on Ponderosa pine. *Canadian Journal of Botany*, **50**, 1943–5.

Wikström, C. (1976). The decay pattern of *Phellinus tremulae* (Bond.) Bond et Borisov in *Populus tremula* L. *European Journal of Forest Pathology*, **6**, 291–301.

9 The place of basidiomycetes in the decay of wood in contact with the ground

J. F. LEVY

Department of Botany, Imperial College, University of London, London SW7 2BB, England

Wood: food source, substrate and environment

To the wood-rotting micro-organisms colonising it, wood consists of a series of conveniently orientated holes surrounded by food. The shape, size and orientation of the holes relate directly to the anatomy of the species of wood being colonised, and differences in the chemical constitution of the cell walls or cell contents may affect the ability of a particular fungus to utilise them as a nutrient source.

Softwoods (gymnosperms) are formed largely by tracheids, with the cell lumen completely enclosed by a cell wall, so that movement of water or micro-organisms must involve passage through a cell wall or pit membrane (Fig. 1*a*). Hardwoods (angiosperms), however, consist of vessels, fibres and parenchyma as well as some tracheids, and here the vessels are often formed from open-ended vessel elements so that both water and micro-organisms can move along the grain without necessarily passing through either a cell wall or pit membrane (Fig. 1*b*,*c*). In addition, the hemicellulose component of the cell wall differs between softwood and hardwood, and in both groups there is a considerable difference between sapwood and heartwood of the same species. Sapwood is likely to contain accumulated nutrients such as starch, residual protoplasts or lipids stored by the tree in the ray parenchyma which is absent in heartwood. The heartwood is likely to have additional materials deposited in the cell walls during the process of heartwood formation. Such materials are known as *extractives* and often provide the dark colouration typical of such woods as mahogany (*Swietenia macrophylla*), teak (*Tectona grandis*), or ebony (*Diospyros* spp.). In some heartwoods these extractives have some degree of toxicity to micro-organisms and may impart a natural durability, as is shown in western red cedar (*Thuja plicata*), teak, or greenheart (*Ocotea rodiaei*).

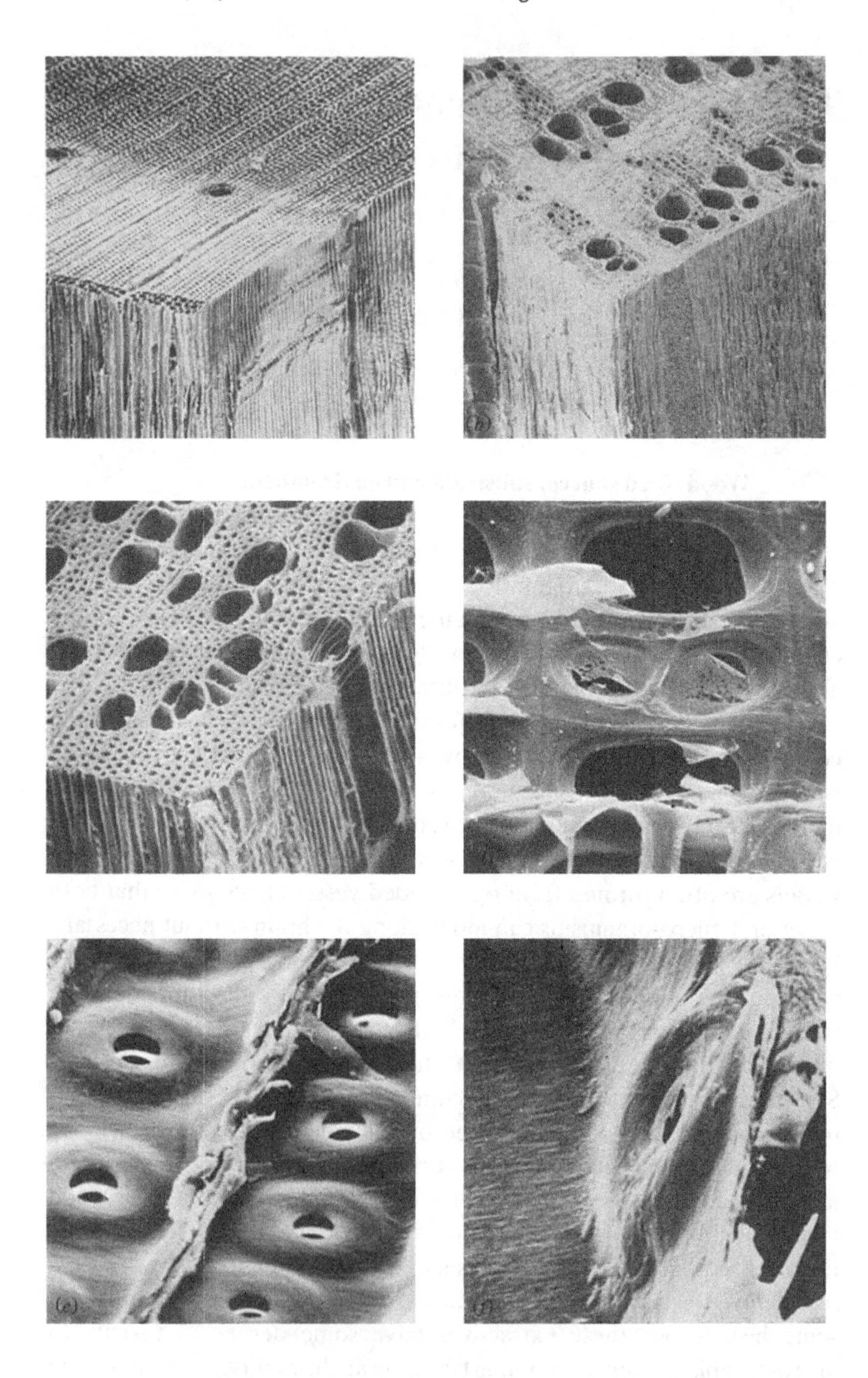

The colonisation sequence of wood by micro-organisms and the development of decay will be affected by these differences as shown in the case of commercial timbers by Dickinson (Chapter 10). Structural and chemical differences constitute part of the ecological factors that will determine the process of decay. Other factors include: the rate of growth of the wood when it was being formed in the tree; the orientation of the grain of the wood with respect to the source of invading organisms and water; the moisture content of the wood, both static and in movement; the presence, movement or accumulation of nutrients; pH and temperature. Such factors give rise to a series of ecological niches which may be filled by a range of micro-organisms depending upon the particular circumstances.

Micro-organisms associated with decay of wood in contact with the ground

A wide range of micro-organisms from many diverse taxonomic groups are readily available to colonise and destroy wood in contact with the ground. Some are specific to a particular wood under certain precise conditions of exposure. Others are more opportunist and may play a different rôle in the decay process under varying conditions. It is not possible to define or restrict each decay type within taxonomic frontiers, since the same organism (e.g. *Phialophora fastigiata*) may, given suitable environmental conditions, fill one or several ecological niches or 'physiological groups' (Clubbe, 1978).

Studies on the micro-organisms colonising wood and on the sequence of events leading to the onset of decay have shown the importance of reclassifying the species involved, irrespective of their taxonomic identity, into a small number of groups relevant to their effect on the wood (Banerjee, 1969; Banerjee & Levy, 1971; Butcher, 1968*a,b*, 1971; Carey, 1980; Clubbe, 1978, 1980; Corbett, 1963; Corbett & Levy, 1963; Cowling, 1965; Greaves, 1966; Käärik, 1967, 1968; Levy, 1968, 1971, 1973, 1975; Merrill & French, 1966). Clubbe (1980) recognised six such

---

Fig. 1.(*a*) Softwood anatomy: Scots pine (*Pinus sylvestris*), × 35. (*b*) Ring-porous hardwood anatomy: oak (*Quercus robur*), × 30. (*c*) Diffuse-porous hardwood anatomy: birch (*Betula pendula*), × 120. (*d*) Breakdown of the membrane of crossfield pits in Scots pine by bacterial action, × 600. (Zainal, 1976*b*.) (*e*) Bordered pits in Scots pine, showing complete breakdown of the pit membrane by bacterial action, × 500. (*f*) Bacteria associated with breakdown of the membrane of an aspirated bordered pit in Scots pine, × 700.

groupings, based on his own observations and other studies published in the literature. They comprised:

Bacteria
'Primary moulds'
'Stainers'
'Soft rots'
Basidiomycetes { White rots / Brown rots }
'Secondary moulds'

which may be defined as follows.

Bacteria

This group includes a range of true bacteria, mainly gram-positive rods, some of which are capable of fixing atmospheric nitrogen. It also includes some actinomycetes such as *Streptomyces* species, which often show a somewhat random distribution. These micro-organisms are usually the first colonisers and may have one or several effects on the wood. (*a*) A progressive breakdown of the membrane of pits in the sapwood is evident (Fig. 1*d–f*). This opens up the wood structure, allows gaseous diffusion to occur, provides an open pathway for water, and gives access to colonising micro-organisms incapable of causing lysis of a wall or pit (Levy, 1973, 1975). (*b*) Fixation of atmospheric nitrogen has been demonstrated to increase progressively from an exposed surface into wood (Levy, Millbank, Dwyer & Baines, 1974; Baines & Millbank, 1976; Baines, Dickinson, Levy & Millbank, 1977). This could provide an additional source of nitrogen, increasing the low level normally found in wood. (*c*) There is antagonism or synergism with fungal colonists (Levy, 1975). (*d*) The presence of bacterial cells and their metabolic products will have changed the internal environment, although no other effect may be observed.

'Primary moulds'

These organisms comprise the first fungal colonists, and can be regarded as akin to Garrett's sugar fungi (Garrett, 1951, 1955). They do not appear to possess enzymes capable of degrading cellulose or lignin and do not seem able to attack the wood cell walls. Their food source is likely to be sugars or simple carbohydrates either present in the ray parenchyma of the sapwood or derived from the soil. They can penetrate into wood only through natural openings, such as end-grain apertures or openings through the walls produced by other organisms, i.e. through pits after bacteria have destroyed the pit membrane. Phycomycetes, ascomycetes, and fungi imperfecti are all represented in this group.

'Stainers'

This group, mainly ascomycetes and fungi imperfecti, is characterised by the pigmentation of the hyphal walls, which gives rise to a discolouration of the sapwood of infected wood. Like the 'primary moulds' their nutrients are primarily the food reserves of the tree stored in the ray parenchyma of the sapwood, or residual sugars and simple carbohydrates present elsewhere in the wood. Hyphae of these organisms are able to penetrate through cell walls in a characteristic manner by means of a fine constriction of the hypha in the wall and resumption of its normal size and shape on emergence into the lumen (Fig. 2*a*); a single hypha may pass through a number of cells in this way (Fig. 2*b*). Several 'stainers' have been shown to be capable of causing soft rot in wood.

'Soft rots'

This group is comprised of ascomycetes and fungi imperfecti. They are the first of the wood-rotting fungi to colonise wood in ground contact. They are grouped together on the basis of forming cavities in the S_2 layer of the cell wall (Käärik, 1974), which is destroyed by the formation of chains of such cavities. Their micromorphology was first illustrated in the middle of the last century, but their economic significance was not realised until nearly a hundred years later. Findlay & Savory first reported their activities in 1950, and Savory (1954*a,b*, 1955) published a series of papers which established the term 'soft rot'. He demonstrated that the fungi were responsible for the decay of the timber fill in water cooling towers. This degradation was previously thought to be due to chemical breakdown and not to be the result of biological action. The soft-rot fungi become established where there is little competition from other fungi and are known to give rise to serious economic losses of timber. Where the basidiomycetes are able to establish dominance, the soft-rot fungi, although present, never become the main causal organisms of wood decay.

Savory showed that soft-rot fungi grew in the S_2 layer of the secondary cell wall of wood cells, forming chains of cavities with pointed ends which were orientated parallel to the cellulose microfibrils in the cell walls (Fig. 2*c*). However, Corbett (1963, 1965) was the first to observe

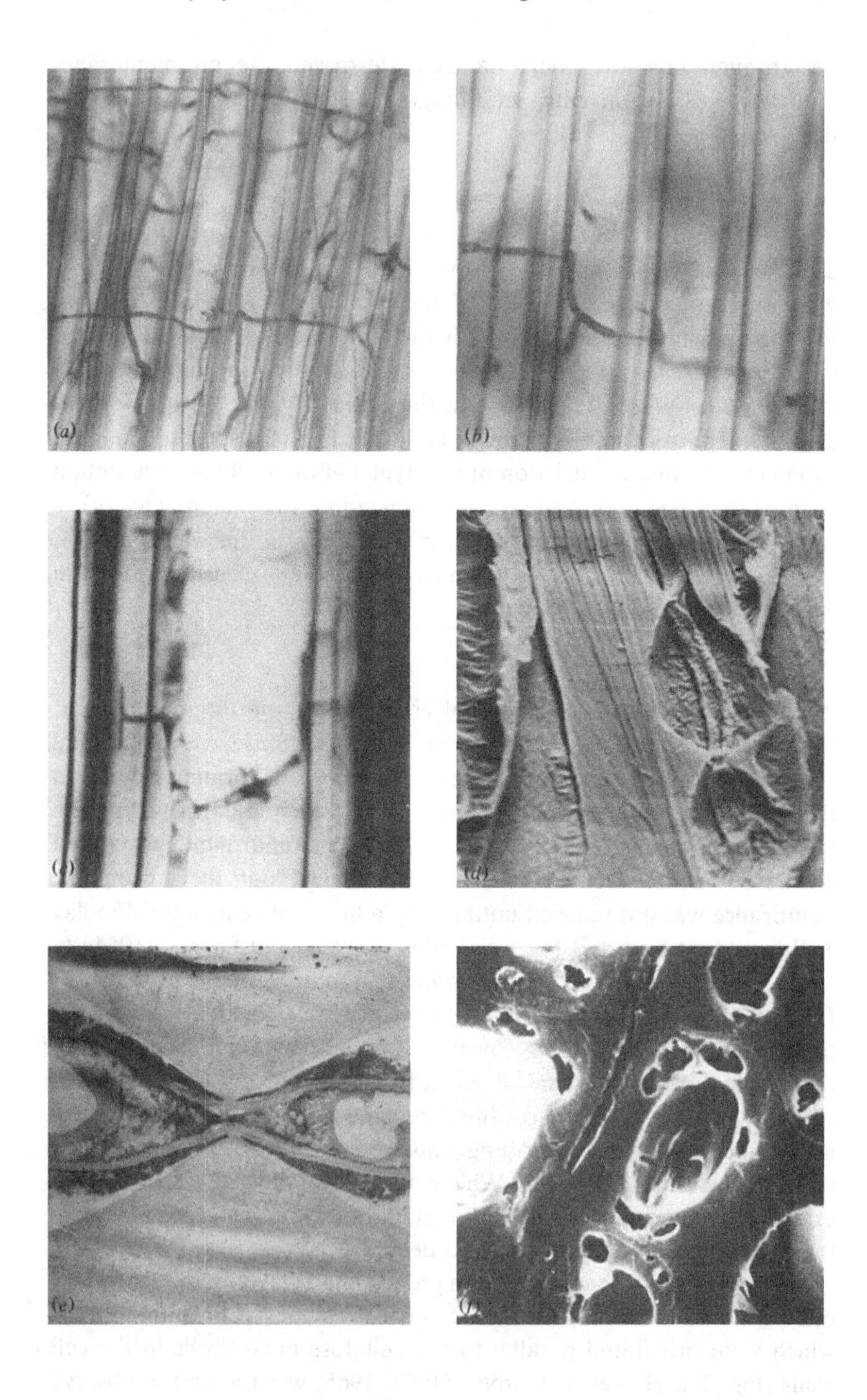
(a)
(b)
(d)
(e)

the way in which the fungal hyphae penetrated the S_2 layer of the wood cell wall. The details are given so that they can be contrasted with the behaviour of basidiomycetes. A short side-branch of a hypha lying in the cell lumen penetrated the S_3 layer perpendicular to the cell wall, and, on reaching the S_2 layer of that wall (or after having passed into the S_2 layer of the adjacent cell wall), the penetration hypha appeared to turn through an angle of 90° to lie parallel with the cellulose microfibrils, and branched at 180° to form what Corbett called a 'T-branch' (Fig. 2*c*). A cavity was eroded around the hyphal branches forming the cross of the T, and the hyphae increased in girth. Fine hyphal strands, termed 'proboscis' hyphae, (Corbett, 1963, 1965; Crossley & Levy, 1977) formed the initial extension growth of the hypha in the cavity.

Leightley (1977), Leightley & Eaton (1977), and M. Hale (personal communication) studied the development of the proboscis hypha by time-lapse photography. A fine hypha penetrates from the distal ends of the cavities in a direction parallel to the cellulose microfibrils. After a period of growth, further extension of the proboscis hypha ceases; the hypha increases in girth and a new cavity forms around it. After some time, extension growth continues by the development of a new proboscis hypha and the process is repeated time and again (Fig. 2*d*). A hypha in a cavity appears to be capable of forming a new T-branch, so setting up a new helical chain of cavities. The cavity itself is straight-edged even when viewed under a transmission electron microscope (Findlay, 1970) (Fig. 2*e,f*). This suggests that the action of the enzyme in forming the cavity is very exact, and that at no time does the enzyme system diffuse freely into the wood cell wall beyond the sharp edge of the cavity. Montgomery (Chapter 3) and Green (1980) suggest a possible reason for this behaviour.

The pointed ends of the cavities have given rise to much speculation and many authors have attempted to explain their formation in relation

---

Fig. 2. (*a*) Hyphae of a 'stainer' fungus in birch (*Betula pendula*), showing constrictions in fibre cell walls, × 150. (*b*) Hyphae of a 'stainer' fungus in birch, showing details of the fine constriction within the cell walls of fibres, × 300. (*c*) T-branch of the soft-rot fungus *Chaetomium globosum* in the S_2 layer of the wall of a tracheid of Scots pine (*Pinus sylvestris*), × 600. (G. Stevens.) (*d*) Chain of cavities caused by a hypha of the soft-rot fungus *Phialophora fastigiata* in birch. × 2500. (Findlay, 1970.) (*e*) Pointed ends of two soft-rot cavities showing the sharpness and straightness of the edge of the cavity and the hypha, × 12 000. (Findlay, 1970.) (*f*) Soft-rot cavities seen in transverse section in birch fibres, × 500. (A. Crossley.)

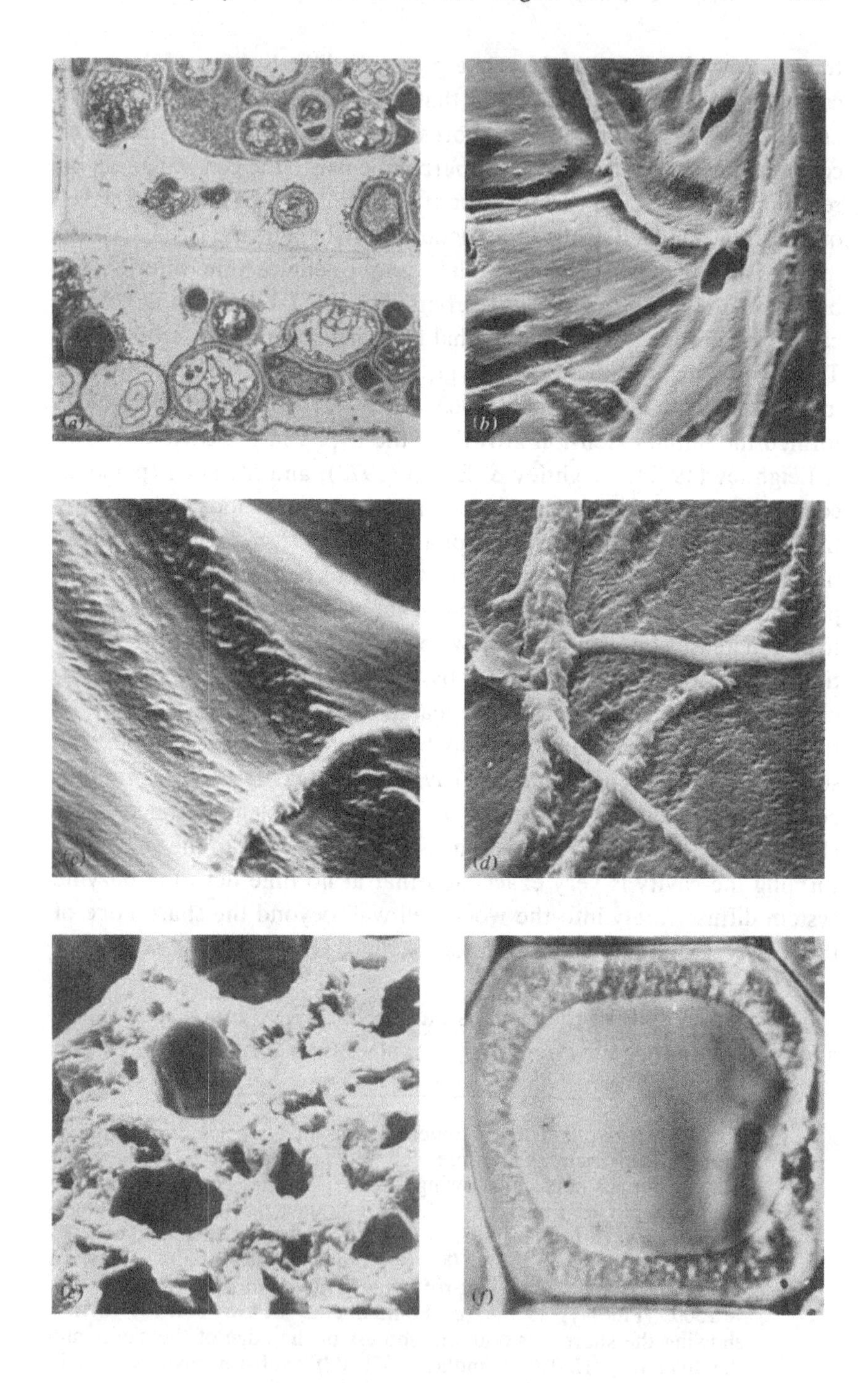
(a)
(b)
(c)
(d)
(e)
(f)

to the fine structure of the cell wall. Preston (1979) suggested a mechanism based on the differential rates of flow of the enzyme through the non-crystalline regions of the cellulose microfibrils.

Whatever the mechanism by which these cavities are formed, the effect on the wood structure can be devastating. Zainal's transmission electron photomicrographs (Zainal, 1976*a,b*) (Fig. 3*a*), show that the entire secondary wall apart from a residue of the S_3 layer can be removed during soft-rot attack on a birch (*Betula*) fibre. Dickinson (Chapter 10) points out the importance of the decay by these organisms when basidiomycetes do not develop.

Basidiomycetes

Until 1950, when Findlay & Savory first reported soft-rot fungi, basidiomycetes were regarded as the only fungi capable of causing severe economic loss of timber. White rots destroy both the holocellulose and lignin in the wood cell walls, bleaching the wood, whereas the brown rots utilise the holocellulose, leaving the lignin of the wood as a brown, brittle, and friable residue (see Hering: Chapter 12; Hintikka: Chapter 13).

Bravery (1971, 1972, 1975 and 1976) published an outstanding series of scanning electron photomicrographs, which illustrated another fundamental difference between the attack by white-rot and brown-rot fungi. His observations were confirmed by Nasroun (1971) with other basidiomycetes. Bravery showed that the hypha of a white-rot fungus penetrated into the cell lumen and lay on the inner surface of the wood cell wall. Lysis of the wall occurred along the hyphal contact, forming a groove or trough with a central ridge on which the hypha rested. As the

---

Fig. 3. (*a*) Cell wall of birch fibre almost completely destroyed by the soft-rot fungus *Phialophora fastigiata*, apart from the middle lamella complex and the S_3 layer, × 7000. (Zainal, 1976*b*.) (*b*) Troughs formed in the cell wall by hyphae of the white-rot fungus *Coriolus versicolor* in Scots pine (*Pinus sylvestris*), × 1000. (Bravery 1972.) (*c*) Part of Fig. 3(*b*) at a higher magnification to show the ends of the cellulose microfibrils visible at one side of the trough. × 3000. (Bravery, 1972.) (*d*) Hyphae of the brown-rot fungus, *Coniophora puteana*, showing that troughs are absent from region round the hyphae, × 3000. (Bravery, 1972.) (*e*) Transverse section of brown-rotted wood to show the destruction of part of the cell wall whilst the S_3 layer remains intact, × 250. (Nasroun, 1971.) (*f*) Transmission electron micrograph of a birch (*Betula pendula*) fibre decayed by the brown-rot fungus *Coniophora puteana*. Note the hypha in the cell lumen. (A. Crossley.)

hypha branched, so new troughs were formed and eventually the troughs coalesced, eroding the wall of the wood cell (Fig. 3*b,c*). The hypha was able to penetrate through the S_3 layer of the cell wall and well into the S_2; if the trough was parallel to the microfibrils its edges were smooth, but if it cut across their orientation the edges were ragged and the ends of the cellulose microfibrils could be seen clearly (Fig. 3*c*). The formation by white-rot fungi of troughs in the cell wall suggests some restriction to the free diffusion of enzyme away from the hypha. Montgomery (Chapter 3), Green (1980) and Green, Dickinson & Levy (1980) give possible explanations for these characteristics.

The brown-rot fungi show a different micromorphology (Bravery 1971; Nasroun, 1971). The hypha again penetrates the lumen and lies on the inner surface of the cell wall. The appearance of both the hypha and the S_3 wall layer change very little, but the other layers are completely altered by breakdown of the holocellulose. The resulting friable residue, with its high lignin content, looks more like expanded polystyrene or foam rubber than the normal smooth texture of unattacked walls (Fig. 3*d–f*). In this case, enzyme movement appears to be unrestricted and the enzyme system seems to diffuse completely through the wall layers, although it does not penetrate through the primary wall–middle lamella complex into adjacent cells (A. Crossley, unpublished) (Fig. 3*e,f*). Montgomery (Chapter 3) and Green (1980) discuss the characteristics of the polysaccharase enzyme system in brown-rot basidiomycetes.

'Secondary moulds'

The final group of organisms are the 'secondary moulds'. This grouping includes all those fungi which do not appear to attack wood itself but possess an active cellulase system, as exemplified by a clearance of ball-milled cellulose in agar culture. Their position in the succession seems to be associated with the appearance and eventual dominance of the decay fungi, particularly the basidiomycetes. The rôle of these 'secondary moulds', predominantly *Trichoderma viride* and *Gliocladium roseum*, is probably one of utilising the cellulose, derived from the breakdown of the wood, which is in excess of the requirements of the decay fungi. This cellulose food source may be a true nutritional excess, or the result of competition between the two groups of organisms for the partially decayed substrate.

Table 1 shows an example of the grouping of organisms isolated from posts in a field experiment.

Table 1. *Species isolated from stakes of* Betula *and* Pinus sylvestris *which had been in the ground for one year.* (After Clubbe, 1980.)

BACTERIA
Various unidentified cultural forms – mainly gram-positive rods
Actinomycetes

'PRIMARY MOULDS'
Arthrinium state of *Apiospora montagnei*
Botrytis cinerea
Cylindrocarpon destructans
Cylindrocarpon state of *Nectria coccinea*
Fusarium sambucinum
Fusarium avenaceum
Fusarium spp.
Mucor hiemalis
Mucor plumbeus
Penicillium cyclopium
Penicillium fennelliae
Penicillium purpurescens
Pestalotia sp.
Trichoderma viride
Verticillium lecanii
Verticillium psalliotae
Yeasts
Zygorhynchus heterogamus
Mycelia sterilia

'STAINERS'
Aphanocladium sp.
Aureobasidium pullulans
Botryodiplodia theobromae
Ceratocystis sp.
Cladosporium herbarum
Coniochaeta subcorticalis
Coniothyrium fuckelii
Drechslera dematioidea
Epicoccum purpurascens
Phialophora fastigiata
Phialophora melinii
Scytalidium sp.
Torula herbarum
Phoma fimeti
Phoma spp.
Xylaria sp.
Mycelia sterilia

'SOFT ROTS'
Coniothyrium fuckelii
Drechslera dematioidea
Epicoccum purpurascens
Humicola fuscoatra
Trichocladium opacum
Phialophora fastigiata
Phialophora melinii
Phialophora spp.
Mycelia sterilia

BASIDIOMYCETES
Sistotrema brinkmannii
Unknown IC/CC/BS7

White Rots
Bjerkandera adusta
Phlebia radiata
Stereum hirsutum
Unknown IC/CC/BS2
Unknown IC/CC/BS4
Unknown IC/CC/BS13
Unknown IC/CC/BS17

'SECONDARY MOULDS'
Acremonium strictum
Gliocladium roseum
Trichoderma viride

Water and nutrients

Corbett (1963, 1965) demonstrated the importance of cellular orientation by showing that fungal hyphae pass quickly through wood along the grain, i.e. from transverse face to transverse face, and more slowly across the grain along the rays, i.e. from tangential face to tangential face. The slowest penetration occurred across both grain and rays, i.e. from radial face to radial face. Here access must involve passage through either cell walls or pits or both. Bravery (1972) pointed out that this was also true for movement of liquids through wood.

A useful example of wood in ground contact is provided by fence posts (Levy, 1968) (Fig. 4). These contain a vertical zonation of microhabitats and yet are basically similar in structure throughout their length. For example, four moisture zones can be found: (*a*) conditions of intermittent wetting and drying; (*b*) permanent low water content; (*c*) permanent high water content, and (*d*) excess water producing anaerobiosis. The variety and activity of the micro-organisms in each zone differ, and this progressively increases the difference between each zone in what is otherwise a relatively uniform structure of void spaces and cell walls. Baines & Levy (1979) demonstrated that a stick of wood with the grain parallel to its long axis will act as a wick if there is a moisture gradient between the two ends. They quantified this wick action by measuring the rate of water movement through sticks of Scots pine (*Pinus sylvestris*) under different rates of evaporation. E. F. Baines (pers. commun.) also showed that, for a post in the ground, the

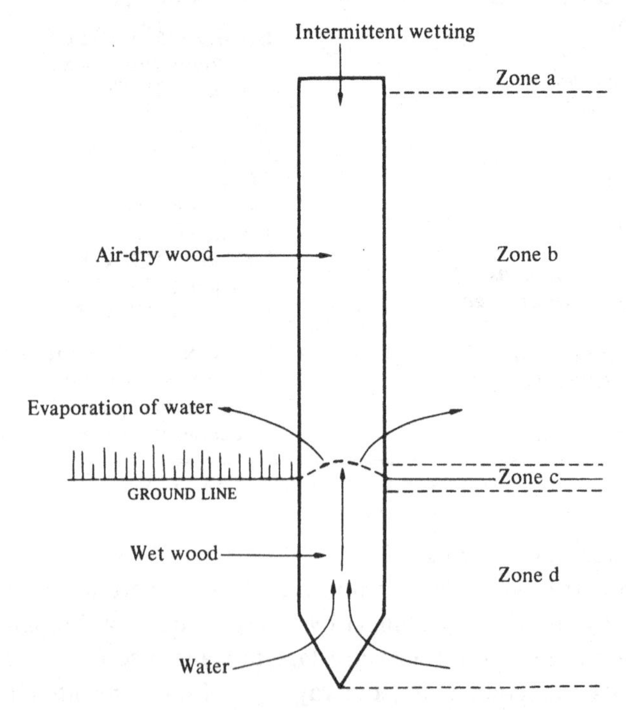

Fig. 4. Water movement in a fence post which produces ecological zones for the colonisation and development of micro-organisms.

equilibrium moisture contents of the four zones along its length will remain constant, but nevertheless water will move from zone (*d*) to zone (*c*) and will evaporate at the boundary between zones (*c*) and (*b*). In this region, materials dissolved in or carried by the water will be deposited where evaporation takes place. This is the ground-line region where the most active decay occurs and any such materials may be nutrients for the decay organisms present. Above the ground line, the movement of air across the surface dries the wood to moisture contents at which decay cannot take place. It is thus only at, or near, the ground line that the conditions are both wet enough and yet sufficiently aerobic to allow fungal growth. It is in this region of the post that biodeterioration will occur. The loss of strength which results is normally caused by decay by basidiomycetes, and is the climax of a series of events which have taken place since the post was first put into the ground. Yet basidiomycetes cannot be isolated for at least the first three months of exposure, (Corbett & Levy, 1963; Butcher, 1971; Käärik, 1967, 1968; Banerjee & Levy, 1971; Clubbe, 1978).

Cowling (1965), discussing the nutritional aspects of wood-inhabiting fungi including the basidiomycetes, pointed out the importance of the low nitrogen content of wood. He also considered the availability of cellulose in cell walls and how far the other wall constituents prevented breakdown of the cellulose. Cowling & Merrill (1966) also discussed nitrogen relationships of wood and its rôle in wood deterioration. Greaves (1966) demonstrated that readily available food reserves in the parenchyma of the rays of the sapwood are also important for the initial colonisation of the wood. In laboratory experiments with monocultures of fungi and bacteria, he showed that, the ray parenchyma was usually the first tissue to be substantially colonised prior to the initiation of decay of the cell walls. If the nutrients are water soluble they can move through wood and be deposited near the surfaces of cut timber in normal drying and seasoning processes (King, Oxley & Long, 1974). Evaporation from the surface causes a movement of water and materials in solution towards the surfaces, and the solutes remain as food sources which encourage scavenging micro-organisms to colonise wood.

Although wood is initially a nitrogen-poor environment for basidiomycetes, the nitrogen content can increase by microbial and physical effects. Sharp & Millbank (1974), Levy *et al.* (1974), Baines & Millbank (1976) and Baines *et al.* (1977) demonstrated that the phenomenon of nitrogen fixation can occur in wood providing it is wet enough. Baines & Millbank (1976) also demonstrated the progressive penetration of nitrogen-fixing organisms into wood from an exposed face, probably via the ray parenchyma. Uju (1979) carried out an investigation of the nitrogen relationships of wood in ground contact, by adding a solution of a soluble inorganic nitrate to sterile soil in which sterile sticks of Scots pine had been half buried. He showed that an increase in nitrogen content took place not only in that half of the stick below soil level but also in the half projecting above the soil level. He subsequently demonstrated this increase in nitrogen in the wood above ground level in a variety of conditions, and produced evidence to show that some of the increase was due to the movement of soluble nitrogen, since it could not be correlated with the presence of any organism (Uju, Baines & Levy, 1980).

Succession and climax of micro-organisms in sapwood stakes

Clubbe (1978, 1980) analysed a field experiment of Scots pine and birch sapwood stakes some of which had been treated with a wood preservative. He emphasised that the identification of individual species did not help to determine patterns of succession amongst the colonising organisms or provide a clue to their importance in the onset and development of decay. However, if the effect of each individual organism on the wood structure was taken into account, it was possible to group them into the categories described earlier in this Chapter and listed in Table 1. It was then possible to see patterns of development in both colonisation and decay.

The results of Clubbe's isolation technique showed the colonisation of birch sapwood stakes at the ground line in the course of a year by the six groups of organism. The bacteria appeared first and were the first colonisers at each depth, but they were isolated with decreasing frequency towards the end of the year. They were followed by the 'primary moulds' which also decreased towards the end of the year. The 'stainers' followed the 'primary moulds' and also showed some decline by the end of a year. The 'soft rots' and basidiomycetes did not appear until three months had elapsed. Both groups had penetrated to 40 mm depth at six months, after which the 'soft rots' declined in frequency of isolation whilst the basidiomycetes were clearly the dominant mycoflora with a very high percentage frequency. The 'secondary moulds' followed the basidiomycetes and built up in importance with them. Thus there appeared to be a clearly defined succession of the mycoflora, with the basidiomycetes forming the climax.

A similar sequence of events was shown to occur in Scots pine stakes, although it took a longer time, with no colonisation to 40 mm depth from the exposed tangential surface for the first nine months. There was some decline in the frequency of isolation of bacteria and 'primary moulds' by the end of a year, whilst the basidiomycetes were clearly increasing and appeared likely to become the climax organisms. The soft-rot fungi were never a major component of the mycoflora, whilst the 'secondary moulds' were building up at the end of the year.

A direct observation technique (Clubbe, 1980) confirmed these observations and showed some soft-rot decay in birch and a very small amount in Scots pine. After eighteen months exposure, considerable basidiomycete decay had occurred in both species at a depth of 3 mm from the exposed tangential surface, with birch showing greater decay. Dickinson (Chapter 10) quotes Clubbe's results from the preservative-treated stakes, where the presence of the preservative eliminated basidiomycetes from the mycoflora, and in their absence the 'soft rots' took over as a climax group of organisms. In treated birch there was clear evidence of soft-rot attack, whilst in Scots pine there was no sign of wall decay although these organisms were isolated with high percentage frequency.

The rôle of basidiomycetes

It is quite clear from the subjective qualitative observations of many researchers, and from the objective quantitative studies of Clubbe, that in normal wood in ground contact the wood-rotting basidiomycetes form the climax of the succession of colonising micro-organisms and are the main cause of decay. However, they first appear late in the succession and it is clear from Clubbe's work (Dickinson: Chapter 10) that even a mild loading of wood preservative eliminates them altogether and allows the soft-rot fungi to become the climax mycoflora. Further work is needed to determine the reasons for both these phenomena, but a number of factors may be concerned:

(*a*) the stake is dry when first put into the ground and may take some time to absorb enough water from the soil to support basidiomycete growth;

(*b*) once the wood gets wet it becomes too wet for the basidiomycetes, and time is required for bacteria to destroy the pit membranes which permits gaseous diffusion and gives rise to less anaerobic conditions;

(*c*) insufficient nitrogen and nutrients may be present until enough time has elapsed for the wick action, by which water moves along the grain, to have built up the nutrient content to a level necessary for the basidiomycetes;

(*d*) spore germination studies (Carey, 1980; J. K. Carey & J. G. Savory, unpublished) have shown that spores of *Gloeophyllum trabeum* will germinate on freshly-felled, green, wood but not on wood that has subsequently been kiln-dried, unless it has been colonised by bacteria and fungi;

(*e*) the fact that soft-rot fungi rapidly form the climax mycoflora when basidiomycetes are unable to colonise suggests a possible interaction between these two groups. Time may be required to enable the basidiomycetes to become dominant.

Conclusions

The wood-rotting basidiomycetes will normally become the climax mycoflora in wood in contact with the ground, and will be the major cause of decay. They do not, however, colonise the wood for at least three months, during which time a succession of organisms occur, some of which can cause partial breakdown of the wood, but normally the only other group of wood-rotting organisms, the soft-rot fungi, play a minor rôle, and are dominated and possibly suppressed by the basidiomycetes. If for any reason the basidiomycetes are eliminated from the mycoflora, the 'soft rots' take on a much more active rôle and become the climax group present.

References

Baines, E. F., Dickinson, D. J., Levy, J. F. & Millbank, J. W. (1977). Wood in ground contact: the nature of the hazard. *Record of the Annual Convention of the British Wood Preserving Association*, 33–45. London.

Baines, E. F. & Levy, J. F. (1979). Movement of water through wood. *Journal of the Institute of Wood Science*, **8**, 109–13.

Baines, E. F. & Millbank, J. W. (1976). Nitrogen fixation in wood in ground contact. *Material und Organismen, Supplement No. 3*, 167–73.

Banerjee, A. K. (1969). Fungi in fence posts: isolation, distribution and interaction of fungal species present and their effect. PhD thesis, University of London.

Banerjee, A. K. & Levy, J. F. (1971). Fungal succession in wooden fence posts. *Material und Organismen*, **6**, 1–25.

Bravery, A. F. (1971). The application of scanning electron microscopy on the study of timber decay. *Journal of the Institute of Wood Science*, **5**, 13–19.

Bravery, A. F. (1972). The effects of certain wood preservatives and their distribution on some fungi causing timber decay. PhD thesis, University of London.

Bravery, A. F. (1975). Micromorphology of decay in preservative treated wood. In *Biological Transformation of Wood by Microorganisms*, ed. W. Liese, pp. 129–42. Berlin: Springer-Verlag.

Bravery, A. F. (1976). Interactions between organic solvent preservatives, wood and decay. *Material und Organismen, Supplement No. 3*, 331–44.
Butcher, J. A. (1968*a*). The ecology of fungi infecting untreated and preservative treated sapwood of *Pinus radiata*. In *Biodeterioration of Materials*, ed. A. H. Walters & J. J. Elphick, pp. 444–59. Amsterdam, London & New York: Elsevier.
Butcher, J. A. (1968*b*). The ecology of fungi infecting untreated sapwood of *Pinus radiata*. *Canadian Journal of Botany*, **46,** 1577–89.
Butcher, J. A. (1971). Techniques for the analysis of fungal floras in wood. *Material und Organismen*, **6,** 209–32.
Carey, J. K. (1980). The mechanisms of infection and decay of window joinery. PhD thesis, University of London.
Clubbe, C. P. (1978). The use of selective media for studying the colonisation of wood in ground contact by micro-organisms. *International Research Group on Wood Preservation, Stockholm*, Document No. IRG/WP/186. Stockholm: IRG Secretariat.
Clubbe, C. P. (1980). Colonisation of wood by micro-organisms. PhD thesis, University of London.
Corbett, N. H. (1963). Anatomical, ecological and physiological studies on microfungi associated with decayed wood. PhD thesis, University of London.
Corbett, N. H. (1965). Micro-morphological studies on the degradation of lignified cell walls by Ascomycetes and Fungi Imperfecti. *Journal of the Institute of Wood Science*, **14,** 18–29.
Corbett, N. H. & Levy, J. F. (1963). Ecological studies on fungi associated with wooden fence posts. *British Wood Preserving Association News Sheet*, No. 27, 1–3; No. 28, 1–10. London.
Cowling, E. B. (1965). Micro-organisms and microbial systems. In *Cellular Ultrastructure of Woody Plants*, ed. W. A. Cote, pp. 341–68. New York: Syracuse University Press.
Cowling, E. B. & Merrill, W. (1966). Nitrogen in wood and its role in wood deterioration. *Canadian Journal of Botany*, **44,** 1533–44.
Crossley, A. & Levy, J. F. (1977). Proboscis hyphae in soft rot cavity formation. *Journal of the Institute of Wood Science*, **7,** 30–3.
Findlay, G. W. D. (1970). Microscopic studies on soft rot in wood. PhD thesis, University of London.
Findlay, W. P. K. & Savory, J. G. (1953). Soft rot. Decomposition of wood by lower fungi. *Proceedings of the 7th International Botanical Congress, Stockholm (1950)*, ed. H. Osvald & E. Åberg, p. 315. Stockholm: Almqvist & Wiksell.
Garrett, S. D. (1951). Ecological groups of soil fungi: a survey of substrate relationships. *New Phytologist*, **50,** 149–66.
Garrett, S. D. (1955). Microbial ecology of the soil. *Transactions of the British Mycological Society*, **38,** 1–9.
Greaves, H. (1966). New concepts of the decay of timber by micro-organisms with reference to the long-term durability of wood from archaeological sites. PhD thesis, University of London.
Green, N. B. (1980). The biochemical basis of wood decay micromorphology. *Journal of the Institute of Wood Science*, **8,** 221–8.
Green, N. B., Dickinson, D. J. & Levy, J. F. (1980). A biochemical explanation for the observed patterns of fungal decay in timber. *International Research Group on Wood Preservation, Stockholm*, Document No. IRG/WP/1111.
Käärik, A. (1967). Colonisation of pine and spruce poles by soil fungi after six months. *Material und Organismen*, **2,** 97–108.

Käärik, A. (1968). Colonisation of pine and spruce poles by soil fungi after twelve and eighteen months. *Material und Organismen*, **3**, 185–98.

Käärik, A. (1974). Sapwood staining fungi. *International Research Group on Wood Preservation, Stockholm*, Document No. IRG/WP/125.

King, B., Oxley, T. A. & Long, K. D. (1974). Soluble nitrogen in wood and its distribution on drying. *Material und Organismen*, **9**, 241–54.

Leightley, L. E. (1977). The mechanism of timber degradation by aquatic fungi. PhD thesis, Portsmouth Polytechnic.

Leightley, L. E. & Eaton, R. A. (1977). Mechanisms of decay of timber by aquatic micro-organisms. *Record of the Annual Convention of the British Wood Preserving Association*, 21–46. London.

Levy, J. F. (1968). Studies on the ecology of fungi in wooden fence posts. In *Biodeterioration of Materials*, ed. A. H. Walters & J. J. Elphick, pp. 424–8. Amsterdam, London & New York: Elsevier.

Levy, J. F. (1971). Further basic studies on the interaction of fungi, wood preservatives and wood. *Record of the Annual Convention of the British Wood Preserving Association*, 63–75. London.

Levy, J. F. (1973). Colonisation of wood by fungi. *British Wood Preserving Association News Sheet*, No. 130, 1–2. London.

Levy, J. F. (1975). Bacteria associated with wood in ground contact. In *Biological Transformation of Wood by Micro-organisms*, ed. W. Liese, pp. 64–73. Berlin: Springer-Verlag.

Levy, J. F., Millbank, J. W., Dwyer, G. & Baines, E. F. (1974). The role of bacteria in wood decay. *Record of the Annual Convention of the British Wood Preserving Association*, 3–13. London.

Merrill, W. & French, D. W. (1966). Colonisation of wood by soil fungi. *Phytopathology*, **56**, 301–3.

Nasroun, T. A. (1971). Mode of changes in the ultrastructure and microstructure of wood attacked by white rots and brown rots. Diploma thesis, Imperial College, University of London.

Preston, R. D. (1979). The shape of soft-rot cavities – a hypothesis. *Wood Science and Technology*, **13**, 155–63.

Savory, J. G. (1954*a*). Damage to wood caused by micro-organisms. *Journal of Applied Bacteriology*, **17**, 213.

Savory, J. G. (1954*b*). Breakdown of timber by Ascomycetes and Fungi Imperfecti. *Annals of Applied Biology*, **41**, 336–47.

Savory, J. G. (1955). Role of microfungi in the decomposition of wood. *Record of the Annual Convention of the British Wood Preserving Association*, 3–35. London.

Sharp, R. F. & Millbank, J. W. (1974). Nitrogen fixation in deteriorating wood. *Experimentia*, **29**, 895–6.

Uju, G. C. (1979). Nitrogen relationships of wood in ground contact. PhD thesis, University of London.

Uju, G. C., Baines, E. F. & Levy, J. F. (1980). Nitrogen uptake by wick action in wood in soil contact. *Journal of the Institute of Wood Science*, **9**, 23–6.

Zainal, A. S. (1976*a*). The soft rot fungi: the effect of lignin. *Material und Organismen, Supplement No. 3*, 121–7.

Zainal, A. S. (1976*b*). Micromorphological studies of soft rot fungi in wood. PhD thesis, University of London.

10 The decay of commercial timbers

D. J. DICKINSON
Department of Botany, Imperial College, University of London, Prince Consort Road, London SW7 2BB, England

Other chapters in this volume discuss the decay by basidiomycetes of living trees and of wood in contact with the ground (Mercer: Chapter 8; Levy: Chapter 9). Here special consideration is given to the rôle of basidiomycetes in the decay of wood utilised by man, i.e. timber. It is impossible to consider the basidiomycetes in isolation as they are often associated with other organisms, but reference to the latter will be made only when they significantly affect the presence or action of the basidiomycetes.

In attempting to remove some of the useful wood from the natural cycle of the forest and utilise it in his constructions, man has learnt, often to his cost, that fungi including many basidiomycetes have to be respected as adversaries. However, wood is still one of the most useful structural materials and in a world of diminishing resources represents a renewable raw material.

Differences between living and dead wood

In living trees the sapwood is considered to be much more resistant to decay than the heartwood. This situation is dramatically changed with the death or felling of the tree and its subsequent conversion for use. The sapwood of all tree species is then subject to microbiological attack. The heartwood resistance varies considerably; the heartwoods of some species are no more resistant than the sapwood, while others will survive for many years in conditions otherwise suitable for decay. The factors which influence the decay resistance of heartwood to fungal decay have been reviewed by Scheffer & Cowling (1966), and the difference in decay resistance between heartwood and sapwood in living trees has been investigated by Shigo (1968). The latter described the host response to injury and invasion as having the effect of cutting off the sapwood present at the time of injury from the newer wood, which remains free of infection and decay. Initially, this reaction is seen as discoloured wood, which eventually may decay due to the action of basidiomycetes. Total decay of the wood present at the time of injury may take up to fifty years. In contrast, sapwood in contact with the ground may be totally decayed in less than five years. The reaction of living trees to decay organisms and its consequences are fully reviewed by Mercer (Chapter 8), and Levy (Chapter 9) discusses the structural features of different types of wood in relation to fungal attack.

Infection of timber in the standing crop

Infection of standing timber by basidiomycetes is of major importance in its subsequent commercial use. Timber is often used with major internal decay, which markedly affects its performance. The infecting organisms may survive, and decay may continue during storage or in service. In timbers of large dimensions decay in storage gives rise to 'hidden decay' which becomes evident only when the timber is resawn (Roff, Cserjesi & Swan, 1974). *Poria carbonica*, for example, causes considerable internal decay in transmission poles in North America. The author has also found this fungus causing severe decay in the structural members of water cooling towers in the UK, and, being a North American fungus, it was apparently present in the timbers prior to construction of the towers. A further example of decay caused by pre-infection by this fungus was the decay of the large flagpole of Douglas fir (*Pseudotsuga menziesii*) in Kew Gardens near London, U.K.

In the past such infections were associated with timber converted from over-mature trees, particularly from North America, but, even in recent forestry practice, commercial thinning and extraction operations in plantations have also contributed greatly to the decay of standing timber. In the UK, control of 'butt-rot' of conifers caused by *Heterobasidion annosum* illustrates how correct practice is the first line of defence in protection. *H. annosum* is the most serious disease in British forests (Anon., 1967). It normally enters conifers through the root, penetrates the heartwood and grows up the stem causing a butt-rot. It can also enter via root wounds caused during extraction of timber, but by far the most important source of infection is via the stumps produced during thinning operations. The fungus colonises the surface of the stump, grows down into the roots and thence into the roots of neighbouring trees. The rate of development of the rot and the age at which it occurs vary greatly between tree species and from site to site. If infection is present throughout the crop at an early age, serious damage can be expected by the time the crop is felled and decay may extend 3 m up the tree with stain up to 6 m in exceptional cases. In British forests, *Phaeolus schweinitzii* and *Armillaria mellea* also cause significant damage to standing timber, but *H. annosum* is responsible for roughly 90% of the decay in conifers (Anon., 1967).

Infection of timber after harvesting

Logs and large timbers may take many months to air-dry, and during this time much internal decay can occur, either continuing from infection in the standing tree or from new infection. Logs are commonly 'water-stored' to prevent post-harvest decay; under these conditions decay fungi cannot survive, and logs can be stored for long periods without fungal attack, although bacteria can degrade the pits. This type of attack by bacteria is described in wood in contact with the ground (Levy: Chapter 9). Apart from water storage, post-harvest degradation of logs is best controlled by careful handling and controlled drying.

The sapwood of sawn timber is readily colonised by a wide range of fungi. The earliest colonisers are not usually basidiomycetes but are fungi that use residual food materials, particularly those stored in rays, and they do not significantly degrade the cell walls of the wood. Many species have been recognised in this early colonisation which results in the phenomenon known as 'blue stain' (Käärik, 1974), but by far the most important organisms in *Pinus* are species of the ascomycete *Ceratocystis*. As pointed out by Levy (Chapter 9), the 'stainers' are among early colonisers which eventually give way to the basidiomycetes. This succession also occurs in converted wood when drying is retarded.

In mills producing large quantities of sawn softwood, some sort of treatment after sawing is normal practice. This treatment need be only superficial provided it is adequate to control the stain during the seasoning period. Modern handling practices such as the use of packaged timber can lead to increases in the stain problem on timber imported into the UK, and also give rise to a substantial increase in incipient decay. A common cause of such decay is *Phlebia gigantea*. Its control is now considered to be as important as the control of 'blue stain', and new chemical treatments for stain must also be effective against decay basidiomycetes. Examples of this problem include the import of *Pinus radiata* from New Zealand (Butcher, 1973) and of green *Pinus pinaster* from Portugal into the UK for the manufacture of packing cases and pallets. Without the use of effective fungicidal treatment such packages would not only be blue-stained, but also severely decayed by *P. gigantea* (D. J. Dickinson, unpublished).

Degradation in service

Decay of timber in buildings

It has been estimated that in 1977 £3 million per week was spent in the UK on repairing damage caused by wood-rotting fungi in buildings (Scobie, 1980). Virtually all of this damage can be attributed to water. The prevention and cure of decay within buildings is therefore initially always linked to understanding and curing such problems as water ingress, rising damp, ineffective sub-floor ventilation, condensation, and with general rainware (i.e. gutters, downpipes, etc.) and plumbing problems. Detailed descriptions of these problems and their cure, although of paramount importance in prevention of decay in buildings, seem inappropriate in the context of this volume and it suffices to stress that wood-destroying fungi, most of which are basidiomycetes, can only establish themselves in buildings if the moisture content of the timber is allowed to increase to a suitable level due to one of these factors.

Decay of timber in buildings is popularly referred to as either 'dry' or 'wet rot'. In the past the term 'dry rot' was loosely used to describe any type of decay in buildings, but the convention nowadays is to restrict this description to the decay caused by the true 'dry rot' fungus, *Serpula lacrimans*. The term 'wet rot' is used to describe decay caused by any other fungus attacking timber in a building, generally under wetter conditions. Visible mycelium is often absent, which presumably led builders in the past to suspect that moisture alone was responsible for the decay, hence the term 'wet rot'. Unfortunately, the presence of mycelium with 'wet rot' fungi often causes great confusion and 'wet rot' organisms are often confused with *S. lacrimans* by inexperienced timber surveyors. It is a matter of great importance to distinguish the two types of decay as the procedures adopted by commercial companies for the eradication of *S. lacrimans* are stringent and costly, whereas relatively simple measures are normally adequate for the eradication of 'wet rot'.

Dry rot. S. lacrimans can be considered to be a truly domesticated fungus as it rarely occurs outside buildings in this country. There have been, and still are, many misconceptions concerning the growth of *S. lacrimans* within buildings. It generally grows in woodwork that is in contact with damp masonry. This takes some time to dry out so that a re-infection of replacement timbers can often take place. The fungus is also able to withstand the alkaline conditions within walls, and it is capable of penetrating and travelling within the walls or between the plaster and brickwork. This can lead to extensive spread of the fungus. *S. lacrimans* is often quoted as being able to colonise dry areas, but generally spread will occur only in wet conditions.

S. lacrimans produces substantial strands behind the advancing hyphal front. The function and physiology of these strands are discussed by Jennings (Chapter 5). It suffices to say here that their function within buildings has often been misunderstood, but they obviously contribute to the survival and success of this organism in an environment where the food source may not be continuous but separated by inert materials. Another popular misconception regarding *S. lacrimans* is that, unlike 'wet rot', it cannot be controlled by drying alone. In practical terms this may well be the case, but it is due to the fact that the fungus may have penetrated the fabric of the building which is virtually impossible to dry out. In reality *S. lacrimans* is more sensitive than the common 'wet rot' organisms to drying out.

When *S. lacrimans* is growing actively in conditions of high humidity as in a wet cellar, it can produce spectacular masses of mycelium which exude drops of water. Under drier conditions the mycelium is compact and reduced to a grey felted 'skin', with characteristic yellow patches and tinges of lilac. It causes a typical 'brown rot' of the wood as described by Levy (Chapter 9). The rotten wood shrinks and splits into cubical pieces with deep cross cracking. The fruit bodies vary from flat pancakes to irregular brackets depending on their position. The basidiospores appear rusty red *en masse*, and are produced in prodigious numbers. They are evident to the naked eye in infected buildings as brick-red dust-like deposits. The appearance of the fruit bodies and the associated spore dust within a building are often the first indications of decay being present.

Wet rot. The decay fungus found most commonly in buildings is *Coniophora puteana*, which is commonly quoted as being responsible for up to 90% of wood decay within buildings. In the early stages of attack by this basidiomycete, darkening of the wood surface occurs, but the attack may seem very similar to that of *S. lacrimans*. Sometimes

Fig. 1. Characteristic dark strands of *Coniophora puteana* on thermal building blocks. (Princes Risborough Laboratory, Crown Copyright.)

cubical cracking is absent and the splitting is largely longitudinal (Anon., 1969*a*). This has led to another popular misconception: that cubical cracking is indicative of decay caused by *S. lacrimans*. *C. puteana* often causes internal decay of wood flooring, leaving a relatively sound skin of timber at the surface. Under most conditions little or no visible mycelium is produced but when present it often produces fine, dark brown-black, strands (Fig. 1). The fruit body is rare in buildings and appears as a thin skin, coloured characteristically olive green to dull olive brown. The spores are olive brown.

Several other wood-rotting basidiomycetes are also found in buildings (Findlay, 1951). The commonest are probably *Fibroporia vaillantii*, which affects wood in a very similar manner to *S. lacrimans*, and *Paxillus panuoides*, which requires very moist conditions and resembles *C. puteana* in the way in which it breaks down the timber. Two other fungi worthy of mention are *Amyloporia xantha* and *Gloeophyllum trabeum*. These fungi are often associated with the decay of timber in roofs. In recent years widespread adoption of flat roof construction has led to many cases of considerable decay in such structures. Leakage and condensation problems lead to wetting of the timbers, but the root cause is often bad design and construction.

Decay of exposed wood in buildings

Most external joinery in the UK is manufactured from softwood, principally Scots pine (*Pinus sylvestris*) and hemlock (*Tsuga*), and most decay problems are associated with mass-produced windows and doors. Although decay in exterior softwood joinery was reported in America in the 1930s (Hubert, 1934), decay in such joinery in the UK was considered to be of consequence only in certain environments such as greenhouses (Cartwright & Findlay, 1958). After the late 1950s, however, a problem became apparent and by the late 1960s and early 1970s had risen to nothing less than a national disaster. This massive increase in decay, especially of window joinery, has been attributed to several factors, but it is generally accepted that the major causes are: (i) a change in the raw material; large amounts of susceptible sapwood are now employed instead of the more durable heartwood, due to the utilisation of smaller trees which contain much lower proportions of heartwood; (ii) design and manufacturing faults which lead to penetration of water into open joints; (iii) a change in paints and primers which are now lead-free. These industrial changes have provided new niches for the basidiomycetes.

The problem has stimulated much work in the UK, principally at the Princes Risborough Laboratory of the Building Research Establishment. Tack (1968) showed in a survey in south-east England that at least half of the window frames examined had one or more of the lower joints moist enough for decay to occur. This survey and subsequent experimental work (Anon., 1969*b*) clearly demonstrated that once moisture had penetrated into joinery, the rate of water loss through the normal three coats of paint was very slow. Other surveys (Beech & Newman, 1975; Soane, 1978) confirmed the incidence of decay in untreated joinery.

At first much of the decay was attributed to *Coniophora puteana*. Subsequent examinations and surveys revealed that this fungus was in fact rare. In window joinery and doors of Canadian hemlock (*Tsuga canadensis*), Soane's survey (1978) yielded a range of basidiomycetes. One of the commonest was the white-rot fungus, *Phellinus contiguus*, followed by *Coriolus versicolor* and strains of *Hyphoderma*, *Hyphodontia*, *Sistotrema brinkmannii* and *Stereum* sp. *Poria placenta* was also isolated as a result of infection of *Tsuga* in Canada (Savory & Carey, 1975). In a systematic study of window frames from various sites, *P. contiguus* was again shown to be abundant and *C. puteana* absent (Savory & Carey, 1976). Moist conditions in joinery components are stabilised by the paint film, which prevents drying. Under such conditions, the basidiomycetes become established and cause extensive decay.

Other organisms must be important in the early stages of this decay of external joinery. Carey (1980) in an extensive survey of exposed and painted components simulating joinery showed that a sequence of micro-organisms colonised the wood. Initially bacteria, 'primary moulds' and 'stainers' are involved, but eventually the decay basidiomycetes appear. This succession is closely analogous to that colonising timber in contact with the ground (Levy: Chapter 9). Several of these early colonisers have been shown to increase the absorbency of the joinery at the joints in a manner similar to the ponding effect described earlier (Carey, 1980). This undoubtedly is a major factor in the colonisation and decay of external joinery by basidiomycetes.

In an attempt to combat the national problem of decay in joinery, the National Housebuilders Council of the UK made preservative treatment mandatory for softwood window joinery and later for doors. The principal fungicide used in these preservatives is tri-*n*-butyl tin oxide (TBTO), which has been shown to impart an initial fungitoxicity to the timber equivalent at least to that of *Pinus* heartwood when applied as a three-minute dip (Baker, Laidlaw, Miller & Savory, 1977). However, recent work (Henshaw *et al.*, 1978) has thrown doubt on the permanence of this fungicide. Current work also indicates that certain fungi, including basidiomycetes and staining fungi isolated from treated wood, are capable of detoxifying this fungicide, which may eventually lead to colonisation by basidiomycetes and subsequent decay (Carey, 1980; D. J. Dickinson, unpublished).

The importance of the establishment of stable wet conditions in joinery has been recognised by the preservative industry and most companies now offer water repellent versions of joinery preservatives. Results of long-term trials in the USA indicate that water-repellent treatment alone can be just as effective as with a fungicide (Feist & Miraz, 1978).

Decay of timber in the ground

The colonisation and subsequent decay of timber in contact with the ground are discussed in depth by Levy (Chapter 9). The decay of such timber by basidiomycetes represents the climax of a succession of micro-organisms. The potential for such decay has been fully appreciated for many years and commercial timbers used in this situation have been either of known durable species or treated with wood preservatives. The first patents for the pressure treatment of timber by creosote were granted in the 1830s and since that time creosote has proved to be a first-class preservative. When used correctly, it has imparted excellent long-term durability to timber in ground contact. Classic examples are the creosoted railway sleepers and transmission poles which have given many years service when correctly treated. Failure of creosoted poles and sleepers has been associated with decay by relatively few species of fungi. The principal fungus involved in this country is *Lentinus lepideus*, which normally occurs on inadequately treated material. In a recent UK survey of some 300 electricity transmission poles of the Eastern Electricity Board (D. J. Dickinson & B. Calver, unpublished), approximately 5% of the poles were found to be affected by decay. In these poles, *L. lepideus* was the dominant organism, and in all cases the decay was associated with inner, untreated, sapwood and heartwood.

Since the last war there has been a steady move towards the use of fixed, water-borne, salt preservatives such as the copper-chrome-arsenates (CCA). In recent years there have been some problems with these preservatives. The difficulties have been loosely described as the 'soft-rot' problem and have been associated principally with hardwoods, particularly transmission poles in the tropics and Australia. In the recent past CCAs have been used primarily with softwood species, principally *Pinus*. With the development of new markets, particularly in Australia, other species including hardwoods such as the eucalypts have been extensively treated. In the early 1970s, it became apparent that these treated hardwood species were not performing as well as their softwood counterparts in Europe (Levy, 1971; Greaves, 1972; Dickinson, 1974). In all cases the decay in the treated sapwood was caused by soft-rot fungi and could not be attributed to the basidiomycetes which were always associated with decay of untreated sapwood or incorrectly treated poles in similar environments.

In an attempt to understand the differences between the performance of hard and softwoods, Dickinson, Sorkhoh & Levy (1976) subjected CCA-treated and untreated *Betula* and *Pinus sylvestris* to monocultures of soft-rot and basidiomycete fungi in laboratory experiments. Their results demonstrated a fundamental difference between decay in treated and untreated wood, and between treated hardwood and treated softwood, at normal commercial levels of CCA. Both species of untreated wood were decayed by soft-rot fungi and basidiomycetes. The treated material, however, performed quite differently. The treated softwood was completely protected, but the treated hardwood was protected against basidiomycetes only, the fibres being attacked by the soft-rot fungi, although the rays were protected. Microdistribution studies of the preservative elements showed considerable differences between the two timber species tested. The treated softwood was evenly treated with the preservative components throughout the tissues, and analysis of the cell-wall layers showed a relatively even distribution across the cell wall, with somewhat higher levels within the S_2 layer. In the hardwoods the preservative components were unevenly distributed, with accumulations in the rays. The S_2 layers of the fibres were poorly treated, but there was an accumulation of preservative on the lumen surfaces, a fact previously noticed in sycamore (*Acer pseudoplatanus*) (Dickinson, 1974). When the way in which basidiomycetes attack the cell wall via the S_3 layer from the lumen is considered (Levy: Chapter 9), it is easy to see how a preservative deposited on the lumen surface might inhibit such fungi. However, in the case of the soft-rot fungal attack of hardwood fibres, once entry into the poorly treated S_2 region of the wall has been gained, decay can proceed normally, since this is the normal mode of soft-rot attack (Levy: Chapter 9). In contrast, softwood with an even distribution of the preservative and good treatment of the S_2 layers of the tracheids was well protected from all fungi.

Clubbe (1980) demonstrated this fundamental difference between untreated and treated timber in field trials of hardwoods and softwoods in contact with the ground. In a controlled three-year trial, differences occurred between untreated and treated stakes. The microbial sequence in the untreated timber (Levy: Chapter 9) showed that the basidiomycetes were the climax colonisers of both the softwood and hardwood, but, after treatment, basidiomycetes failed to colonise both types of wood and the soft-rot fungi assumed greater importance. This work substantiated the field observations already reviewed, showing that decay due to soft rot occurred in the hardwoods followed by similar decay in the treated softwoods at a much later date. Further confirmation of these trends has been obtained in a current international field trial involving fifty-eight species of timber. Untreated controls have been invaded by basidiomycetes, and treated hardwoods have begun to show signs of soft rot at the lower treatment levels (Dickinson & Levy,

1980). Treated timber in the ground is therefore an inhospitable habitat for basidiomycetes. The soft-rot fungi assume a greater importance and are not displaced by the basidiomycetes (Levy: Chapter 9) but proceed to decay the treated wood slowly.

Conclusions

Commercial timbers, therefore, present various habitats for basidiomycete colonisation and decomposition. In some cases the environment is too severe and the soft-rot fungi or even the bacteria assume greater importance, basidiomycetes failing to colonise the substrate. In others, bad practice by the users of timber and neglect of buildings provide ideal situations for rapid colonisation and decay by basidiomycetes.

References

Anon. (1967). *Fomes annosus. Forestry Commission, U.K.*, Leaflet, No. 5. London: Her Majesty's Stationery Office.

Anon. (1969*a*). *Decay in building: recognition, prevention and cure.* Technical Note, No. 44. Princes Risborough Laboratory, Building Research Establishment, UK.

Anon. (1969*b*). *Loss of moisture from painted wood.* Technical Note, No. 34. Princes Risborough Laboratory, Building Research Establishment, UK.

Baker, J. M., Laidlaw, R. A., Miller, E. R. & Savory, J. G. (1977). Research in wood preservation at Princes Risborough Laboratory 1975 and 1976. *Record of the Annual Convention of the British Wood Preserving Association*, 3–27. London.

Beech, J. C. & Newman, P. L. (1975). *A survey of decay, insect attack and moisture contents in timber components in Scottish houses.* Current Paper, No. 97/75, Princes Risborough Laboratory, Building Research Establishment, UK.

Butcher, J. A. (1973). Laboratory screening trials of new prophylactic chemicals against sap stain and decay in sawn timber. *Material und Organismen*, **8**, 51–70.

Carey, J. K. (1980). The mechanisms of infection and decay of window joinery. PhD thesis, University of London.

Cartwright, K. St G. & Findlay, W. P. K. (1958). *Decay of Timber and its Prevention*, 2nd edn. London: Her Majesty's Stationery Office.

Clubbe, C. P. (1980). Colonisation of wood by micro-organisms. PhD thesis, University of London.

Dickinson, D. J. (1974). The microdistribution of copper-chrome-arsenate in *Acer pseudoplatanus* and *Eucalyptus maculata. Material und Organismen*, **9**, 21–3.

Dickinson, D. J. & Levy, J. F. (1980). Preliminary results of the field experiment to determine the performance of preservative-treated hardwood with particular reference to soft rot. *Report of Working Group 3, International Research Group on Wood Preservation, Stockholm.*

Dickinson, D. J., Sorkhoh, N. A. A. & Levy, J. F. (1976). The effect of the microdistribution of wood preservatives on the performance of treated wood. *Record of The Annual Convention of the British Wood Preserving Association*, 25–40. London.

Feist, W. C. & Miraz, E. A. (1978). Protecting mill work with water repellants. *Forest Products Journal*, **28**, 31–5.

Findlay, W. P. K. (1951). A note on fungi of less common occurrence in houses. *Transactions of the British Mycological Society*, **34**, 35–7.

Greaves, H. (1972). Structural distribution of chemical components in preservative-treated wood by energy dispersive X-ray analysis. *Material und Organismen*, **7**, 277–86.

Henshaw, B. G., Laidlaw, R. A. Orsler, R. J., Carey, J. K. & Savory, J. G. (1978). The permanence of tributyl tin oxide in timber. *Record of the Annual Convention of the British Wood Preserving Association*, 19–29. London.

Hubert, E. E. (1934). The protection of jointed wood products against decay and stain. *University of Idaho Bulletin*, No. 29, 1–36.

Käärik, A. (1974). Sapwood staining fungi. *International Research Group on Wood Preservation, Stockholm,* Document No. IRG/WP/125. Stockholm: IRG Secretariat.

Levy, J. F. (1971). Long-term trials of wood preservatives in fence posts. *Timber Trades*, **227**, 71–3.

Roff, J. W., Cserjesi, A. J. & Swan, G. W. (1974). Prevention of sap stain and mold in packaged timber. *Department of the Environment, Canadian Forestry Service*, Publication No. 1325.

Savory, J. G. & Carey, J. K. (1975). Why preserve external doors? *Timber Trades Journal*, **295**, 12–15.

Savory, J. G. & Carey, J. K. (1976). Laboratory assessment of the toxic efficiency of joinery preservative treatments. *Record of the Annual Convention of the British Wood Preservation Association*, 3–21.

Scobie, D. (1980). *Timber Trades Journal*, **312**, 21.

Sheffer, T. C. & Cowling, E. B. (1966). Natural resistance of wood to microbial deterioration. *Annual Review of Phytopathology*, **4**, 147–70.

Shigo, A. L. (1968). Discoloration and decay in hardwoods following inoculations with hymenomycetes. *Phytopathology*, **58**, 1493–8.

Soane, G. E. (1978). The incidence of decay in external doors. *The British Wood Preservation Association, News Sheet*, No. 151.

Tack, C. H. (1968). Window joinery in service. *Building*, **214**, 135–6.

11 Decomposition by basidiomycetes in aquatic environments

E. B. GARETH JONES

Department of Biological Sciences, Portsmouth Polytechnic, Portsmouth PO1 2DY, England

Introduction

In aquatic environments, substratum decomposition is the result of the activities of a variety of organisms adapted or tolerant of conditions prevailing in a fluid medium. Materials whose decomposition in aquatic habitats has been studied include: leaves (Kaushik & Hynes, 1971; Barlocher & Kendrick, 1974; Descals, Nawawi & Webster, 1976; Ingold, 1976), mangroves (Newell, 1976), seaweeds (Haythorn, Jones & Harrison, 1980), sewage (Bridge Cooke, 1976) and wood (Jones, 1968, 1971; Jones, Kühne, Trussell & Turner, 1972).

In this chapter I have confined my observations to decomposition of materials (especially wood) in aquatic habitats by basidiomycetes. However, fungi imperfecti and ascomycetes are the most frequently encountered fungi on wood in aquatic habitats (Jones, Turner, Furtado & Kühne, 1976; Eaton, 1976; Ingold, 1976; Sanders & Anderson, 1979). Microfungi appear to be able to withstand the saturated conditions prevailing in habitats such as water-cooling towers far better than basidiomycetes (Levy: Chapter 9). Microfungi cause soft-rot decay of wood in such habitats (Eaton, 1976) and their ability to penetrate the S_3 layer of the secondary wall and grow within the S_2 layer may confer on them a competitive advantage over the basidiomycetes (E. B. G. Jones, 1981). This aspect will be considered in greater detail later in this chapter.

The succession of organisms on submerged wood is similar to that described by Levy (Chapter 9) for wood in ground contact. Early colonisation of such wood is by bacteria and actinomycetes (Holt, Jones & Furtado, 1979), followed by micro-fungi (Jones, 1963, 1968; Eaton & Jones, 1971*a,b*; Irvine, 1974; D. Kane, unpublished), whereas basidiomycetes appear later in the succession (Byrne & Jones, 1974). In freshwater habitats, nematodes and oligochaete worms also play a rôle in the decomposition of wood (Jones, 1981) in that they tunnel the wood, exposing deeper layers for further colonisation by fungi. In marine environments, crustacean and molluscan wood borers are extremely active in the decay of timber (Becker, 1971; Jones, Turner, Furtado & Kühne, 1976).

Taxonomy and distribution

Aquatic basidiomycetes are taxonomically unrelated, belonging to such diverse groups as the Ustilaginales, Polyporales, Tremellales and Gasteromycetes (Tables 1, 2).

Marine species

Marine basidiomycetes are neither numerous (Table 1) nor common in occurrence. Of ninety-eight test blocks submerged at six test sites (Byrne & Jones, 1974) only nine test blocks were colonised by basidiomycetes: *Digitatispora marina* (six) and *Nia vibrissa* (three). Rees, Johnson & Jones (1979) found no basidiomycetes on driftwood collected from sand dunes in Jutland, Denmark. However, heterobasidiomycetous yeasts have been repeatedly isolated from seawater, but rarely associated with any particular substratum. They are able to derive the necessary nutrients from seawater (van Uden & Fell, 1968) and do not appear to be parasites of plants, unlike their terrestrial ancestors (Fell, 1976).

Of the three marine homobasidiomycetes known, *N. vibrissa* is the most frequently collected species. The pale-pink fruit bodies generally appear on substrates after a period of incubation (Fig. 1*c*). J. Kohlmeyer & E. Kohlmeyer (1964), Fazzani & Jones (1977) and Leightley & Eaton (1979) have shown that the fruit bodies are covered by tightly interlaced appendages, the tips of which are curved or bifurcate (Fig. 2*a*). The basidia are globose and deliquesce to release the appendaged basidiospores (Fig. 2*b*). *N. vibrissa* has been found growing on mangroves, driftwood, test panels (Leightley & Eaton, 1979) and feathers (G. Rees, unpublished).

Fruit bodies of *Digitatispora marina* are inconspicuous and easily overlooked, or confused with bacterial colonies on wood. It has a resupinate fruit body with tetraradiate basidiospores (Fig. 1*a,b*). *Halocyphina villosa* was described as an imperfect fungus (Kohlmeyer, J. & Kohlmeyer, E., 1965), but Ginns & Malloch (1977) suggested it

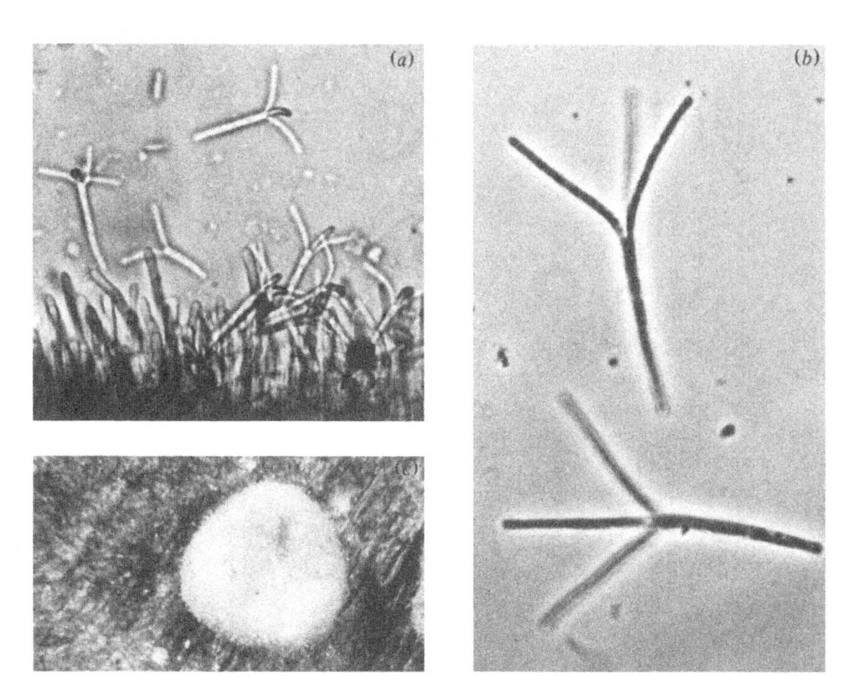

Fig. 1. (*a, b*) *Digitatispora marina:* (*a*) Section through hymenium showing basidia and basidiospores, × 250; (*b*) Tetraradiate hyaline basidiospores, × 700. (*c*) *Nia vibrissa*: fruit body on wood, × 6.

was a basidiomycete ('Cyphellaceae'/Polyporales). It has been reported from mangroves, *Spartina* (Kohlmeyer, J. & Kohlmeyer, E., 1965), submerged wood and branches of *Tamarix gallica* (Kohlmeyer, J. & Kohlmeyer, E., 1977). Unlike *N. vibrissa* and *D. marina*, the basidiospores of *H. villosa* are not appendaged (Fig. 3*g*). Kendrick & Watling (1979) regard the basidiospores of *D. marina* and *N. vibrissa* as stauroconidia, the propagules having specialised appendages. However, Doguet (1962, 1963, 1967, 1969) and J. Kohlmeyer & E. Kohlmeyer (1971) are of the opinion that both fungi have basidiospores.

Eighteen marine-occurring heterobasidiomycetous yeasts have been isolated (Table 1) and referred to three genera: *Leucosporidium, Rhodosporidium* and *Sporobolomyces*. They are tentatively assigned to the Sporobolomycetaceae, Sporobolomycetales (von Arx, 1979). Moore (1980) has proposed the family Sporidiaceae (Sporidiales, Sporidiomycetes, Ustomycota) for the genera *Rhodosporidium* and *Leucosporidium* while the genus *Sporobolomyces* is referred to the

Table 1. *Summary of the characteristics of marine basidiomycetes*

Classification	Teleomorph	Anamorph	Clamp connections	Dolipore	Substrate	Apobasidium	Passive release of basidiospores	Utilisation of cellulose	Utilisation of cellobiose	References
Hymenomycetes										
Polyporales										
Corticiaceae	*Digitatispora marina*	–	+	+	Wood	+	+	0	0	3
'Cyphellaceae'	*Halocyphina villosa*	–	+	0	Mangroves, wood	–	–/+*	0	0	1,2
Gasteromycetes										
Melanogastrales										
Melanogastraceae	*Nia vibrissa*	–	+	+	Wood, *Spartina*	+	+	+	+	4
Teliomycetes										
Ustilaginales										
Tilletiaceae	*Melanotaenium ruppiae*	–	–	–	*Ruppia maritima*	0	0	0	0	5
Heterobasidiomycetous yeasts										
Sporidiales[a]										
Sporidiaceae	*Leucosporidium antarcticum*	–	+	0	Seawater	+	+	0	–	6
	L. scottii	*Candida scottii*	+	0	Seawater	+	+	0	+	6
	Rhodosporidium bisporidii	–	+	0	Seawater	+	+	0	+	7
	R. capitatum	–	–	0	Seawater	+	+	0	+	7
	R. dacryoidum	–	+	0	Seawater	+	+	0	Latent –/+	7
	R. diobovatum	*Rhodotorula* sp.	+	0	Seawater	+	+	0	+	8
	R. infirmo-miniatum	*Cryptococcus infirmo-miniatus*	+ (few)	0	Seawater, Zooplankton	+	+	0	+	7
	R. malvinellum	–	+	0	Seawater	+	+	0	+	8
	R. sphaerocarpum	*Rhodotorula glutinis*	+	0	Seawater	+	+	0	+	9
	R. toruloides	–	+	0	Seawater, animals, oil slicks	+	+	0	0	9

Sporboolomycetales										
Sporobolomycetaceae	*Sporobolomyces hispanicus*	–	0	0	Seawater	0	–	0	–	10
	S. odorus	–	0	0	Seawater	0	–	0	–	10
	S. pararoseus	–	0	0	Seawater	0	–	0	+	10
	S. roseus	–	0	0	Seawater Oil slicks	0	–	0	+ (– rarely)	10
	S. salmonicolor	–	0	0	Seawater	0	–	0	–	10
Malasseziales										
Malasseziaceae	–	*Sterigmatomyces halophilus*	0	0	Seawater	0	0	0	Weak	11
	–	*S. indicus*	0	0	Seawater	0	0	0	Weak	12
Not referable to order ? Deuteromycotina	–	*Sympodiomyces parvus*	0	0	Seawater	0	0	0	+	13

References: 1, Kohlmeyer & Kohlmeyer (1977); 2, Ginns & Malloch (1977); 3, Brooks (1975); 4, Leightley & Eaton (1977); 5, Feldmann (1959); 6, Fell, Statzell, Hunter & Phaff (1969); 7, Fell, Hunter & Tallman (1973); 8, Newell & Hunter (1970); 9, Newell & Fell (1970); 10, Fell (1976); 11, Fell (1966); 12, Fell (1970); 13, Fell & Statzell (1971).

+ = presence of the character; – = absence of the character; 0 = not examined or tested; –/+ = results not clear, or conflicting; * released forcibly at first and then passively out of the fruit body.

[a] An *apobasidium* in this order is one in which the spores are not forcibly discharged; a symmetrical arrangement of the spores on the sterigmata is included only in the more restricted definition.

Table 2. *Summary of the characteristics of freshwater basidiomycetes*

Classification	Teleomorph	Anamorph	Clamp connections	Dolipore	Substrate	Apobasidium	Passive release of basidiospores	Utilisation of cellulose	Utilisation of cellobiose	References
Gasteromycetes Hymenogastrales Limnoperdaceae	*Limnoperdon incarnatum*	–	+	0	Hardwood twigs, marshwater	+	+	+?	+?	1
Hymenomycetes Polyporales Corticiaceae	*Leptosporomyces galzinii*	*Tricladium* sp.	+	0	Wood or bark	–	–	0	–	2
		Taeniospora gracilis?								
	Bulbillomyces farinosus	*Aegerita candida*	+	0	Wood	–	–	0	0	
	Subulicystidium longisporum	*Aegerita tortuosa*	+	0	Wood	–	–	0	0	
	Sistotrema sp. cf. *oblongisporum*	*Ingoldiella hamata*	+	0	Wood	–	–	0	–	3
Tremellales	*Xenolachne flagellifera*	–	+	0	Logs	+	+	0	0	4
Not referable to orders or families	–	*Dendrosporomyces prolifer*	–	+	Freshwater foam	0	0	0	0	5
	–	*Ingoldiella fibulata*	+	0	Foam	0	0	0	0	6
	–	*Varicosporium splendens*	–	+	Foam	0	0	0	0	7
	–	*Tricladium malaysianum*	–	+	Foam	0	0	0	0	8
	Sev. 286*	–	+	0	Wood	0	+	+	+	9

Teliomycetes										
Ustilaginales										
Filobasidiaceae	*Rogersiomyces okefenokeensis*	–	+	0	Decaying leaves	+	+	0	0	10
Heterobasidiomycetous yeasts										
Malasseziales										
Malasseziaceae	–	*Vanrija aquatica* (syn. *Candida aquatica*)	–	0	Foam, *Equisetum fluviatile*	–	–	0	0	11, 12, 13

References: 1, Escobar, McCabe & Harpel (1976); 2, Nawawi, Descals & Webster (1977); 3, Nawawi (1977); 4, Rogers (1947); 5, Nawawi, Webster & Davey (1977); 6, Nawawi (1973*b*); 7, Nawawi (1973*a*), 8, Nawawi (1974); 9, Leightley & Eaton (1977); 10, Crane & Schoknecht (1978); 11, Jones & Slooff (1966); 12, Moore (1980); 13, Webster & Davey (1975).

+ = presence of the character; – = absence of the character; 0 = not examined or tested; * = undescribed.

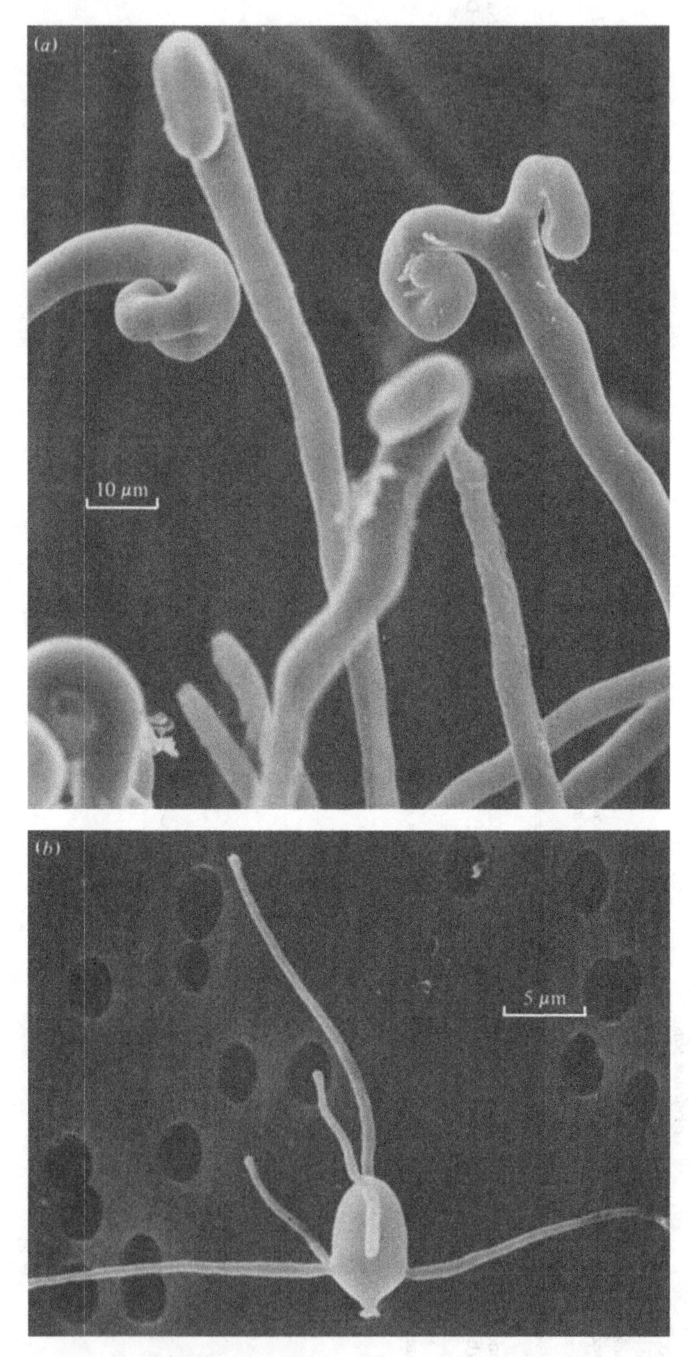

Fig. 2. *Nia vibrissa*: (*a*) bifurcate fruit body appendages; (*b*) basidiospore with one polar appendage and four equatorial appendages.

Sporidiobolaceae. Heterobasidiomycetous yeasts have been isolated from estuarine and inshore waters (van Uden, 1967) and from oceanic regions (Fell, 1976).

The different spore types encountered in the heterobasidiomycetous yeasts are not appendaged and do not appear to be adapted for an aquatic habitat. Fell (1976) and his co-workers (Fell, Hunter & Tallman, 1973) have described the heterothallic life cycles of these fungi and the different sexual compatibility systems. Most of these yeasts have been isolated from seawater or plankton samples, and only rarely have they been reported from specific substrata (van Uden & Fell, 1968). Although widely distributed in oceanic waters, they are present in low numbers (5–10 litre^{-1}), while in estuarine and coastal waters yeast cell counts of 118–1228 litre^{-1} (van Uden, 1967) are not uncommon.

Freshwater species

Table 2 lists the freshwater basidiomycetes described, and the substrata on which they have been found growing. Three homobasidiomycetes are known: *Limnoperdon incarnatum*, *Leptosporomyces galzinii* and *Sistotrema* sp. (cf. *oblongisporum*). All produce non-appendaged basidiospores in semi-aquatic to terrestrial habitats, usually on decaying wood, bark, moss, ferns, leaves or conifer needles. It is the anamorph that is found in aquatic habitats and in *L. galzinii* and *S. oblongisporum* tetraradiate and branched (*Ingoldiella hamata*) spores have been observed, usually growing on decaying leaves. The spores have clamp connections (Fig. 3*a,e*).

The amphibious nature of these fungi, with the teleomorph largely terrestrial and the anamorph aquatic, is intriguing. However, Bandoni (1972, 1974) has reported a number of aquatic Hyphomycetes on leaves in woods well away from streams and rivers. *L. incarnatum* is known only from one collection, while *L. galzinii* and *S. oblongisporum* (*I. hamata*) have been collected more frequently (Nawawi, 1977).

Three heterobasidiomycetes have been found in aquatic or semi-aquatic habitats: *Xenolachne flagellifera, Subulicystidium longisporum* and *Bulbillomyces farinosus*. All have unbranched basidiospores while the anamorph *Aegerita* has spores that trap air bubbles like the aero-aquatic fungi (Fischer, 1977). Eriksson & Ryvarden (1976) suggest that the diaspore or bulbil of the *Aegerita* state is a floating unit for dispersal in water.

A number of fungi collected in freshwater foam have conidia with clamp connections or dolipore septa: *Dendrosporomyces prolifer, Ing-*

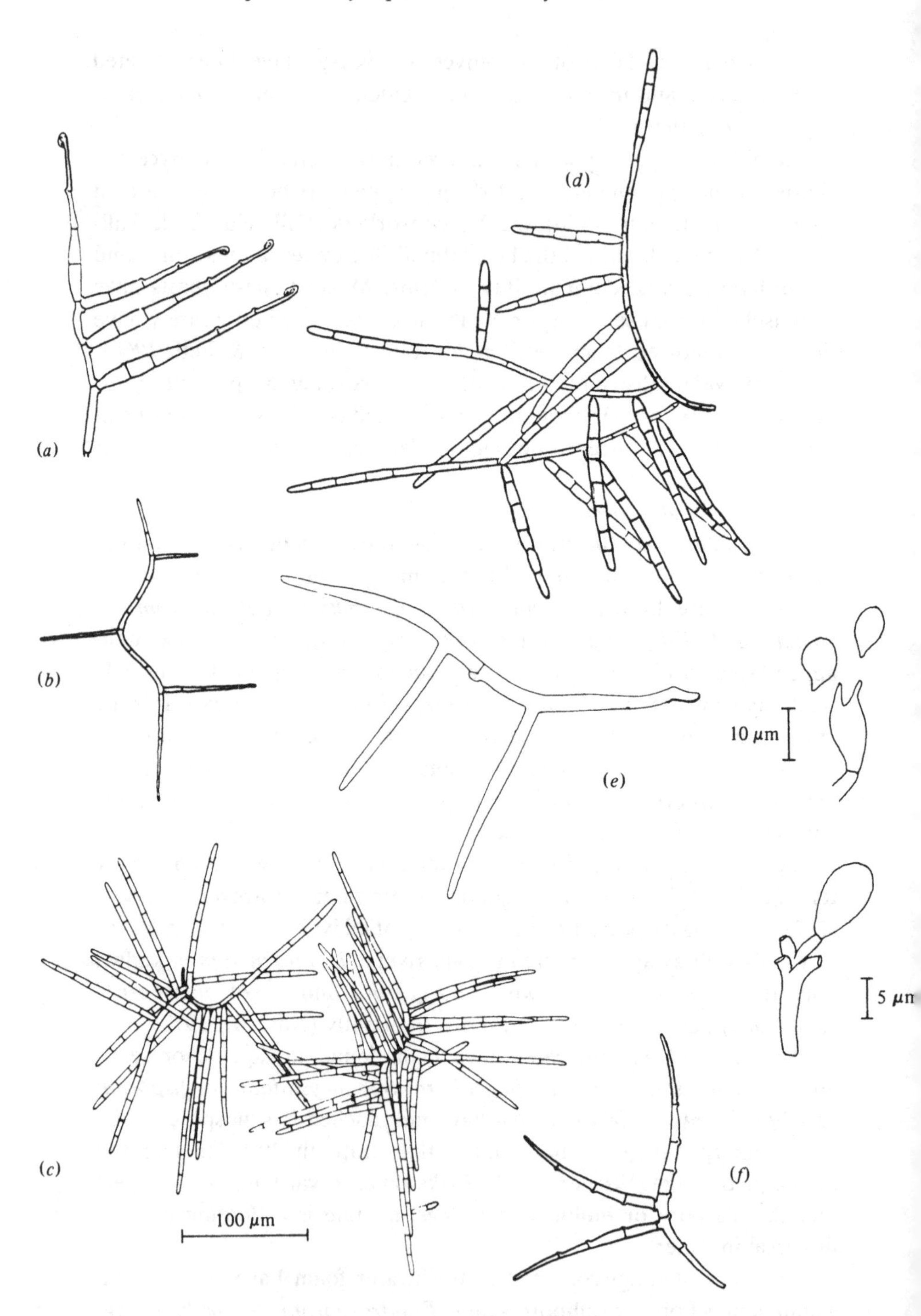
(a)
(b)
(c)
(d)
(e)
(f)
10 μm
5 μm
100 μm

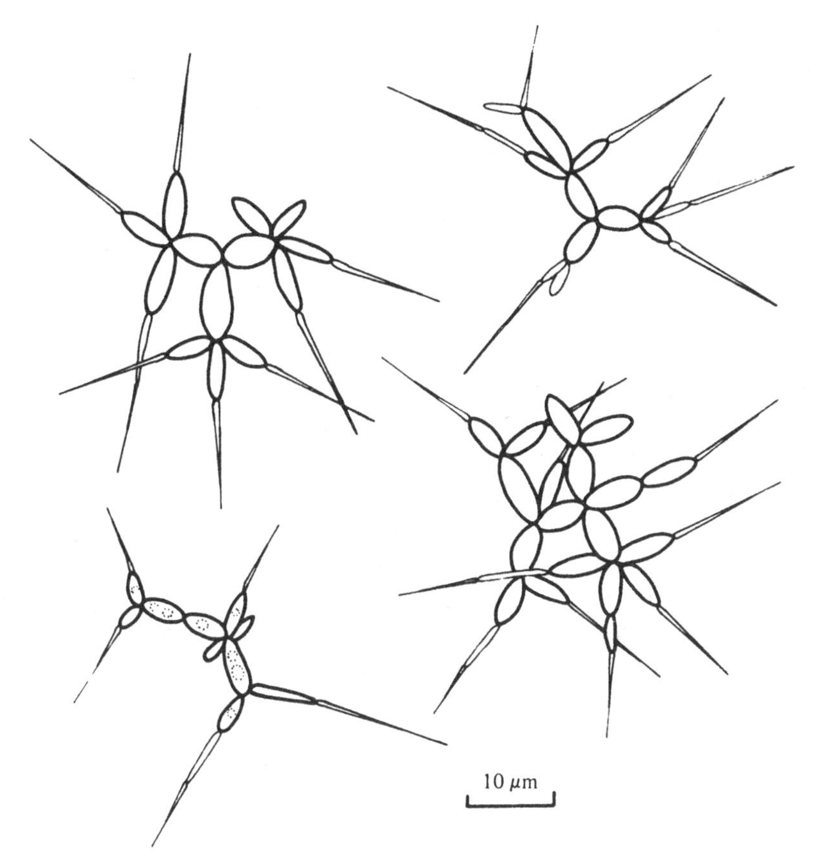

Fig. 4. *Vanrija aquatica*: tetraradiate groupings of cells.

oldiella fibulata, *Varicosporium splendens*, *Tricladium malaysianum*, and the conidia are illustrated in Fig. 3*a–f*. Kendrick & Watling (1979) refer to the conidia of the anamorphs *D. prolifer*, *I. fibulata*, *I. hamata*, *T. malaysianum* and *Taeniospora gracilis* as stauroconidia with modified hyphal systems. Further species await description, as Nawawi (1977) and Ingold (1959) have observed other tetraradiate spores bearing clamp connections in their foam samples.

Crane & Schoknecht (1978) described *Rogersiomyces okefenokeensis*, an interesting member of the Filobasidiaceae, growing on decaying

---

Fig. 3. Spores of freshwater and marine fungi: (*a*) *Ingoldiella hamata*; (*b*) *Tricladium malaysianum*; (*c*) *Dendrosporomyces prolifer*; (*d*) *Varicosporium splendens*; (*e*) *Leptosporomyces galzinii*; (*f*) *Ingoldiella fibulata*; (*g*) *Halocyphina villosa*; (*h*) *Rogersiomyces okefenokeensis*.

leaves. The basidiospores are borne on apobasidia and are released passively (Fig. 3*h*). E. B. G. Jones (unpublished), Irvine (1974) and Leightley & Eaton (1977) have reported an unidentified basidiomycete growing on test wood blocks submerged in freshwater rivers and in water-cooling towers.

Freshwater basidiomycetous yeasts have received little attention (Bridge Cooke, 1976). A species (see Fig. 4) frequently encountered in freshwater habitats is *Vanrija aquatica* (syn. *Candida aquatica*) first described from freshwater foam collected at Malham Tarn by Jones & Slooff (1966). More recently it has been collected by Ingold (1975) and Webster & Davey (1975) in foam and culms of *Equisetum fluviatile* (respectively). Moore (1980) suggests that the teleomorph of *V. aquatica* may be a basidiomycete.

Although apparently tolerant of the conditions, none of the fungi discussed can be regarded as common in aquatic habitats (Byrne & Jones, 1974). Nearly half of the species listed in Tables 1 and 2 are known only from water or phytoplankton samples, but decaying leaves and timber are the substrata that have yielded the greatest numbers, although this may reflect sampling procedures.

Spore dispersal

Marine basidiomycetes (with the exception of *Halocyphina villosa* and the yeast genus *Sporobolomyces*) release their basidiospores passively, and sterigmata have not been observed. In *Nia vibrissa*, the basidia deliquesce and disintegrate, leaving the basidiospores surrounded by mucilage (Leightley & Eaton, 1979). They are released from the fruit body by breakdown of the peridial wall. Fazzani & Jones (1977) have suggested that the peridial wall breaks down due to bacterial action, while Ingold (1971) was of the opinion that the mucilage inside the fruit body swells, rupturing the wall and releasing the spores into the surrounding water. Fazzani & Jones (1977) suggested that the fruit bodies were dislodged from the substratum and dispersed in seawater – the fruit body acting as a dispersal unit. The tips of the fruit-body appendages are bifurcate, as shown in Fig. 2(*a*), and may help to entangle or attach the floating fruit bodies to suitable substrata. In *Digitatispora marina* the basidiospores are released into the surrounding seawater and their tetraradiate nature may aid in flotation.

Forcible discharge of basidiospores has not been observed in the yeast genera *Leucosporidium* and *Rhodosporidium* (Fell, Statzell, Hunter & Phaff, 1969; Newell & Hunter, 1970). However, Ginns & Malloch

(1977) suggested that the basidiospores in *H. villosa* are forcibly discharged, pushed up the funnel-shaped fruit body, the spores accumulating in a mass at the aperture (Kohlmeyer, J. & Kohlmeyer, E., 1977). They are subsequently washed away by water currents.

Our knowledge of spore dispersal in aquatic fungi is fragmentary as little experimental or quantitative work has been undertaken (Webster, 1959; Jones, 1973). Rees (1980) determined spore sedimentation rates for eighteen marine fungi, as well as spore density and degree of appendage cover. He concluded that the appendages of marine fungal spores are probably the most important single factor controlling the sedimentation rate. Removal of the appendages of the basidiospores of *N. vibrissa* by mild sonication resulted in more rapid settling, 0.280 mm s^{-1} as opposed to 0.115 mm s^{-1}.

Both passive and forcible release of basidiospores is reported for freshwater basidiomycetes. Passive release of spores has been reported for *Limnoperdon incarnatum* (Escobar, McCabe & Harpel, 1976), *Xenolachne flagellifera* (Rogers, 1947) and *Rogersiomyces okefenokeensis* (Crane & Schoknecht, 1978). No anamorphs have been described for these fungi and the details of spore release are fragmentary. In *L. incarnatum*, the spores have a beaked pedicel which arises as a sterigmatic extension. Escobar, McCabe & Harpel (1976) reported that fruit bodies float on water, which may aid spore dispersal. Cultures submerged in water develop hyphae that extend to the surface and then form floating fruit bodies. The peridium complex has dendrophyses. In *X. flagellifera* and *R. okefenokeensis* apobasidia are produced, both fungi sporulating under moist or wet weather conditions.

Forcible release of basidiospores has been reported for the teleomorphs of *Leptosporomyces galzinii* and *Sistotrema* sp. (Nawawi, Descals & Webster, 1977; A. Nawawi, unpublished), whereas the anamorphs of these fungi release the conidia passively. Similarly *Subulicystidium longisporum* and *Bulbillomyces farinosus* produce basidiospores in terrestrial habitats, but on wood near streams the *Aegerita* state is produced. The spores are hydrofuge, like aero-aquatic fungi (*Helicoon, Spirosphaera, Helicodendron*), and the entrapment of air may aid in their dispersal (Kendrick, 1979). However, little experimental work has been done on the spore dispersal of these fungi.

Decay of wood

The rôle of aquatic basidiomycetes in the decomposition of organic materials has been studied by Leightley & Eaton (1977, 1979,

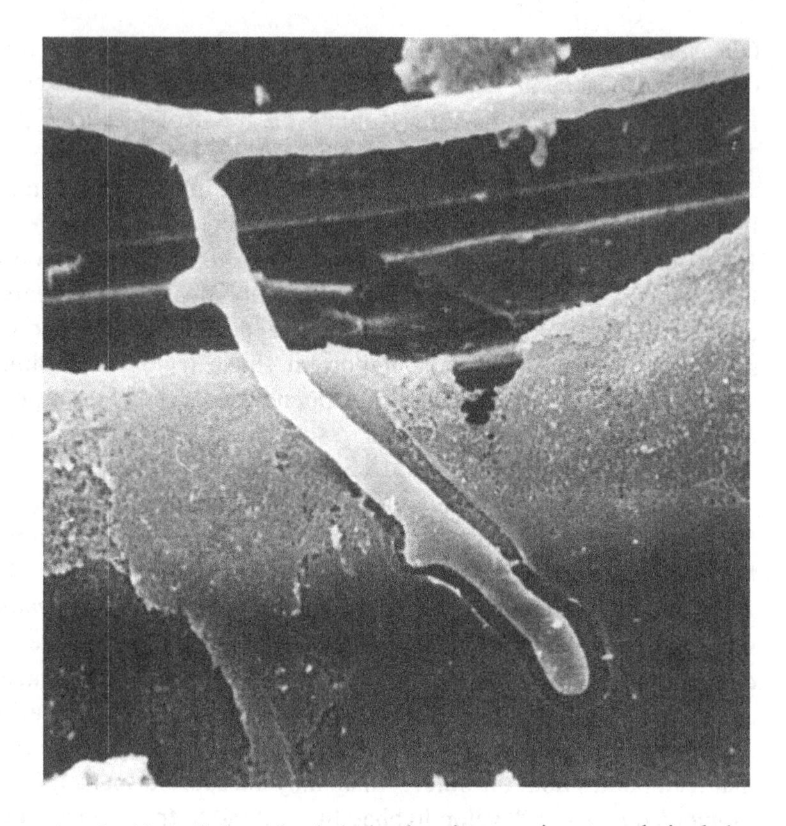

Fig. 5. *Nia vibrissa*: hyphal tip forming erosion trough in balsa (*Ochroma*) wood. Scanning electron micrograph, × 8600.

1980) who investigated the ability of *Nia vibrissa* and an unidentified species to decay wood. Two aspects were investigated: production of enzymes which decay wood (using the method of Rautela & Cowling, 1966), and the ability of the fungi to break down the wall structure of wood cells.

Both fungi were shown to produce cellulase, xylanase and mannanase when grown in a tube containing the carbohydrates: cellulose (as Whatman CF 11 cellulose powder swollen in 95% *o*-phosphoric acid); larch xylan (Koch-Light Laboratory Ltd) and glucomannan (prepared from Scots pine (*Pinus sylvestris*) and containing 78.2% xylose, 1.2% arabinose). Phenol oxidase activity was confirmed by the Käärik drop test (Käärik, 1965), and the Bavendamm test (Bavendamm, 1928) was positive for gallic and tannic acids utilisation. These tests confirm the white-rot decay characteristic of these fungi. Of the two species, SEV

286 was the most active, causing significant weight loss of beech (*Fagus*) and Scots pine test blocks (28.1% and 11.4% respectively after three months exposure to the fungus). When the fungi were grown on balsa (*Ochroma lagopus*) and beech test blocks, or beech and Scots pine veneers (Leightley & Eaton, 1976), the hyphae penetrated the cell walls, passing from cell to cell, and lysis of the cell wall was observed adjacent to hyphal tips (Fig. 5). The production of discrete erosion channels is considered diagnostic for white-rot fungi (Liese, 1970; Bravery, 1971). White-rot fungi are able to utilise not only the cellulose but also the lignin (Levy: Chapter 9).

Heterobasidiomycetous aquatic yeasts have been shown to assimilate a wide range of carbon compounds, including starch, mannitol, xylose and arabinose. Some are able to assimilate cellobiose (*Leucosporidium scottii, Rhodosporidium bisporidii, R. capitatum, R. diobovatum, R. malvinellum, R. infirmo-miniatum* and *Sporobolomyces pararoseus*), although cellulose does not appear to be utilised.

These observations indicate that aquatic basidiomycetes, although only a few have been described, are able to utilise a wide range of substrates, and at least two are able to cause decay of wood.

Adaptation to aquatic conditions

A number of reasons can be advanced as to why so few basidiomycetes are to be found in aquatic habitats, and they include the following. Firstly, large fruit bodies are impractical in fast-moving streams, and they would need tough, pliable, thalli (as in the fucoid marine algae) to withstand the physical forces in these aquatic environments. Soft or putrescent fruit bodies would need some protection from bacterial action, such as the unwettable wall of *Nia vibrissa* fruit bodies (Fazzani & Jones, 1977).

Secondly, suitable substrates for colonisation may be absent in aquatic habitats. Leaf fall in the autumn yields an abundance of material for colonisation by aquatic Hyphomycetes in streams (Iqbal & Webster, 1977), but in lakes and oceans lack of substrates for colonisation may be a very real problem. This may explain why test panels submerged in the sea are sometimes poorly colonised (Jones, 1963; Jones, Kühne, Trussell & Turner, 1972). However, Turner (1973) has shown that, even at great depths in the oceans, wood is available for colonisation by bacteria, fungi and marine borers.

Thirdly, forcible release of basidiospores in water would be wasteful. Ingold (1971) has suggested that spores would be projected for only a short distance, and that this would be of limited advantage to the fungus. It is interesting that fungi listed in Tables 1 and 2 as having forcible release of basidiospores all have teleomorphs that are semi-aquatic or amphibious, while the anamorphs are aquatic. Passive release of basidiospores (*N. vibrissa, Digitatispora marina, Limnoperdon incarnatum* and the heterobasidiomycete yeasts) is restricted to those producing teleomorphs in aquatic habitats. Webster & Descals (1979) considered this topic and posed the question: 'Do the basidiospores colonise leaves before they fall into streams?' Indeed, are the basidiospores (or the ascospores of other aquatic Hyphomycetes with a teleomorph ascomycete) responsible for the wider dispersal of these fungi? This may explain why so-called 'aquatic' Hyphomycetes have been recovered from terrestrial situations (Bandoni, 1972, 1974; Jones, 1981). Crane & Schoknecht (1978) suggested that angiocarpic development or an adaptation to aquatic habitats were two plausible explanations for the loss of forcible discharge of spores from the basidia, and for the modification of the structures that make discharge possible. This is true for most of the basidiomycetes listed in Tables 1 and 2. *Halocyphina villosa* and *L. galzinii* may be regarded as inter-tidal or semi-aquatic species, while the anamorphs of *Subulicystidium longisporum* and *Bulbillomyces farinosus* may have affinities with the aero-aquatic fungi (Fischer, 1977). In many marine ascomycetes, active release of the spores has also been lost, asci deliquescing in the centrum of the perithecia and spores being released passively in a mass of mucilage. This mode of release occurs in the Spathulosporales (Kohlmeyer, 1973), Plectascales and Halosphaeriaceae (Jones, 1976) and in marine lichens (A. Fletcher, 1975, personal communication). Webster & Descals (1979) reported a similar tendency in freshwater ascomycetes, e.g. *Massarina* sp. and *Nectria lugdunensis*. Similarly, the development of apobasidia has emerged in a number of groups, e.g. Gasteromycetes, Filobasidiaceae, Cyphellaceae, Corticiaceae and Ustilaginaceae (Watling: Chapter 1).

Finally, release of extracellular enzymes may be a positive disadvantage for basidiomycetes in aquatic habitats. Diffusible enzymes would soon be diluted or washed away from the substrate. This may explain why brown-rot fungi, with readily diffusible enzymes (Montgomery: Chapter 3), have not been reported from aquatic environments. Leightley & Eaton (1979) considered that enzyme diffusion to the cell wall surface was restricted in *N. vibrissa* and an unidentified species, and suggested that the slime layer observed at the tip of the hyphae might aid in adhesion and enzyme transport. Similar mucilaginous layers around wood-degrading fungal hyphae have been observed by Schmid & Liese (1966) and Maret (1972). Holt & Jones (1978) postulated a similar system for aquatic wood-decaying bacteria. One reason why soft-rot fungi (members of the Ascomycotina and Deuteromycotina) are so successful in aquatic conditions, such as water-cooling towers (Eaton & Jones, 1971*a*,*b*; Jones, 1972; Eaton, 1976) and wood in ground contact (Levy: Chapter 9), is that the hyphae of these fungi are within the S_2 layer of the secondary wood cell wall. Here, the area for enzyme release is restricted, and the loss of enzymes to the surrounding water is much reduced. Jones (1981) has shown that the aquatic Hyphomycetes: *Anguillospora crassa, A. longissima, Tricladium splendens* and *Lunulospora curvula* cause soft-rot decay of wood. The highest weight loss of 24% was obtained with *A. longissima* on balsa wood. *Heliscus lugdunensis, Lemonniera aquatica* and *Varicosporium elodeae* were able to utilise cellulose but did not cause soft-rot decay of wood. It would be interesting to know if the anamorphs *Dendrosporomyces prolifer, Ingoldiella fibulata, Varicosporium splendens* and *Tricladium malaysianum* were able to decay wood, and if so, the type of decay.

Summary

Feldman (1959) described the first marine basidiomycete. In the last decade more have been reported as workers examined their collections more critically and used more advanced techniques for their study. Improved isolation procedures, greater success in obtaining sporulating cultures under laboratory conditions and the use of benomyl in isolating media have all aided their study. Further investigations may prove that aquatic basidiomycetes are indeed more prevalent than we believe. However, as discussed above, the success of basidiomycetes in the aquatic environment may be limited by lack of substrates for colonisation, physical factors limiting the size of fruit bodies, passive discharge of basidiospores and loss of enzymes by diffusion into the surrounding water. The soft-rot fungi are able to overcome many of these problems, especially the loss of enzymes, by growing within the S_2 layer of the wood cell-wall.

Over thirty aquatic basidiomycetes are listed in Tables 1 and 2. Many are able to utilise cellobiose (Fell, Statzell, Hunter & Phaff, 1969), while Leightley & Eaton (1977) have shown that two basidiomycetes are able to utilise cellulose and lignin. However, further studies are required to establish their frequency of occurrence, the materials they grow on in nature, and their rôle in the decomposition of materials in aquatic habitats.

Acknowledgements. I am grateful to the following for permission to reproduce their figures or micrographs: Drs L. E. Leightley, R. A. Eaton (Fig. 5), A. Nawawi, E. Descals and Professor J. Webster (Fig. 3*a–f*), and to Mr E. Hawton for photographic assistance.

References

Bandoni, R. J. (1972). Terrestrial occurrence of some aquatic Hyphomycetes. *Canadian Journal of Botany*, **50**, 2283–8.

Bandoni, R. J. (1974). Mycological observations on the aqueous films covering decaying leaves and other litter. *Transactions Mycological Society of Japan*, **15**, 309–15.

Bärlocher, F. & Kendrick, B. (1974). Dynamics of the fungal population on leaves in a stream. *Journal of Ecology*, **62**, 761–91.

Bavendamm, W. (1928). Über das Vorkommen und den Nachweis von Oxydasen bei holzzerstörenden Pilzen. *Zeitschrift für Pflanzenkrankheiten*, **38**, 257–76.

Becker, G. (1971). On the biology, physiology and ecology of marine wood-boring Crustaceans. In *Marine Borers, Fungi and Fouling Organisms of Wood*, ed. E. B. G. Jones & S. K. Eltringham, pp. 303–26. Paris: Organisation for Economic Co-operation and Development.

Bravery, A. F. (1971). The application of scanning electron microscopy in the study of timber decay. *Journal of the Institute of Wood Science*, **5**, 13–19.

Bridge Cooke, W. (1976). Fungi in sewage. In *Recent Advances in Aquatic Mycology*, ed. E. B. G. Jones, pp. 389–434. London: Elek Science.

Brooks, R. D. (1975). The presence of dolipore septa in *Nia vibrissa* and *Digitatispora marina*. *Mycologia*, **67**, 172–4.

Byrne, P. J. & Jones, E. B. G. (1974). Lignicolous marine fungi. *Veröffentlichungen der Instituten für Meeresforschung in Bremerhaven*, Supplement, **5**, 301–20.

Crane, J. L. & Schoknecht, J. D. (1978). *Rogersiomyces*, a new genus in the Filobasidiaceae (Homobasidiomycetes) from an aquatic habitat. *American Journal of Botany*, **65**, 902–6.

Descals, C. E., Nawai, A. & Webster, J. (1976). Developmental studies in *Actinospora* and three similar aquatic hyphomycetes. *Transactions of the British Mycological Society*, **67**, 207–22.

Doguet, G. (1962). *Digitatispora marina* n.g., n.sp., Basidiomycète marin. *Compte rendu Hebdomadaire des séances l'Académie des Sciences, Série D*, **254**, 4336–8.

Doguet, G. (1963). Basidiospores anormales chez le *Digitatispora marina*. *Bulletin de la Société mycologique de France*, **78**, 283–90.

Doguet, G. (1967). *Nia vibrissa* Moore et Meyers, remarquable Basidiomycète marin. *Compte rendu Hebdomadaire des séances l'Académie des Sciences, Série D*, **265**, 1780–3.

Doguet, G. (1969). *Nia vibrissa* Moore et Meyers, Gasteromycète marin. II. Développement des carpophores et des basides. *Bulletin de la Société Mycologique de France*, **84**, 93–104.

Eaton, R. A. (1976). Cooling tower fungi. In *Recent Advances in Aquatic Mycology*, ed. E. B. G. Jones, pp. 359–87. London: Elek Science.

Eaton, R. A. & Jones, E. B. G. (1971*a*). The biodeterioration of timber in water cooling towers. I. Fungal ecology and the decay of wood at Connah's Quay and Ince. *Material und Organismen*, **6**, 51–80.

Eaton, R. A. & Jones, E. B. G. (1971*b*). The biodeterioration of timber in water cooling towers. II. Fungi growing on wood in different positions in water cooling towers. *Material und Organismen*, **6**, 81–92.

Eriksson, J. & Ryvarden, L. (1976). *The Corticiaceae of North Europe*, vol. 4, *Hyphodermella-Mycoacia*. Oslo: Fungiflora.

Escobar, G. A., McCabe, D. E. & Harpel, C. W. (1976). *Limnoperdon*, a floating gasteromycete isolated from marshes. *Mycologia*, **68**, 874–80.

Fazzani, K. & Jones, E. B. G. (1977). Spore release and dispersal in marine and brackish water fungi. *Material und Organismen*, **12**, 235–48.

Feldmann, G. (1959). Une ustilaginale marine, parasite du *Ruppia maritima* L. *Revue Général Botanique*, **66**, 35–8.

Fell, J. W. (1966). *Sterigmatomyces*, a new fungal genus from marine areas. *Antonie van Leeuwenhoek, Journal of Microbiology and Serology*, **32**, 99–104.

Fell, J. W. (1970). In *The Yeasts*, ed. J. Lodder, p. 1232. Amsterdam: North-Holland.

Fell, J. W. (1976). Yeasts in oceanic regions. In *Recent Advances in Aquatic Mycology*, ed. E. B. G. Jones, pp. 93–124. London: Elek Science.

Fell, J. W., Hunter, I. L. & Tallman, A. S. (1973). Marine basidiomycetous yeasts (*Rhodosporidium* spp.n.) with tetrapolar and multiple allelic bipolar mating systems. *Canadian Journal of Microbiology*, **19**, 643–57.

Fell, J. W. & Statzell, A. C. (1971). *Sympodiomyces* gen. n., a yeast-like organism from southern marine waters. *Antonie van Leeuwenhoek, Journal of Microbiology and Serology*, **37**, 359–67.

Fell, J. W., Statzell, A. C., Hunter, J. L. & Phaff, H. J. (1969). *Leucosporidium* gen. n., the heterobasidiomycetous stage of several yeasts of the genus *Candida*. *Antonie van Leeuwenhoek, Journal of Microbiology and Serology*, **35**, 433–63.

Fischer, D. J. (1977). New methods of detecting and studying saprophytic behaviour of aero-aquatic hyphomycetes from stagnant water. *Transactions of the British Mycological Society*, **68**, 407–11.

Fletcher, A. (1975). Key for the identification of British marine and maritime lichens. I. Siliceous rocky shore species. *Lichenologist*, **7**, 1–52.

Ginns, J. & Malloch, D. (1977). *Halocyphina*, a marine Basidiomycete (Aphyllophorales). *Mycologia*, **69**, 53–8.

Haythorn, J., Jones, E. B. G. & Harrison, J. L. (1980). Observations on marine algicolous fungi including the new hyphomycete *Sigmoidea marina* sp. nov. *Transactions of the British Mycological Society*, **74**, 615–23.

Holt, D. M. & Jones, E. B. G. (1978). Bacterial cavity formation in delignified wood. *Material und Organismen*, **13**, 13–30.

Holt, D. M., Jones, E. B. G. & Furtado, S. E. J. (1979). Bacterial breakdown of wood in aquatic habitats. *British Wood Preserving Association Record Annual Convention*, 13–24.

Ingold, C. T. (1959). Aquatic spora of Omo Forest, Nigeria. *Transactions of the British Mycological Society*, **42**, 479–85.

Ingold, C. T. (1971). *Fungal spores: their liberation and dispersal*. Oxford: Clarendon Press.

Ingold, C. T. (1975). *An illustrated guide to aquatic and water-borne hyphomycetes*. Science Publication No. 30. Ambleside: Freshwater Biological Association.

Ingold, C. T. (1976). The morphology and biology of freshwater fungi, excluding phycomycetes. In *Recent Advances in Aquatic Mycology*, ed. E. B. G. Jones, pp. 335–57. London: Elek Science.

Iqbal, S. H. & Webster, J. (1977). Aquatic hyphomycete spora of some Dartmoor streams. *Transactions of the British Mycological Society*, **69**, 233–41.
Irvine, J. (1974). An investigation of the factors affecting the biodeterioration of treated timber in aquatic habitats. PhD Thesis, Portsmouth Polytechnic, UK.
Jones, E. B. G. (1963). Observations on the fungal succession on wood test blocks submerged in the sea. *Journal of the Institute of Wood Science*, **11**, 14–23.
Jones, E. B. G. (1968). The distribution of marine fungi on wood submerged in the sea. In *Biodeterioration of Materials*, ed. J. J. Elphick & A. H. Walters, pp. 460–85. Amsterdam: Elsevier.
Jones, E. B. G. (1971). The ecology and rotting ability of marine fungi. In *Marine Borers, Fungi and Fouling Organisms of Wood*, ed. E. B. G. Jones & S. K. Eltringham, pp. 237–58. Paris: Organisation for Economic Cooperation and Development.
Jones, E. B. G. (1972). The decay of timber in aquatic environments. *British Wood Preserving Association: Record of Annual Convention*, 31–49.
Jones, E. B. G. (1973). Marine fungi: spore dispersal, settlement and colonization of timber. *Proceedings of the 3rd International Congress of Marine Corrosion and Fouling*, ed. R. F. Acker, B. F. Brown, J. R. DePalma & W. P. Iverson, pp. 640–7. Evanston, Illinois: Northwestern University Press.
Jones, E. B. G. (1976). Lignicolous and algicolous fungi. In *Recent Advances in Aquatic Mycology*, ed. E. B. G. Jones, pp. 1–49. London: Elek Science.
Jones, E. B. G. (1981). Observations on the ecology of lignicolous aquatic hyphomycetes. In *The Fungal Community: Its Organization and Role in the Ecosystem*, ed. D. T. Wicklow & G. C. Carroll, pp. 731–42. New York: Dekker.
Jones, E. B. G., Kühne, H., Trussell, P. & Turner, R. D. (1972). Results of an international cooperative research programme on the biodeterioration of timber submerged in the sea. *Material and Organismen*, **7**, 93–118.
Jones, E. B. G. & Slooff, W. C. (1966). *Candida aquatica* sp. n. isolated from water scums. *Antonie van Leeuwenhoek, Journal of Microbiology and Serology*, **32**, 223–8.
Jones, E. B. G., Turner, R. D., Furtado, S. E. J. & Kühne, H. (1976). Marine biodeterogenic organisms. I. Lignicolous fungi and bacteria and the wood boring molluscs and crustacea. *International Biodeterioration Bulletin*, **12**, 120–34.
Käärik, A. (1965). The identification of the mycelia of wood decay fungi by their oxidation reactions with phenolic compounds. *Studia Forestalia Suecica*, No 31.
Kaushik, N. K. & Hynes, H. B. N. (1971). The fate of the dead leaves that fall into streams. *Archiven für Hydrobiologie*, **68**, 465–515.
Kendrick, B. (1979). Dialogue on the ecology of anamorphs and teleomorphs. In *The Whole Fungus*, ed. B. Kendrick, vol. 2, pp. 623–34. Ottawa: National Museum of Natural Science, National Museums of Canada, and the Kananaskis Foundation.
Kendrick, B. & Watling, R. (1979). Mitospores in Basidiomycetes. In *The Whole Fungus*, ed. B. Kendrick, vol. 2, pp. 473–545. Ottawa: National Museum of Natural Science, National Museums of Canada, and the Kananaskis Foundation.
Kohlmeyer, J. (1973). Spathulosporales, a new order and possible missing link between Laboulbeniales and Pyrenomycetes. *Mycologia*, **65**, 614–47.
Kohlmeyer, J. & Kohlmeyer, E. (1964–9). *Icones Fungorum Maris*, vol. 1, Tables 1–90. Weinheim: J. Cramer.
Kohlmeyer, J. & Kohlmeyer, E. (1965). New marine fungi from mangroves and trees along eroding shorelines. *Nova Hedwigia*, **9**, 89–104.
Kohlmeyer, J. & Kohlmeyer, E. (1971). *Synoptic plates of higher marine fungi*, 3rd edn. Lehre: Cramer.

Kohlmeyer, J. & Kohlmeyer, E. (1977). Bermuda marine fungi. *Transactions of the British Mycological Society*, **68**, 207–19.
Leightley, L. E. & Eaton, R. A. (1976). A method for studying the growth of soft rot fungi in wood veneers. *International Biodeterioration Bulletin*, **12**, 44–8.
Leightley, L. E. & Eaton, R. A. (1977). Mechanisms of decay of timber by aquatic micro-organisms. *British Wood Preserving Association. Record Annual Convention*, 221–50.
Leightley, L. E. & Eaton, R. A. (1979). *Nia vibrissa* – a marine white rot fungus. *Transactions of the British Mycological Society*, **73**, 35–40.
Leightley, L. E. & Eaton, R. A. (1980). Micromorphology of wood decay by marine microorganisms. In *Biodeterioration, Proceedings of the 4th International Symposium*, ed. T. A. Oxley, D. Allsopp & G. Becker, pp. 83–8. London: Pitman Publications Ltd and the Biodeterioration Society.
Liese, W. (1970). Ultrastructural aspects of woody tissue disintegration. *Annual Review of Phytopathology*, **8**, 231–58.
Maret, R. (1972). Chimie et morphologie submicroscopique des parois cellulaires de l'Ascomycète *Chaetomium globosum*. *Archiv für Mikrobiologie*, **81**, 68–90.
Moore, R. T. (1980). Taxonomic proposals for the classification of marine yeasts and other yeast-like fungi including the smuts. *Botanica Marina*, **23**, 361–73.
Nawawi, A. (1973*a*). A new species of *Varicosporium* from Malaysia. *Nova Hedwigia*, **24**, 39–43.
Nawawi, A. (1973*b*). Two clamp-bearing aquatic fungi from Malaysia. *Transactions of the British Mycological Society*, **61**, 521–8.
Nawawi, A. (1974). Two new *Tricladium* species. *Transactions of the British Mycological Society*, **63**, 267–72.
Nawawi, A. (1977). Conidial Aquatic Basidiomycetes (Abstract). *Second International Mycological Congress, Tampa*, ed. H. E. Bigelow & E. G. Simmons, vol. 2, p. 249.
Nawawi, A., Descals, E. & Webster, J. (1977). *Leptosporomyces galzinii*, the basidial state of a clamped branched conidium from fresh water. *Transactions of the British Mycological Society*, **68**, 31–6.
Nawawi, A., Webster, J. & Davey, R. A. (1977). *Dendrosporomyces prolifer* gen. et sp. nov., a basidiomycete with branched conidia. *Transactions of the British Mycological Society*, **68**, 59–63.
Newell, S. Y. (1976). Mangrove fungi: the succession in the mycoflora of Red mangrove (*Rhizophora mangle* L.) seedlings. In *Recent Advances in Aquatic Mycology*, ed. E. B. G. Jones, pp. 51–91. London: Elek Science.
Newell, S. Y. & Fell, J. W. (1970). The perfect form of a marine-occurring yeast of the genus *Rhodotorula*. *Mycologia*, **62**, 272–81.
Newell, S. Y. & Hunter, I. L. (1970). *Rhodosporidium diobovatum* sp. n., the perfect form of an asporogenous yeast (*Rhodotorula* sp.). *Journal of Bacteriology*, **104**, 503–8.
Rautela, G. S. & Cowling, E. B. (1966). Simple cultural test for relative cellulolytic activity of fungi. *Applied Microbiology*, **14**, 892–8.
Rees, G. (1980). Factors affecting the sedimentation rate of marine fungal spores. *Botanica Marina*, **23**, 375–85.
Rees, G., Johnson, R. G. & Jones, E. B. G. (1979). Lignicolous marine fungi of Danish sand dunes. *Transactions of the British Mycological Society*, **72**, 99–106.
Rogers, D. P. (1947). A new gymnocarpus heterobasidiomycete with gasteromycetous basidia. *Mycologia*, **39**, 556–64.
Sanders, P. F. & Anderson, J. M. (1979). Colonization of wood blocks by aquatic hyphomycetes. *Transactions of the British Mycological Society*, **73**, 103–7.

Schmid, R. & Liese, W. (1966). Elektronmikroskopische Beobachtungen on hyphen von holzpilsen. *Material und Organismen*, **1**, 250–9.
Turner, R. D. (1973). Deepwater wood-boring mollusks. *Proceedings of the 3rd International Congress of Marine Corrosion and Fouling*, ed. R. F. Acker, B. F. Brown, J. R. DePalma & W. P. Iverson, pp. 836–41. Evanston, Illinois: Northwestern University Press.
van Uden, N. (1967). Occurrence and origin of yeasts in estuaries. In *Estuaries*, ed. G. H. Lauft, Publication No. 83, pp. 306–10. Washington, DC: American Association for the Advancement of Science.
van Uden, H. & Fell, J. W. (1968). Marine yeasts. In *Advances in the Microbiology of the Sea*, ed. M. Droop & E. J. F. Wood, pp. 167–201. London and New York: Academic Press.
von Arx, J. A. (1979). Propagation in the yeasts and yeast-like fungi. In *The Whole Fungus*, ed. B. Kendrick, vol. 2, pp. 555–71. National Museum of Natural Science, National Museums of Canada, and the Kananaskis Foundation.
Webster, J. (1959). Experiments with spores of aquatic hyphomycetes. I. Sedimentation and impaction on smooth surfaces. *Annals of Botany*, N.S., **23**, 595–611.
Webster, J. & Davey, R. A. (1975). Sedimentation rates and trapping efficiency of cells of *Candida aquatica. Transactions of the British Mycological Society*, **64**, 437–40.
Webster, J. & Descals, E. (1979). The teleomorphs of water-borne Hyphomycetes from fresh water. In *The Whole Fungus*, ed. B. Kendrick, vol. 2, pp. 419–51. Ottawa: National Museum of Natural Science, National Museums of Canada, and the Kananaskis Foundation.

12 Decomposing activity of basidiomycetes in forest litter

T. F. HERING

Department of Physiology and Environmental Studies, School of Agriculture, University of Nottingham, Sutton Bonington, Loughborough LE12 5RD, England

In discussion on the decomposition of forest litter, it is always admitted that basidiomycetes play an important rôle. In general, we cannot yet express this importance by means of a numerical estimate of the proportion of litter materials that they process, and this review will cover work that illustrates their importance in qualitative terms. It will draw on fairly old as well as recent work, for new work on the decomposer basidiomycetes appears at a relatively slow rate.

Mull and mor

In temperate countries the amount of litter reaching the forest floor each year varies widely, with a mean value of about 2 t ha^{-1} dry weight. Decomposition is a long process, and the half-life for the carbon mineralisation of a litter sample may be two years or longer, depending on conditions. In nature the process always involves animals, fungi and bacteria, and we can recognise two very distinctive patterns of decay, leading to very different kinds of soil profile, although there are many intermediate types. In mull, soil animals are active in the fresh litter, and there is a peak of microbial activity later, whereas in mor, the litter is initially unpalatable to animals, so a primary phase of microbial decay precedes attack by animals. In typical mull soils the litter is quickly removed from the soil surface by animals, especially earthworms, so that for most of the year the soil surface is almost bare of litter, being a crumbly mass of worm casts, in which organic and inorganic matter are intimately mixed. The animals divide up the plant matter finely, but they make little impact on macromolecules such as cellulose and lignin; these are incorporated in the soil in casts and faecal pellets, where they become available for attack by micro-organisms. The evidence that the

Table 1. *Total numbers and fresh weights of decomposer agarics collected from permanent plots totalling 300 m² in each of two woodlands in Cumbria, England.* In the period September–November of three successive years, a total of twenty-five visits were made to each site. (From Hering, 1967)

	Mull site		Mor site	
	Number	Fresh weight (g)	Number	Fresh weight (g)
Clitocybe langei sensu Singer			41	30
Collybia peronata			50	139
Cystoderma amianthinum			17	10
Laccaria amethystea			34	43
L. laccata	44	12	352	171
Marasmius epiphyllus	190	2		
M. ramealis	31	1		
Mycena galericulata	28	27	203	210
M. galopus	226	36	292	34
M. metata	171	10		
M. polygramma	24	33		
M. speirea	211	4		
M. sanguinolenta	4	1	14	1
Psathyrella squamosa	4	1	62	85
Total	933	127	1065	723

higher fungi are involved in this process comes from the fact that their fruit bodies emerge from this layer of the soil. Many small toadstools, such as species of *Lepiota* and *Pluteus*, that are described in floras as growing on damp rich soil in woods, may be decomposing this finely divided, buried, material. However, there has been very little detailed study of such fungi to confirm this. Table 1 shows the results of comparable surveys on mull and mor sites in the same region. Mor had a much greater weight of fruit bodies, and comparative equality of numbers was reached only because of the presence on the mull site of very small toadstools such as *Mycena speirea* and *Marasmius epiphyllus*, fruiting on twigs and stalks lying on the soil surface. This material is unpalatable to earthworms, and harbours a distinctive component of the toadstool flora. It seems likely that low fruit-body production is a general feature of mull soils, though they are not often surveyed. On the evidence of fruit bodies – and that is all we have to go on as yet – the rôle of basidiomycetes in mull could be a fairly minor one.

In a mor profile, the litter is not sufficiently palatable to be rapidly removed by animals, so it lies on the soil surface until it is covered by new litter, and the mineral soil is permanently hidden by a series of organic layers, traditionally described as:

- L intact dead leaves;
- F broken up, but still recognisably composed of leaf pieces;
- H dark, powdery, organic material.

There is a very obvious connection between decomposer basidiomycetes and the F-layer. Often fruit bodies can be seen to be 'rooted' in the F-layer material, with hyphal systems radiating out across the leaf pieces. Mycelial strands are very frequent too, and often show clamp connections, as does the surface mycelium. This in itself does not conclusively show that these organisms are carrying out decomposition, as the mycelia of mycorrhizal basidiomycetes must be abundant in the same habitat. However, bleaching of litter is generally considered to be a sign of basidiomycete activity (Hintikka, 1970; Chapter 13) and in some cases this is very obvious, as when large bleached patches are found beneath troops of fruit bodies of *Collybia peronata* in *Fagus* litter.

We know much less about the activities of basidiomycetes in the H-layer. None of the lines of evidence given above is very helpful with H-layer material, and isolation studies indicate a greatly increased rôle for heavy-sporing moulds such as *Trichoderma* and *Penicillium*. When F-layer material becomes H-layer material, the greater part of it has been eaten by small animals, and the litter has become essentially a mass of animal pellets. The initial unpalatability of fresh litter has been progressively lost in the F-layer stage. One cause of unpalatability is the presence in the original plant matter of phenolic compounds, which are more likely to occur in the litter of acid-loving plants, and on acidic, nutrient-poor, sites. Many workers, including Satchell and Lowe (1967) have shown that such compounds have a deterrent effect on detritus-feeding animals. Phenolics may be lost by simple leaching, but microbial activity is probably more important. As Lindeberg (1948) and others have shown, litter-inhabiting basidiomycetes are strong producers of polyphenol oxidases; in the long run this property may be responsible both for the bleaching of litter and also for its increased palatability.

Another factor related to palatability is the state of the nitrogen present. In mor litter, nitrogen is scarce anyway, and C:N ratios of around 100 are common. Many litter fungi, including basidiomycetes, can tolerate this condition, and produce mycelium with a much narrower ratio than the material that they feed on. For this reason, leaves permeated by mycelium should be more nutritious to soil animals than

Table 2. *The ability of various species of basidiomycete to decompose, in pure culture, synthetic complexes formed from protein (either gelatin or leaf protein from* Sambucus*) by reaction with phenolic-containing extracts from various plants.* The cultures were incubated for 7 months. (After Handley, 1954)

	Source of phenolics			
	Quercus	*Pinus*	*Acer*	*Calluna*
Collybia butyracea	++	++	++	++
Hirschioporus abietinus	++	++	++	++
Lenzites betulina	++	++	+	+
Coriolus versicolor	++	+	++	+
Collybia dryophila	Nt	+	++	Nt
Lentinus lepideus	Nt	0	++	Nt
Mycena pura	Nt	0	++	Nt

0 No change.
+ Complex partly decomposed.
++ Complex completely decomposed.
Nt Not tested.

raw litter. In addition, much of the nitrogen in mor litter is in the form of phenolic–protein complexes (Handley, 1954) which seem to be of low digestibility to animals. Handley studied the decomposition of these complexes by synthesising them in the laboratory and testing them with pure cultures; a small selection of his data appears in Table 2. Basidiomycetes, including wood decomposers and also litter fungi such as *Collybia butyracea*, were capable of decomposing the complexes produced from a variety of plant phenolics. Some of his assessments suggested that particular fungi were more active when the phenolics were derived from certain kinds of plant, so the compounds could exercise selectivity over fungal decomposers. This may be so, but the conditions of Handley's tests were too artificial for such conclusions to be properly drawn. Basidiomycetes have access here to a nitrogen source that seems to be closed to animals, and perhaps to some other micro-organisms as well.

The increase in palatability and reduction of C:N ratio, which occur during fungal decomposition of F-layer material, are the basic reason why a prolonged phase of basidiomycete decomposition comes to an end. When basidiomycetes have sufficiently permeated and altered the material, soil animals invade and destroy the substrate.

Not all basidiomycetes that fruit on the forest floor are decomposers. It is well known that typical mycorrhizal fungi, such as species of the genera *Amanita* and *Boletus*, show negligible decomposing activity when cultures are incubated with litter materials. It is generally useful to think of decomposers and mycorrhizal fungi as being two watertight, mutually exclusive classes, both spreading their mycelium in the same material but essentially using it for different purposes: the decomposers as a source of energy, and the mycorrhiza-formers as a source of inorganic elements. The only really satisfactory evidence for a mycorrhizal rôle is the synthesis of mycorrhizas with seedlings in pure culture, and the only satisfactory evidence for decomposing ability is the incubation of pure cultures with sterile litter resulting in a significant loss of weight. In these ways, many species have now been claimed as one or the other, but many are still unclaimed – the smaller toadstools in the Entolomataceae, for instance, seem to be a largely uncharted group. Some higher fungi are probably active in both capacities. *Laccaria laccata* was shown by Mikola (1956) to be able to decompose litter in pure culture, while several authors, including Suggs & Grand (1972), have demonstrated mycorrhiza formation with various trees. Since several genera of agarics contain both mycorrhiza-formers and decomposers, it seems possible for a fungus to evolve from one way of life to the other, and one should reasonably expect to find some species with both capacities; indeed Lyr (1963) showed that some typical mycorrhizal fungi like *Amanita citrina* and *Suillus variegatus* produced small yields of cellulase and xylanase, and could play a small part in the destruction of litter materials.

Decomposition studies with pure cultures

By pure-culture methods, quite a long list of species have been shown to be litter-decomposers – most of these are in the genera *Mycena, Marasmius, Collybia* and *Clitocybe* (Lindeberg, 1944, 1947; Hering, 1967). This confirms the impression given by these species in the forest, where they are associated with local patches of bleaching of mor litter, and mycelium can be followed out from the toadstool to pieces of decomposing leaf. On similar grounds, I think that others, including certainly species of *Psathyrella* and *Tubaria*, are likely to be litter-decomposers, but there has been no laboratory study of them. This kind of evidence cannot be gained easily in mull, where the mycelial connections of both mycorrhiza-formers and presumed decomposers are less obvious than in the organic horizons of a mor. Watling (Chapter 1) discusses in detail possible ecological groupings of basidiomycete taxa.

Figure 1 shows some results from a pure-culture decomposition study on *Quercus* litter. The two basidiomycetes tested caused substantially greater dry-weight losses than the non-basidial litter fungi; moreover, they showed every sign that these rates could be maintained over long periods, whereas moulds like *Trichoderma viride* gave an early flush of decomposition and sporing and then stagnated. Similar results were obtained in a study by de Boois (1976). Assuming an exponential rate of dry-weight loss, Hering's (1967) data gave an estimate of the half-life of litter decomposed by *Mycena galopus* of about 600 days (or – 0.1% day^{-1}). This estimate was made at temperatures approximating to mean annual outdoor temperatures (12–15 °C). Higher temperatures give increased rates, and Lindeberg (1947), with *Fagus* litter incubated at 25 °C, obtained data which yielded estimates of 200–250 days (approx. – 0.3% day^{-1}) for this and several other species.

Much the greater part of the dry matter of litter is in the form of cell-wall macromolecules; estimates of the percentage dry matter made up by lignin, cellulose and hemicellulose vary from 70% to over 80%. Virtually all the basidiomycetes studied by Lindeberg were able to decompose lignin as well as cellulose and hemicellulose, so they were physiologically akin to the white-rot fungi of wood. Most of them caused bleaching of the litter in pure culture, and this effect may be correlated with lignin-decomposing ability, as it is among wood-rotting fungi

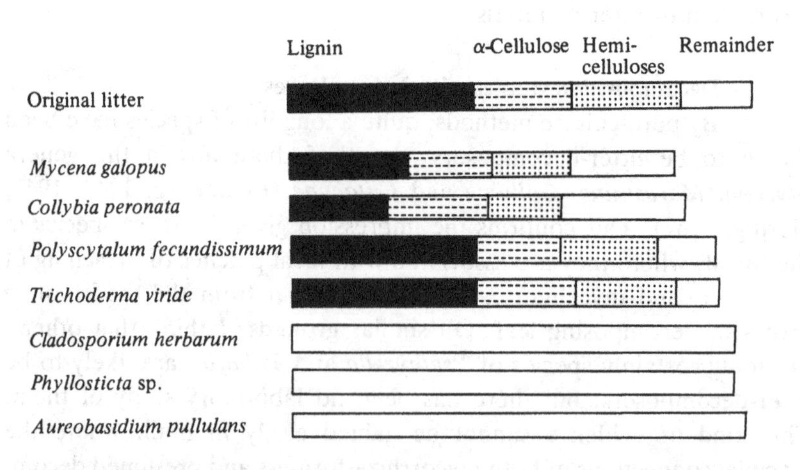

Fig. 1. Changes in dry weight and chemical composition of sterilised *Quercus* leaf litter incubated with single cultures of fungi for 6 months at 12–15 °C. For the last three species, only dry-weight changes are given. Data from Hering (1967).

(Davidson, Campbell & Blaisdell, 1938). It was very common to find that lignin disappears from the litter faster than cellulose. The very small degradation of cellulose by *Collybia peronata* in Fig. 1 confirmed similar results obtained with this fungus by both Lindeberg (1944) and Harris (1945). There is every reason to think that basidiomycetes are the main agents of lignin destruction in mor litter, but it is probably premature to claim, as is sometimes done, that they are the sole agents. Most claims for the involvement of other kinds of micro-organisms are based on tests with 'extracted lignin', and may be unreliable as a guide to what occurs in nature. Where intact forest litter has been incubated with pure cultures, lignin decomposition has been convincingly recorded only with basidiomycetes, though Haider & Schetters (1967) demonstrated losses of lignin in straw, caused by imperfect fungi that were probably non-basidiomycetes. There is still some doubt about basidiomycetes that do not bleach litter – are they lignin-decomposers or not? Mikola (1956) recorded that *Laccaria laccata* was one such fungus but did not carry out chemical analyses of litter. Chastukhin (1962) also found a non-bleaching type of decay with *Lepiota procera*; however, Lindeberg (1947) did find that this species degraded lignin.

Studies in glass flasks do not, of course, indicate how much decomposition is carried out in nature by the same species, when competing with other decomposers. Interesting data (Fig. 2) were obtained by Saitô (1957), who studied the weight changes in *Fagus* leaves undergoing a bleaching type of decay on the forest floor, almost certainly caused by basidiomycetes. The chemical changes were similar to those

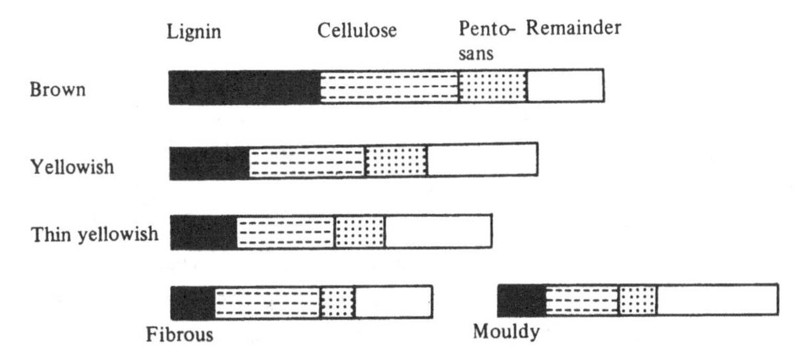

Fig. 2. Changes in dry weight and chemical composition of *Fagus* leaves undergoing a bleaching type of decomposition in a Japanese forest. Dry-weight changes from leaf-area measurements on weighed samples. The terms 'brown', etc. denote successive layers in the soil profile. Data from Saitô (1957).

occurring in pure-culture tests, with lignin disappearing faster than cellulose. The leaves had lost 30–40% of their initial dry weight by the time they reached the condition described by Saitô as 'fibrous' or 'mouldy'; both these terms probably indicated the end of an exclusive basidiomycete phase, with massive invasion by animals and non-basidiomycete fungi. In pure-culture studies these competitors are kept out, and much higher weight losses can be found with basidiomycetes alone. Chastukhin (1962) carried out prolonged incubations, and took account of the fact that in glass flasks the salts released by decomposition cannot escape, and will accumulate until they become toxic. His samples were washed in sterile water, removing both salts and soluble organic matter. Several species caused dry-weight losses of over 70% in two years (82% with *Collybia dryophila*). This work suggested that these fungi were able to destroy all kinds of macromolecule present in litter; however, in nature the substrate is usually eaten by soil animals when less than half of the dry matter has been lost.

Pure-culture studies show that fungi that bleach litter also acidify it; the white rots of litter seem to differ from those of wood in this respect. In studies on *Mycena galopus* (Hering, 1967) there was a fall of pH in 6 months from 4.0 to 3.2, and similar changes were recorded for many fungi by Mikola and Lindeberg. The fall in pH may bring an ecological advantage, since the fungus generates soluble compounds, including sugars, by hydrolysis, and a low pH probably has the effect of restricting bacterial exploitation of these compounds. Kühnelt (1963) pointed out that litter bleached and acidified by the activity of *Clitocybe infundibuliformis* was very poor in most kinds of soil animals, although a few species preferred it. By delaying animal attack, such effects may help to prolong the survival of the fungus.

The experience of collectors shows that most species of decomposer basidiomycetes are relatively specific to certain kinds of litter. For example, *Collybia dryophila* prefers broad-leaved litter, and *Marasmius androsaceus* prefers conifer needles. Little is known of any biochemical background to such differences. In a few cases, such as *Marasmius hudsonii* which is strictly limited to leaves of *Ilex aquifolium*, it may well be that the fungus is partly parasitic, colonising the leaves before they reach the ground. But in pure cultures many species do not show strong preferences, as Table 3, from Mikola (1956), shows. All the fungi he tested decayed *Populus tremula* litter about twice as fast as *Pinus* needles; this was equally true of the needle-fungus, *M. androsaceus.* Likewise, there seemed to be no chemical factor in pine needles to

Table 3. *Percentage weight losses in sterilised samples of leaf litter incubated with basidiomycetes for 107 days.* (From Mikola, 1956)

	Pinus needles	*Populus* leaves
Collybia dryophila	17.2	46.3
Micromphale perforans	15.0	25.8
Mycena galopus	14.7	27.0
Collybia confluens	14.0	19.2
Marasmius androsaceus	13.4	31.8
Laccaria laccata	5.9	13.2

prevent attack by *Collybia dryophila*, nor even to slow its growth detectably. In nature, when *M. androsaceus* grows on mixed leaves and needles, fruit bodies can sometimes be found attached to *Quercus* leaves, but it does not seem to be able to maintain itself indefinitely on purely broad-leaved litter. The natural distribution of the various species probably reflects the result of long-continued competition between species.

We thus have a reasonable, if partial, picture of the activities of decomposer basidiomycetes growing on litter. We can also make rough estimates of the amount of decomposition that they carry out on the litter. For Scandinavian forests, an estimate can be gained from the work of Hintikka (1970) who coined the term 'white-rot humus' for a material predominantly attacked by basidiomycetes (see Hintikka; Chapter 13). Coverage of the forest floor by this humus varied from low levels to about 10% of the litter present. As an estimate of basidiomycete involvement, this is subject to at least two qualifications. Firstly, the total throughput of the bleaching process may be greater than the proportion of bleached litter visible at any one time. Secondly, some basidiomycete species cause little bleaching in the forest. We can certainly expect decomposer basidiomycetes to make a significant contribution wherever they occur, and they do occur very widely.

Fruiting as an index of activity

We can make the last statement of the previous section confidently because of the occurrence of fruit bodies. They are useful indicators of the complement of species present, and of the distribution of a given species across a site. Counts or weighings of fruit bodies, however, can never be better than a rough guide to the relative importance of the various species. The production of fruit bodies never utilises more than a small fraction of the energy contained in the litter that gives rise to them. For comparison, one should consider that 100 kg of fresh mushroom compost (including water) can yield something of the order of 25 kg of mushrooms within 8 weeks (Atkins, 1974). Taking a whole year's harvest, the equivalent yield from 100 kg of woodland litter seems to be at best 0.2–0.5 kg, i.e. of the order of a hundred times less. This disparity would not be greatly different if dry-weight figures were used. If mycorrhizal species were included, the yield of the forest floor would be much higher, but these species are drawing on carbon sources not included in the litter fall. It seems probable that decomposer basidiomycetes assimilate far more carbon than they ever use in producing fruit bodies; the main reason for this may be the limitation on fruiting imposed by the weather.

Fruiting behaviour is probably not a good guide to the period of activity of the fungus; in Britain, strands and apparently actively growing basidiomycete mycelium may be found at any time of the year, but, in general, fruiting of soil and litter types is restricted to the summer and autumn. Figure 3 shows the seasonal fruiting profiles of

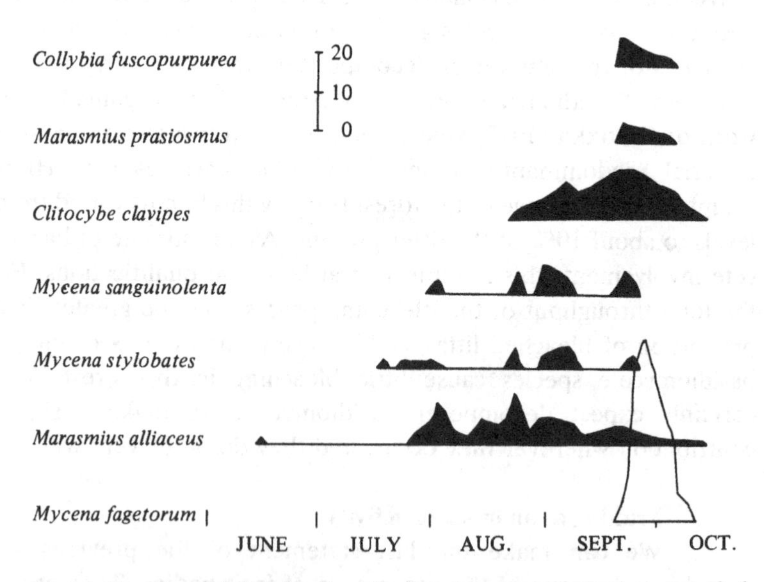

Fig. 3. Numbers of fruit bodies of some decomposer agarics, recorded every 2–3 days during the fruiting season in a plot of 400 m^2 in a *Fagus* forest at Rabsztyn, Poland. Data from Guminska (1962).

some decomposers in a Polish woodland of *Fagus* (Guminska, 1962). Three of the species showed a series of peaks and troughs of fruiting in response to the weather; the effect of a wet spell followed by a dry one in late August can be clearly seen. These decomposers probably resemble the cultivated mushroom, in that they can fruit at any time of the year, given a suitable temperature and soil moisture. As their fruit bodies are small, forming primordia at relatively shallow depths, they are markedly affected by fluctuations in moisture, while the large *Clitocybe clavipes*, arising deeper, is less affected. Its seasonal curve resembles that shown by many mycorrhizal fungi. Three other species, typified by *Mycena fagetorum*, were limited to an autumn season regardless of the weather. Several of the *Mycena* species studied by Hintikka (1963) over five successive seasons showed a similar limitation (Fig. 4). This could represent a biological advantage for these fungi, since, in autumn, conditions are damper, permitting the easier establishment of new colonies from spores, which may also be able to colonise the current year's litter, before or after it falls. It seems that this group of decomposers is 'taking a cue' for fruiting from some feature of the environment, such as frosts or changes in daylength. For the present we can only guess at the nature of this cue, which needs to be explored experimentally.

We have, from observation and experiment, ample evidence to confirm what was said about litter-decomposing basidiomycetes in the opening sentence – their rôle is a very important one. In forests, they

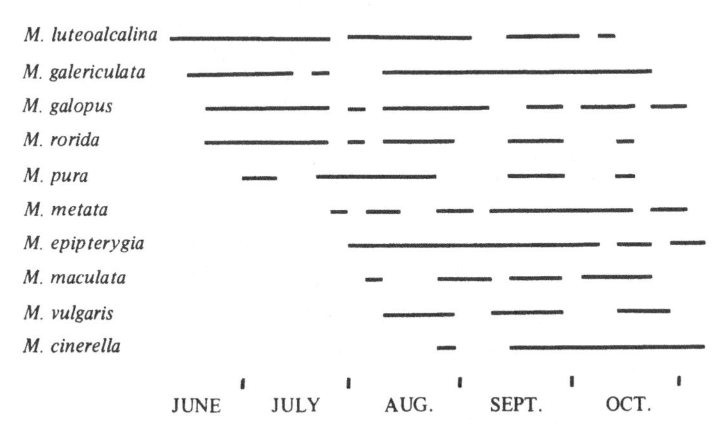

Fig. 4. Fruiting of *Mycena* species on the forest floor in Finland. The lines show 5-day periods during which the given species was found fruiting at least once during 5 years of observations. Data from Hintikka (1963).

are almost ubiquitous; when tested, they decompose litter materials at a steady rate over long periods, and they seem to make a vigorous attack upon some compounds, such as lignin, that would decompose slowly or not at all in their absence. A great deal more remains to be learned about their rôle, and about the specific attributes that cause a given species to be confined to certain materials or to certain soils. To put the present state of our knowledge into proportion, one should realise that only about fifty species of litter-decomposing basidiomycetes have ever been subject to any sort of experimental investigation, while the number of non-mycorrhizal basidiomycete species in the forest floor, in Britain alone, is probably in excess of 500.

References

Atkins, F. C. (1974). *Guide to mushroom growing*. London: Faber.

Chastukhin, V. Ya. (1962). Razlozhenie lesnogo opada chistymi kul'turami bazidial'nykh gribov. *Uchenye zapiski Leningradskogo ordena Lenina gosudarstvennogo universiteta No. 313 Seria Biologicheskykh Nauk,* **49,** 3–22.

Davidson, R. W., Campbell, W. A. & Blaisdell, J. (1938). Differentiation of wood-decaying fungi by their reactions on gallic or tannic acid medium. *Journal of Agricultural Research*, **57,** 683–95.

de Boois, H. M. (1976). Fungal development on oak leaf litter and decomposition potentialities of some fungal species. *Revue d'Ecologie et de Biologie du Sol*, **13,** 437–48.

Guminska, B. (1962). Mikoflora lasow bukowych Rabsztyna i Maciejowej, Studium florystyczno-ekologiczne. *Monographiae Botanicae*, **13,** 3–85.

Haider, K. & Schetters, C. (1967). Mesophile ligninabbauender Pilze in Ackerböden und ihr Einfluss auf die Bildung von Humusstoffen. In *Progress in Soil Biology* ed. O. Graff & J. E. Satchell, pp. 425–39. Amsterdam: North-Holland.

Handley, W. R. C. (1954). Mull and mor formation in relation to forest soil. *Forestry Commission Bulletin*, No. 23. London: Her Majesty's Stationery Office.

Harris, G. C. M. (1945). Chemical changes in beech litter due to infection by *Marasmius peronatus* (Bolt.) Fr. *Annals of Applied Biology*, **32,** 38–9.

Hering, T. F. (1966). The terricolous higher fungi of four Lake District woodlands. *Transactions of the British Mycological Society*, **49,** 369–83.

Hering, T. F. (1967). Fungal decomposition of oak leaf litter. *Transactions of the British Mycological Society*, **50,** 267–73.

Hintikka, V. (1963). Studies in the genus *Mycena* in Finland. *Karstenia*, **6–7,** 77–87.

Hintikka, V. (1970). Studies on white-rot humus formed by higher fungi in forest soil. *Communicationes Instituti Forestalis Fenniae*, **69,** 4–68.

Kühnelt, W. (1963). Über den Einfluss des Mycels von *Clitocybe infundibuliformis* auf die Streufauna. In *Soil Organisms*, ed. J. Doeksen & J. van der Drift, pp. 281–8. Amsterdam: North-Holland.

Lindeberg, G. (1944). Über die Physiologie ligninabbauender Bodenhymenomyceten. *Symbolae botanicae Upsalinenses*, **8,** 1–183.

Lindeberg, G. (1947). On the decomposition of lignin and cellulose in litter caused by soil-inhabiting Hymenomycetes. *Arkiv für Botanik*, **33a**, 1–16.

Lindeberg, G. (1948). On the occurrence of polyphenol oxidases in soil-inhabiting Basidiomycetes. *Physiologia Plantarum*, **1,** 196–205.

Lyr, H. (1963). Zur Frage des Streuabbaues durch ektotrophe Mykorrhizapilze. In *Mykorrhiza*, ed. W. Rawald & H. Lyr, pp. 123–46. Jena: G. Fischer.

Mikola, P. (1956). Studies on the decomposition of forest litter by basidiomycetes. *Communicationes Instituti Forestalis Fenniae*, **69**, 4–48.

Saitô, T. (1957). Chemical changes in beech litter under microbiological decomposition. *Ecological Review, Sendai*, **14**, 209–16.

Satchell, J. E. & Lowe, D. G. (1967). Selection of leaf litter by *Lumbricus terrestris*. In *Progress in Soil Biology*, ed. O. Graff & J. E. Satchell, pp. 102–19. Amsterdam: North-Holland.

Suggs, E. G. & Grand, L. F. (1972). Formation of mycorrhizae in monoxenic culture by pond pine (*Pinus serotina*). *Canadian Journal of Botany*, **50**, 1003–7.

13 The colonisation of litter and wood by basidiomycetes in Finnish forests

V. HINTIKKA

Institute of General Botany, University of Helsinki, Viikki, SF-00710 Helsinki 71, Finland

Colonisation of litter

Although basidiomycete mycelia are active in the deeper layers of forest humus (Jensen, 1974), in Finnish conditions their activity is most easily seen in the uppermost litter layers in late autumn (Hintikka, 1964). Recently-fallen leaves of *Betula, Populus tremula* and *Quercus* are soon invaded by whitish or reddish fan-like mycelia, which bear clamp connections. These mycelia are fairly easy to isolate in aseptic culture. If a piece of mycelium, or of leaf colonised by the mycelium, is incubated on malt agar at 5 °C, the fungus grows rapidly and is usually free of bacteria. Exactly which groups these psychrotolerant mycelia belong to is not known, but some are probably agarics such as *Mycena galopus*. It is interesting that such fungi often produce a profuse aerial mycelium differentiated into strand-like structures in the cold (5 °C) but not at room temperature. This could be interpreted as an adaptation for colonising fresh litter from below. In the litter layer such mycelial strands can be observed growing straight upwards (Hintikka, 1964). Some of these fungi decompose lignin, leaving bleached patches on the uppermost litter in the spring. The decomposition of newly fallen litter appears to be significant in late autumn and even in winter despite low temperatures because the leaves are continuously moist, in contrast to the warmer conditions in summer when the litter dries out for long periods. Microbiological activity can take place even under a snow cover (Havas & Mäenpää, 1972), and in subarctic conditions the weight loss of leaf litter seems to depend on moisture rather than on temperature (Kärenlampi, 1971).

The decomposition of litter falling on the forest floor is a complex process in which, in addition to fungi, millions of bacteria and many small animals participate. The exact rôle of each group of organisms is not seen as readily as in the decay of wood. This is particularly true of the mycelia of soil and litter basidiomycetes, which are difficult to isolate and to assess in terms of the amount of active mycelium. In our studies (Hintikka, 1970*a*), the starting-point was the regular occurrence of bleached litter and humus, 'white-rot humus', around the fruit bodies of certain lignin-decomposing basidiomycetes. When the litter layer is thick and composed of leaves, white-rot humus is invariably seen around and near the bases of fruit bodies, the boundaries between it and the darker litter being very sharp. In conifer needle or bark litter the limits are often less sharp. White-rot humus is found frequently under *Ramaria ochraceo-virens, Clavariadelphus ligula, Clitocybe clavipes, C. infundibuliformis, C. odora, Collybia butyracea, C. confluens, C. dryophila, C. peronata, Mycena pura* and *Agaricus silvicola*, and under the ascomycetes *Spathularia flavida* and *Cudonia confusa*.

The occurrence of white-rot humus causes considerable small-scale variation in many properties of decomposing litter and humus, especially in deciduous forests with a ground flora of grasses and herbs, where it is more abundant than in dry coniferous forests. The following properties seem to be the most important. White-rot humus is more acid than the surrounding humus, the pH value being lower by about 0.5 to 1.5 units. The amounts of both mineral and total nitrogen are higher, and so is the rate of respiration as measured with a Warburg respirometer. The content of exchangeable calcium is lower, probably owing to the presence of calcium oxalate, and the capacity of the humus to oxidise phenolic compounds is higher.

Although the formation of white-rot humus in forest soils may involve the activities of many other organisms, e.g. secondary saprophytic sugar fungi (Hintikka, 1971*b*), the lignin-decomposing fungi probably produce the dominant effects, as they bring about similar changes when they decompose litter in pure cultures. For a more detailed discussion of this see Hering (1967, and Chapter 12).

Finnish forests also harbour several basidiomycete species that produce in the humus layers near their fruit bodies visible effects which differ from those of white-rot fungi. The most conspicuous example is *Hydnellum ferrugineum*, (Hintikka & Näykki, 1967). Its mycelium is readily visible, tough and mat-like, and extends over several square metres in dry pine forest on sandy soils. Sites occupied by the mycelium differ from the surroundings in their vegetation; dwarf-shrubs are lacking, and lichens and mosses dominate above the mycelial mat, which is impenetrable by water and causes the soil to be drier than that beyond the mycelium. This *Hydnellum* is probably a mycorrhizal species, although it has not yet been possible to obtain the fungus in pure culture and synthesise mycorrhizas. In spite of this, it evidently brings about a decrease in the content of organic substances in the humus. Moreover, when the mycelium grows on stones, the iron oxide coating dissolves in the very sites where the mycelium is visible, and the more easily weathered schists are softened. Many other basidiomycetes are probably active in the humus layers of these forests, but their mycelia are less obvious and often undetected unless they are fruiting.

Comparison of aspects of the physiology of wood- and litter-decomposing basidiomycetes

The decomposers of leaf litter and wood are usually considered separately, although the decomposition processes are similar and white rot in litter resembles that in wood. This has concealed some of the physiological differences between these decomposers. The basidiomycete communities of the two types of substrate are more distinct than

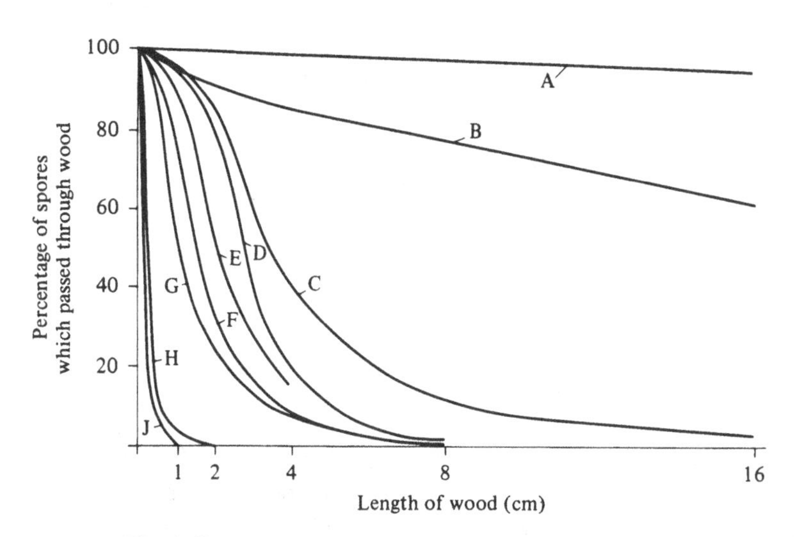

Fig. 1. Percentage of spores of *Armillaria mellea* which passed through a piece of wood the diameter of which was 3 cm and the length of which is indicated on the horizontal axis. A, *Quercus robur*; B, *Fraxinus excelsior*; C, *Populus tremula*; D, *Acer platanoides*; E, *Corylus avellana*; F and G, *Tilia cordata*; H, *Betula pubescens*, and J, *Alnus glutinosa, Pinus sylvestris* and *Picea abies*. (After Hintikka, 1973.)

Table 1. *The effect of carbon dioxide concentration on the germination of basidiospores of lignicolous basidiomycetes.* Percentage germination of spores on Hagem's agar in air and in different CO_2/air mixtures; each value is the mean of two replicate counts of 1000 spores

		Carbon dioxide concentration (%)			
Species	Time (h)	0.03	1	5	20
Coprinus micaceus	17	63.2	70.6**	69.5**	63.0
	24	64.3	79.7***	81.6***	69.3*
Fomitopsis pinicola	16	60.4	92.7***	95.8***	91.8***
Galerina mutabilis	12	30.0	61.0***	49.7***	13.4
	18	64.8	75.1***	77.9***	39.9
Gymnopilus penetrans	24	0.3	1.2*	1.7**	1.6**
	41	0.3	0.9	2.6***	2.4***
Hirschioporus abietinus	42	10.9	18.6***	16.1***	12.1
Hypholoma capnoides	6	20.6	25.8**	39.2***	9.8
	12	70.5	75.8**	74.9*	64.5
H. fasciculare	12	11.6	16.3**	10.9	0.7
	18	16.7	23.8***	16.2	7.7
H. sublateritium	12	64.4	72.5***	65.3	10.6
	24	78.6	82.3*	66.9	62.1
Pholiota aurivella	41	0.0	1.4***	13.3***	11.6***
	47	0.3	3.7***	29.0***	29.6***
P. squarrosa	122	0.8	1.0	3.0***	5.4***
Piptoporus betulinus	21	31.2	39.3***	36.4*	18.5
Polyporus squamosus	66	11.4	62.9***	21.7***	2.6
	90	19.1	—	27.4***	18.6
Spongipellis borealis	24	0.6	1.5*	0.9	0.1
	33	2.6	12.3***	19.6***	7.8***
	44	7.2	36.9***	42.3***	29.0***

***$P < 0.001$; **$P < 0.01$; *$P < 0.05$ (statistically significant stimulation compared with germination in air with a CO_2 concentration of 0.03%). –, not determined.

the communities of microfungi, but the reasons are usually unknown. Why does *Mycena galopus*, for example, grow on leaf litter and twigs but not on tree trunks? It seems that the sheer bulk and density of much woody debris is crucial. Unlike the superficial microfungi, many basidiomycetes can penetrate wood, but to survive they must tolerate the extreme environment of the interior. Some of these extreme conditions are discussed here and related to our observations on the growth of certain lignicolous and non-lignicolous basidiomycetes.

If we start from the invasion of wood by spores, we see that the spores of wood-decomposing fungi are as a rule much smaller (usually < 15 μm in diameter) than the diameter of the conducting elements of wood (30–500 μm). Thus spores may find their way into vessels and tracheids, and the longer these conducting elements are, the deeper the spores will penetrate. In an experiment with spore suspensions of *Armillaria mellea* (Hintikka, 1973), the spores passed almost freely through the wood of *Quercus* and *Fraxinus*, but a piece of conifer wood 1 cm long filtered off all the spores (Fig. 1). This demonstrates one effect the anatomy of wood can have on basidiomycete colonisation.

Effects of carbon dioxide and aeration

Spores can make their way into narrow vessels where aeration is restricted and where carbon dioxide may accumulate. Carbon dioxide stimulates spore germination in many fungi (Cochrane, 1958; Fries, 1966; Tabak & Cooke, 1968; Schánĕl, 1976), including wood-decomposers, and Table 1 lists the species in which a statistically significant stimulation of germination on Hagem agar was observed (Hintikka, 1970*b*). Stimulation seemed to be independent of concomitant changes in pH, as it occurred at all pH values.

The deeper the mycelium penetrates into the wood, the more restricted the aeration becomes. When the gas in living or dead wood is analysed, the concentration of carbon dioxide is found to be much higher than in the air. Determinations made by McDougall & Working (1933), Chase (1934), Thacker & Good (1952), Jensen (1967, 1969), and Novikov & Solovev (1973) indicated that both living and decomposing wood commonly contain carbon dioxide concentrations of 10–20%. Carrodus & Triffett (1975) found that the gas in living stems of *Acacia* is almost pure carbon dioxide. In Finnish conditions, decomposed wood of deciduous trees contains slightly more carbon dioxide than coniferous wood (Hintikka & Korhonen, 1970), probably owing to its more rapid decomposition. In forest humus layers, on the other hand, carbon dioxide hardly accumulates at all in normal conditions (Romell, 1928; Brierley, 1955), maximum values being around 1%. Accordingly, the lignicolous (i.e. wood-decomposing) species should be better adapted to growing in higher carbon dioxide concentrations than the species that decompose litter.

Hintikka & Korhonen (1970) determined the carbon dioxide tolerances of thirty soil or litter inhabiters and fifty lignicolous species.

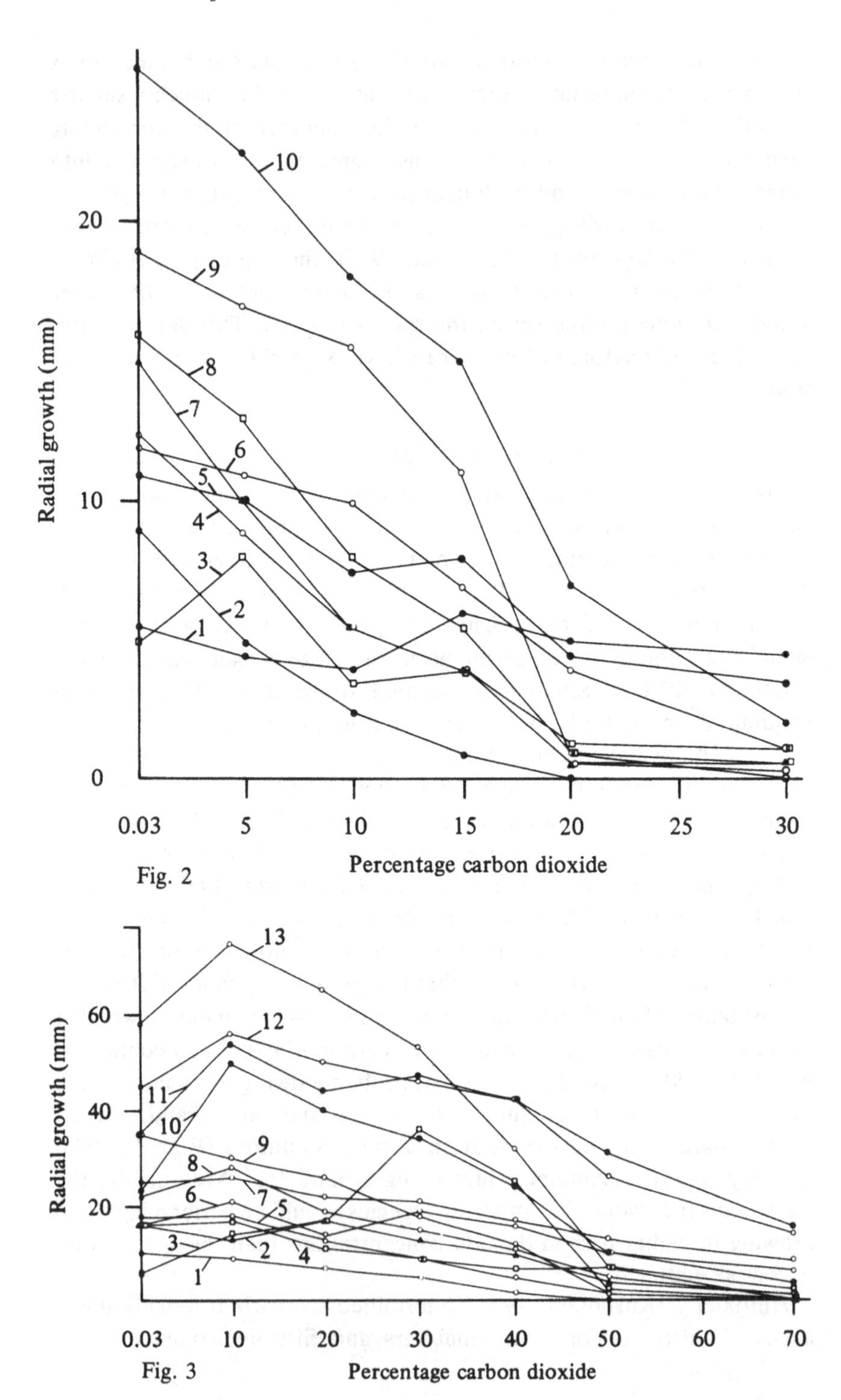
Radial growth (mm)
20
10
0
0.03
5
10
15
20
25
30
Percentage carbon dioxide
1
2
3
4
5
6
7
8
9
10

Fig. 2

Radial growth (mm)
60
40
20
0.03
10
20
30
40
50
60
70
Percentage carbon dioxide
1
2
3
4
5
6
7
8
9
10
11
12
13

Fig. 3

Inoculated Hagem agar plates were placed in desiccators in a partial vacuum; measured amounts of carbon dioxide were introduced and normal pressure reinstated. Fig. 2 shows the radial growth of ten non-lignicolous species; the maximum concentration of carbon dioxide they could tolerate was around 20–30%. Lignicolous species in general had higher tolerances, and some cultures grew well in air containing 70% CO_2 (Fig. 3). This method, however, does not cover all the conditions which could occur naturally. When mycelium produces carbon dioxide it takes up an approximately equal amount of oxygen, and even at 20% CO_2 oxygen may be absent. In this experiment, with a 70% CO_2 concentration, the atmosphere initially contained only 6% O_2. However, several wood-decomposing fungi are known to need only small amounts of oxygen, e.g. *Heterobasidion annosum* (Gundersen, 1961). Schánĕl (1976) investigated the changes in oxygen and carbon dioxide concentrations when wood decomposers were grown in closed flasks, and found that the carbon dioxide concentration sometimes rose to 40%.

At lower concentrations of carbon dioxide, i.e. around 10%, the growth of lignicolous species in our experiments was markedly stimulated (Fig. 3); a similar effect was found by Zycha (1937), and Schánĕl (1976). The latter also found that 10% CO_2 enhanced the production of laccase and peroxidases in *Pleurotus*, and affected the respiratory quotient, which was below 0.2. Carbon dioxide is evidently an important controlling factor in the ecology and physiology of wood-decomposing basidiomycetes.

Because accumulation of carbon dioxide in wood indicates oxygen-deficient conditions, the tolerance of both lignicolous and soil or litter

---

Fig. 2. The effect of carbon dioxide concentration in CO_2/air mixtures on the radial growth (mm) of soil- and litter-inhabiting basidiomycetes after 10 days on Hagem agar. 1, *Psathyrella candolleana*; 2, *Clitocybe infundibuliformis*; 3, *Clitopilus prunulus*; 4, *Mycena clavicularis*; 5, *Lepiota clypeolaria*; 6, *Mycena sanguinolenta*; 7, *Suillus bovinus*; 8, *Collybia confluens*; 9, *Mycena epipterygia*; 10, *Marasmius androsaceus*. (After Hintikka & Korhonen, 1970.)

Fig. 3. The effect of carbon dioxide concentration in CO_2/air mixtures on the radial growth (mm) of lignicolous basidiomycetes after 10 days on Hagem agar at room temperature. 1, *Mycena haematopus*; 2, *Hirschioporus abietinus*; 3, *Fomes fomentarius*; 4, *Inonotus radiatus*; 5, *Hypholoma capnoides*; 6, *Stereum sanguinolentum*; 7, *Pholiota squarrosa*; 8, *Hypholoma sublateritium*; 9, *Lentinus lepideus*; 10, *Piptoporus betulinus*; 11, *Lenzites betulina*; 12, *Fomitopsis pinicola*; 13, *Heterobasidion annosum*. (After Hintikka & Korhonen, 1970.)

Table 2. *The effect of ammonium acetate on the relative radial growth rate of lignicolous and non-lignicolous basidiomycetes, where 100 represents the growth rate on 1% malt agar* (pH 4.9–5.4)

Ammonium acetate in the medium (%)	Relative growth rate							
	0	*0.01*	*0.025*	*0.05*	*0.075*	*0.10*	*0.25*	*0.50*
Non-lignicolous species								
Mycena epipterygia	100	120	10	0	0	0	—	—
Clitocybe clavipes	100	120	—	0	0	0	—	—
C. odora	100	100	33	0	0	0	—	—
Micromphale perforans	100	100	14	0	0	0	—	—
Collybia confluens	100	93	7	0	0	0	—	—
Mycena galopus	100	83	10	0	0	0	—	—
Collybia peronata	100	78	22	0	0	0	—	—
Marasmius prasiosmus	100	62	0	0	0	0	—	—
Collybia butyracea	100	33	11	0	0	0	—	—
Suillus variegatus	100	0	—	0	0	0	—	—
Clitopilus prunulus	100	+	—	0	0	0	—	—
Lignicolous species								
Laetiporus sulphureus	100	—	—	—	—	88	74	26
Daedalea quercina	100	—	—	—	—	—	59	9
Pycnoporus cinnabarinus	100	—	—	—	—	57	33	0
Fomitopsis pinicola	100	—	—	—	—	74	25	0
Polyporus brumalis	100	—	—	—	—	91	20	0
Inonotus rheades	100	—	157	143	129	14	0	0
Phellinus pini	100	—	129	129	114	43	0	0
Fomes fomentarius	100	147	—	124	105	59	0	0
Phellinus igniarius	100	130	—	100	50	10	0	0
Coriolus zonatus	100	82	—	75	63	63	0	0
Heterobasidion annosum	100	—	100	65	27	11	0	0
Galerina mutabilis	100	—	89	5	0	0	0	0

— Not determined
+ Slight growth

inhabiting species was investigated in relation to certain products of anaerobic respiration, namely: ethanol, methanol, lactate, formate, acetate and propionate (Hintikka, 1969). These were added to 1% malt extract agar in closed flasks, and radial growth was measured. Acids were neutralised with ammonium hydroxide to pH 4.9–5.5, and all concentrations had approximately the same pH value. There was not much difference in their tolerance of alcohols, but wood decomposers could grow at slightly higher concentrations. There was, however, a distinct difference between the two groups in their tolerance of acetate

(Table 2); wood-decomposing species commonly tolerated concentrations of 0.1–0.2%, whereas soil/litter inhabiting species stopped growing at 0.025–0.05%. Similar differences occurred on formate and propionate, but the reactions to lactate were less distinct. *Quercus* wood contains acetic acid in such high concentrations that it may corrode metals (Packman, 1960), and it should be noted that in nature only the most tolerant species grow on this wood. Physiologically, as G. Lindeberg & M. Lindeberg (1974), Milward (1976) and Thornton, Robinson & French (1977) have pointed out, short-chain aliphatic acids are toxic to fungi, but the physiological background to this difference is not clear. In lower concentrations they may stimulate wood-decomposing species (Suolahti, 1951; Glasare, 1970).

Effects of terpenes and phenols

Coniferous wood differs from wood of deciduous trees in its content of volatile monoterpenes. It is probable that in living coniferous wood the gas phase is saturated with terpene vapours. Monoterpenes are toxic to fungi (Cobb, Krstic, Zavarin & Barber, 1968; Hintikka, 1970*c*; DeGroot 1972; Krupa, Andersson & Marx, 1973; Väisälä, 1974, 1978; Chou & Zabkiewicz, 1976; Schuck, 1977). As pointed out by Schuck, their insolubility in water causes some methodological difficulties, because the hyphae which grow from the inoculum into the agar are more exposed to terpene vapours than hyphae deeper in the agar, and evidently not all results are comparable. In one experiment (Hintikka, 1970*c*) various Hymenomycetes were grown on 1% malt extract agar in closed desiccators into which measured amounts of terpenes were pipetted, and the air was circulated with a propeller. There was a definite difference between species growing in nature on coniferous wood and those on hardwood. Many species which inhabit conifers, especially those which occur on dead tissues of living hosts (perthophytes), were able to grow in saturated atmospheres of several terpenes. Fungi of hardwoods were much more sensitive to small amounts of terpenes in the air (Table 3).

Several authors (Cochrane, 1958; Popoff, Theander & Johansson, 1975) have found that fungi differ in their tolerance of phenols, which occur naturally in wood, appear as degradation products of lignin, or are synthesised by fungi. This was shown in one experiment by measuring the radial growth of forty-six wood-decomposing fungi on Hagem agar to which various phenols had been added (Hintikka, 1971*a*). When the agar contained *p*-cresol, the most tolerant species were *Fomitopsis*

Table 3. *The effect of terpenes on radial growth (mm) of lignicolous basidiomycetes grown on 1% malt extract agar for 10 days at room temperature.* Each value obtained from one Petri-dish culture

	α-Pinene						β-Pinene						Camphene						Δ^3-Carene					
Concentration of terpenes (ml l^{-1} air)	0	0.001	0.0025	0.005	0.01	0.025	0	0.001	0.0025	0.005	0.01	0.025	0	0.0025	0.005	0.01	0.025	conc.	0	0.001	0.005	0.01	0.025	conc.
Species growing mainly in coniferous wood																								
Fomitopsis pinicola	25	33	17	14	13	3	25	20	16	5	3	3	26	24	4	5	2	1	28	10	7	2	0	2
Gloeophyllum sepiarium	23	23	18	10	10	+	20	17	18	2	2	2	24	21	3	2	0	0	25	20	6	4	—	1
Heterobasidion annosum	44	45	20	20	20	11	45	19	18	13	13	6	45	45	20	14	10	8	45	43	20	9	5	—
Hirschioporus abietinus	16	23	20	9	8	3	17	18	16	2	1	1	16	13	2	1	1	0	21	17	3	2	2	1
Lentinus lepideus	24	20	17	14	13	7	24	17	13	7	3	0	26	26	13	11	4	4	22	22	0	5	0	0
Phellinus pini	8	8	10	8	8	5	8	10	8	5	4	4	10	9	5	3	2	1	9	7	5	5	5	5
Stereum sanguinolentum	37	14	11	9	8	4	14	8	5	4	2	2	26	7	3	3	1	1	10	7	3	2	3	1
Species growing mainly in wood of deciduous trees																								
Coriolus hirsutus	36	23	9	1	1	0	35	18	10	0	0	0	38	9	0	0	0	0	37	2	0	0	0	0
Fomes fomentarius	21	6	1	0	0	0	27	6	4	0	0	0	29	10	0	0	0	0	28	0	+	0	+	0
Galerina mutabilis	15	11	7	1	—	0	27	16	13	1	+	0	19	7	0	0	0	0	24	12	1	+	0	0
Grifola frondosa	25	2	0	0	0	0	23	10	0	0	0	0	30	0	0	0	0	0	30	0	0	0	0	0
Inonotus rheades	6	7	7	+	+	0	8	8	7	0	0	0	7	+	0	0	0	0	8	1	0	0	0	0
Laetiporus sulphureus	28	8	3	+	+	0	30	20	6	0	0	0	17	3	0	0	0	0	28	0	0	0	0	0
Piptoporus betulinus	28	25	8	+	+	0	32	20	7	0	0	0	25	1	0	0	0	0	34	1	0	0	+	0
Pycnoporus cinnabarinus	23	17	2	0	0	0	23	14	1	0	0	0	25	1	0	0	0	0	25	13	0	0	0	0

—, Not determined; +, slight growth.

pinicola, Inonotus rheades, Lentinus lepideus and *Phellinus pini*. When ferulic acid was present, *Gloeophyllum sepiarium, Phellinus pini* and *Stereum sanguinolentum* grew in the highest concentrations tested. When tannic acid was added, *Daedalea quercina, Fomitopsis pinicola, Galerina mutabilis, Laetiporus sulphureus, Phellinus igniarius* and *Piptoporus betulinus* grew best (most of the species of this group can grow in nature on *Quercus* wood). On agar containing vanillin, the least tolerant species were *Heterobasidion annosum, Laetiporus sulphureus* and *Fomes fomentarius*. Thus, in these conditions, different species showed definite differences in tolerance to different phenols. However, it should be noted that the toxicity of phenols is dependent on pH (Cochrane, 1958; Cruickshank & Perrin, 1964). Popoff *et al.* (1975) found that the toxicity of pyrocatechol and 4-methylcatechol to *Heterobasidion annosum* decreases with increasing pH. The pH dependence of the inhibitory effects of phenols varies among the enzymes produced by *H. annosum* (Johansson, Popoff & Theander, 1976). In our experiments (V. Hintikka, unpublished) we found that this species was definitely more sensitive to certain phenolics when it was growing on acid media.

Conclusions

In Finnish forests there appear to be two distinct groups of decomposer basidiomycetes, those occurring on wood and those characteristic of leaf litter. The experiments described here suggest that lignicolous species are better adapted than litter-decomposing basidiomycetes to growing where both aeration is restricted and various volatile substances (e.g. short-chain fatty acids) may accumulate. In addition, the restriction of lignicolous fungi to certain tree species may well be related to the distribution of different monoterpenes and phenols. To test the validity of these hypotheses, a more comprehensive picture of the occurrence of such substances in nature, as well as of their effects on fungal physiology, is needed.

References

Brierley, J. K. (1955). Seasonal fluctuations in the oxygen and carbon dioxide concentrations in beech litter with reference to the salt uptake of beech mycorrhizas. *Journal of Ecology*, **43**, 404–8.

Carrodus, B. B. & Triffett, A. C. K. (1975). Analysis of composition of respiratory gases in woody stems by mass spectrometry. *New Phytologist*, **74**, 243–6.

Chase, W. W. (1934). The composition, quantity and physiological significance of gases in tree stems. *University of Minnesota Agricultural Experiment Station, Technical Bulletin*, No. 99, 1–51.

Chou, C. K. S. & Zabkiewicz, J. A. (1976). Toxicity of monoterpenes from *Pinus radiata* cortical oleoresin to *Diplodia pinea* spores. *European Journal of Forest Pathology*, **6**, 354–9.

Cobb, F. W. Jr., Krstic, M., Zavarin, E. & Barber, H. W., Jr. (1968). Inhibitory effects of volatile oleoresin components on *Fomes annosus* and four *Ceratocystis* species. *Phytopathology*, **58**, 1327–35.

Cochrane, V. W. (1958). *Physiology of fungi*. London & New York: Chapman & Hall.

Cruickshank, I. A. M. & Perrin, D. K. (1964). Pathological functions of phenolic compounds in plants. In *Biochemistry of Phenolic Compounds*, ed. J. B. Harborne, pp. 511–44. London & New York: Academic Press.

DeGroot, R. C. (1972). Growth of wood-inhabiting fungi in saturated atmospheres of monoterpenoids. *Mycologia*, **64**, 863–70.

Fries, N. (1966). Chemical factors in the germination of spores of Basidiomycetes. In *The Fungus Spore*, ed. M. F. Madelin, pp. 189–99. Colston Research Society. London: Butterworth.

Glasare, P. (1970). Volatile compounds from *Pinus sylvestris* stimulating growth of wood-rotting fungi. *Archiv für Microbiologie*, **72**, 333–43.

Gundersen, K. (1961). Growth of *Fomes annosus* under reduced oxygen pressure and the effect of carbon dioxide. *Nature, London*, **190**, 649.

Havas, P. & Mäenpää, E. (1972). Evolution of carbon dioxide at the floor of a *Hylocomium Myrtillus* type spruce forest. *Aquilo Series Botanica*, **11**, 4–22.

Hering, T. F. (1967). Fungal decomposition of oak leaf litter. *Transactions of the British Mycological Society*, **50**, 267–73.

Hintikka, V. (1964). Psychrophilic basidiomycetes decomposing forest litter under winter conditions. *Communicationes Instituti Forestalis Fenniae*, **29**:2, 1–20.

Hintikka, V. (1969). Acetic acid tolerance in wood- and litter-decomposing Hymenomycetes. *Karstenia*, **10**, 177–83.

Hintikka, V. (1970*a*). Studies on white-rot humus formed by higher fungi in forest soils. *Communicationes Instituti Forestalis Fenniae*, **69**:2, 1–68.

Hintikka, V. (1970*b*). Stimulation of spore germination of wood-decomposing Hymenomycetes by carbon dioxide. *Karstenia*, **11**, 23–7.

Hintikka, V. (1970*c*). Selective effect of terpenes on wood-decomposing Hymenomycetes. *Karstenia*, **11**, 28–32.

Hintikka, V. (1971*a*). Tolerance of some wood-decomposing Basidiomycetes to aromatic compounds related to lignin degradation. *Karstenia*, **12**, 46–52.

Hintikka, V. (1971*b*). *Mucor oblongisporus* as a psychrophilic secondary sugar fungus. *Karstenia*, **12**, 59–65.

Hintikka, V. (1973). Passive entry of spores into wood. *Karstenia*, **13**, 5–8.

Hintikka, V. & Korhonen, K. (1970). Effects of carbon dioxide on the growth of lignicolous and soil-inhabiting Hymenomycetes. *Communicationes Instituti Forestalis Fenniae*, **69**:5, 1–29.

Hintikka, V. & Näykki, O. (1967). Notes on the effects of *Hydnellum ferrugineum* (Fr.) Karst. on forest soil and vegetation. *Communicationes Instituti Forestalis Fenniae*, **62**:2, 1–22.

Jensen, F. K. (1967). Oxygen and carbon dioxide affect the growth of wood-decaying fungi. *Forest Science*, **13**, 384–9.

Jensen, F. K. (1969). Oxygen and carbon dioxide concentrations in sound and decaying red oak trees. *Forest Science*, **15**, 246–51.

Jensen, V. (1974). Decomposition of Angiosperm tree leaf litter. In *Biology of Plant Litter Decomposition*, ed. by C. H. Dickinson & G. J. F. Pugh. London & New York: Academic Press.

Johansson, M., Popoff, T. & Theander, O. (1976). Effect of spruce rot constituents on extracellular enzymes of *Fomes annosus*. *Physiologia Plantarum*, **37**, 275–82.

Kärenlampi, L. (1971). Weight loss of leaf litter on forest soil surface in relation to weather at Kevo Station, Finnish Lapland. *Report of the Kevo Subarctic Research Station*, **8**, 101–3.

Krupa, S., Andersson, J. & Marx, D. H (1973). Studies on ectomycorrhizae of pine. IV. Volatile organic compounds in mycorrhizal and nonmycorrhizal root systems of *Pinus echinata* Mill. *European Journal of Forest Pathology*, **3**, 194–200.

Lindeberg, G. & Lindeberg, M. (1974). Effect of short chain fatty acids on the growth of some mycorrhizal and saprophytic Hymenomycetes. *Archiv für Microbiologie*, **101**, 109–14.

McDougall, D. T. & Working, E. B. (1933). The pneumatic system of plants, especially trees. *Carnegie Institute of Washington Publication*, **441**, 1–87.

Milward, Z. (1976). Further experiments to determine the toxicity of propionic acid to fungi infesting stored grain. *Transactions of the British Mycological Society*, **66**, 319–24.

Novikov, M. A. & Solovev, V. A. (1973). Diurnal changes of the composition of the gas in the wood of spruce trees infested with some decomposing fungi. *Vestnik Akademii nauk SSSR, Ser. Biol.*, **5**, 52–5.

Packman, D. F. (1960). The acidity of wood. *Holzforschung*, **14**, 178–83.

Popoff, T., Theander, O. & Johansson, M. (1975). Changes in sapwood of roots of Norway spruce, attacked by *Fomes annosus*. II. *Physiologia Plantarum*, **34**, 347–56.

Romell, L.-G. (1928). Markluftsanalyser och markluftning. *Meddelanden Statens Skogsförsöksanstalt*, **24**, 67–80.

Schánĕl, L. (1976). Role of carbon dioxide in growth and decaying activity of wood-rotting fungi. *Folia Facultatis Scientiarum Naturalium Universitatis Purkynianae Brunensis*, **17**, Biol. 54, op. 6, 5–54.

Schuck, H. J. (1977). The effect of monoterpenes on the mycelial growth of *Fomes annosus* (Fr.) Cooke. *European Journal of Forest Pathology*, **7**, 374–84.

Suolahti, O. (1951). Über eine das Wachstum von Fäulnispilzen beschleunigende chemische Fernwirkung von Holz. Diss. Helsinki.

Tabak, H. H. & Cooke, W. B. (1968). The effects of gaseous environments on the growth and metabolism of fungi. *Botanical Review*, **34**, 126–252.

Thacker, D. G. & Good, H. M. (1952). The composition of air in trunks of sugar maple in relation to decay. *Canadian Journal of Botany*, **30**, 475–85.

Thornton, J. D., Robinson, P. J. & French, J. R. J. (1977). Toxicity of aliphatic acids against soft rot organisms and *Gloeophyllum trabeum*. *International Biodeterioration Bulletin*, **13**, 108–11.

Väisälä, L. (1974). Effects of terpene compounds on the growth of wood-decomposing fungi. *Annales Botanici Fennici*, **11**, 275–8.

Väisälä, L. (1978). Effects of terpene compounds on the growth and peroxidase activity of *Phellinus pini*. *Annales Botanici Fennici*, **15**, 131–7.

Zycha, H. (1937). Über das Wachstum zweier holzzerstörender Pilze und ihr Verhältnis zur Kohlensäure. *Zentralblatt für Bakteriologie, Parasitenkunde und Infektionskrankheiten*, **97**, 222–44.

14 Biomass and nutrient cycling by decomposer basidiomycetes

J. C. FRANKLAND
Institute of Terrestrial Ecology, Merlewood Research Station, Grange-over-Sands, Cumbria LA11 6JU, England

Half a century ago Waksman was questioning the relationship between soil fertility and microbial activity. Since then much has been discovered about the influence of mycorrhizal basidiomycetes on plant growth and nutrient uptake, but we still know little about the involvement in mineral cycling of those basidiomycetes which are primarily decomposers of dead organic matter. Their supreme importance in the breakdown of cellulose and lignin is amplified in other chapters. This review discusses from an ecological viewpoint their biomass and influence on the cycling of nutrient elements other than carbon, particularly in temperate woodlands.

Biomass of decomposer basidiomycetes in litter and soil

On etymological grounds the term 'biomass' should be restricted to living matter, but frequent misuse has led to confusion. In this chapter it signifies, therefore, the total standing crop of an organism unless specified as being only living or dead, and 'production' is the biomass produced over a certain period of time.

The decomposer basidiomycetes rarely figure in lists of fungi isolated from litter or soil, even when their fruit bodies are abundant, and we still cannot obtain an accurate measurement of their total biomass in an ecosystem. We can, however, attempt to apportion biomass, within some ecosystems, to the various groups of decomposer and consumer organisms, and can obtain, as will be shown, some idea of the relative magnitude of the basidiomycete biomass.

It is an accepted fact that decomposition by heterotrophs is important in the release of nutrients for green plants and that heterotrophic metabolism forms a significant part of the total metabolism of a terrestrial ecosystem. Reichle (1977) estimated that heterotrophic metabolism (g C m^{-2} yr^{-1}) ranged from 34 to 57% of the total metabolism in forest, prairie and tundra ecosystems. Micro-organisms in general have a relatively high rate of production and turnover but a small living biomass; nevertheless the major proportion of the living biomass of the heterotrophic community of a temperate forest floor is microbial. Reichle, for example, attributed only 7% of the living heterotrophic biomass in a temperate *Liriodendron* forest in Tennessee to the pulmonates, nematodes and arthropods but 85% to the microflora. In a few ecosystems, the apportionment of biomass has been extended to the constituent groups of the microflora. For example, in the International Biological Programme analysis of Meathop Wood, an English deciduous woodland with a thin accumulation of litter and a mull humus on limestone, 96% of the microbial biomass in the leaf litter layer and soil (excluding woody remains and roots) was found to belong to the actinomycetes and bacteria, and only 4% to the fungi (Gray, Hissett & Duxbury, 1973; Frankland, 1975*a*,*b*, and unpublished). In terms of living biomass, which is more relevant to activity than this total biomass, the fungi had a much higher status, constituting 66% of the microflora. Attempts to subdivide microbial biomass usually cease at this point.

However, the type and distribution of fungal mycelium in a soil profile can be clues to the abundance of basidiomycetes, even if their species are rarely distinguishable unless fruiting. The distribution of the fungal biomass in the soil of Meathop Wood, estimated from hyphal length and classified as belonging to basidiomycetes or microfungi according to the presence or absence of clamp connections, is shown in Table 1, although it is recognised that the basidiomycete component may have been underestimated because the frequency of clamp formation varies between species. A relatively large quantity of non-basidiomycete mycelium (kg ha^{-1}) occurred in the lowest soil horizons. However, this reflected the sheer bulk of the subsoil. Fungal mycelium of all types was most concentrated (g g^{-1} substrate) in the thin organic L and (Oh + Ah) surface horizons, and basidiomycete mycelium, excluding as far as possible that of mycorrhizal and pathogenic species on and in living roots, was almost confined to this area of the profile, dominating the mycoflora during the decomposition of the cell walls of plant debris. As many as 89% of the dead *Quercus* leaves, which formed the principal component of the L horizon, were found by direct observation to be colonised by basidiomycetes 18 months after leaf-fall. Basidiomycete mycelium was also abundant on other leaf types in this

Table 1. *Comparison of the distribution of the biomass of basidiomycetes (kg ha^{-1} dry wt) with that of other microbial decomposers in the floor of a temperate deciduous woodland with mull humus (Meathop Wood, Cumbria, UK)*

Substrate or horizon		Basidiomycetes Living	Basidiomycetes Total	Other fungi Living	Other fungi Total	Bacteria and actinomycetes Living	Bacteria and actinomycetes Total
Woody debris		30.5	216.9	7.3	34.7	2.6	601.6
L		3.1	8.7	0.5	4.1		
(Oh + Ah)	depth	8.9	31.7	3.4	12.9	37.3	8433.3
A	36 cm	<1.0	<1.0	26.4	97.5		
B		<1.0	<1.0	31.4	155.6		
Dead roots		228.0	1628.5	65.1	325.7	8.0	1851.1
Total		271.5	1886.8	134.1	630.5	47.9	10 886.0

woodland, and the abundance increased with age of the litter (Fig. 1).

It is more difficult to classify hyphae accurately in litter after it has been homogenised as in the classic technique for measuring hyphal length (Jones & Mollison, 1948), but by using this method and converting length to biomass (see Frankland, Lindley & Swift, 1978) it was estimated that at least 68 and 71% respectively of the total fungal biomass in the L and (Oh + Ah) horizons of Meathop Wood (excluding roots) was of the basidiomycete type, closely resembling that seen on intact leaves. On this basis, basidiomycetes represented approximately one-eighth of the total fungal biomass in the litter/soil profile. The living biomass was also measured, assuming that hyphae with cell contents were alive (Frankland, 1975*b*). By this criterion, 36 and 28% respectively of the basidiomycete mycelium in the two organic horizons was living, i.e. approximately one-sixth of the living fungal biomass in the litter and soil of this woodland belonged to this group of fungi. The microfungi appeared to be represented by many more species, but basidiomycete mycelia characteristically are less ephemeral and accumulate more biomass than those of the microfungi.

Basidiomycetes can form therefore a significant proportion of the microbial biomass of a woodland soil, but their ecological importance in biomass terms becomes much more obvious if account is taken of the large quantities of fungal mycelium in dead wood and dead roots. In Meathop Wood, these substrates were often packed with mycelium, and many dead tree branches and roots contained virtually a 'pure culture'

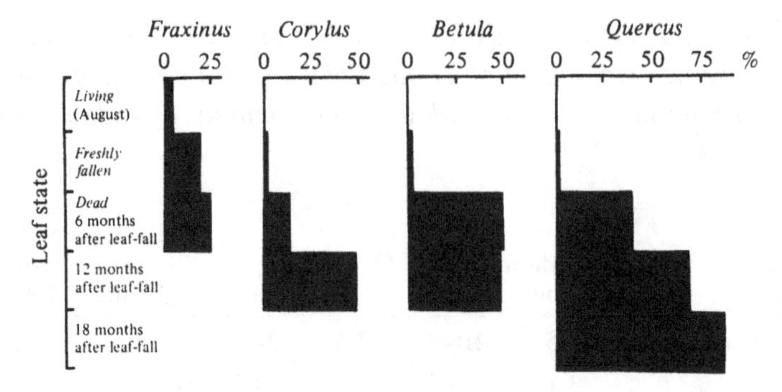

Fig. 1. Number of leaves (%) colonised by basidiomycetes (Meathop Wood, Cumbria, UK).

of a basidiomycete, such as *Stereum hirsutum* or *Armillaria mellea*. The total and living biomass of basidiomycetes and other fungi in the dead wood (M. J. Swift, personal communication) and roots are shown in Table 1. The mycelial content of roots was estimated from the weight of dead roots (J. E. Satchell, unpublished) by assuming that the fungal content was the same as in dead wood. According to these data on decomposer fungi, over 80% of all the mycelium and over 90% of basidiomycete mycelium on and in the forest floor was located in dead wood and dead roots, which suggests that research workers' past preoccupation with leaf litter in decomposition studies of woodlands may well have been misplaced.

The biomass of bacteria and actinomycetes is also included in Table 1, showing that they formed the major component of dead microbial material, but that basidiomycetes contributed at least 60% (272 kg ha^{-1}) of all living microbial biomass. However, all these estimates must be tentative until better methods of detecting these fungi are developed.

The annual production of the fruit bodies of the litter-inhabiting basidiomycetes of Meathop Wood was calculated from Hering's records (1966) to be only 1 kg ha^{-1} dry wt at most, so in this woodland the reproductive biomass of the basidiomycetes appears to be small in comparison with the estimates of their vegetative mycelium.

There is meagre information on the quantity of basidiomycete mycelium in other soil types, but the estimates for the calcareous mull in Meathop Wood are probably at the low end of the scale for temperate woodlands. Fungal mycelium in general is less concentrated in mull with a thin accumulation of litter than in mor with a well-developed litter

Table 2. *The proportion of clamp-bearing hyphae and dark pigmented hyphae in the profile of a Swedish forest podsol expressed as a percentage of the total length of fungal hyphae in each horizon: mean values, (n = 6).* (From Bååth & Söderström, 1977)

Horizon	Clamp-bearing hyphae	Dark pigmented hyphae
A_{00}	2	21
A_{01}	7	19
A_{02}	13*	14
A_2	7	15
B	5	27

* Significantly higher than in the other horizons, $P < 0.05$ (Bååth & Söderström, 1979).

layer. Witkamp (1960) confirmed this and found that, in autumn, 91% and 100% respectively of the mycelium in the upper 2 cm of the mineral soil of a calcareous mull and a mor under temperate oak forest was of the basidiomycete type. The total biomass of basidiomycete mycelium in a well-developed iron podsol under mature *Pinus sylvestris* in central Sweden (Bååth & Söderström, 1977, 1979) was higher than in the Meathop mull, although the length of hyphae with clamp connections as a percentage of the total hyphal length was somewhat lower. The highest proportion of basidiomycete hyphae ($P < 0.05$) was again near the surface in an organic A_{02} (H) horizon (Table 2).

Even less is known about the amount of basidiomycete mycelium in the soil of other climatic zones, but Flanagan & Scarborough (1974) reported that hyphae with clamp connections accounted for at least one-third of the total fungal biomass in some soils of the American tundra. Again, hyphae of mycorrhizal species may have been included with those of decomposer species; a lower proportion of basidiomycete mycelium was recorded in tundra soil lacking ectomycorrhizas by Miller & Laursen (1974). The biomass of saprophytic basidiomycetes might be expected to be greatest in tropical rain forests, which produce about three times as much organic matter as the most productive woodlands of the temperate region. This is unlikely, however, because the accumulation of litter is very small; on the other hand, higher temperatures and wet conditions result in speedy growth and decay, so mycelial production is probably much greater than in temperate woodlands.

Whereas biomass is the key to the capture of plant nutrients, the turn-over of this biomass is all-important in nutrient release, but as yet we cannot measure the annual production of basidiomycete mycelium in a natural soil. The only production estimate appears to be one obtained by the author for a single decomposer, *Mycena galopus*, growing on *Quercus* litter in Meathop Wood (Frankland, 1975*a*; see Frontispiece). It was determined by measuring changes in the biomass and decay rate of the mycelium on litter with a distinctive white rot, and was about twice the maximum biomass. The ratios of vegetative production : vegetative biomass : fruit body production for this species were approximately 10:5:1. If these ratios were extended to the other fruiting basidiomycetes of the litter, the mycelial production and biomass of the basidiomycetes in the L horizon became at least 10 kg ha^{-1} yr^{-1} and 5 kg ha^{-1} respectively. This biomass figure is of the same order of magnitude as the estimate obtained from hyphal length, but it is doubtful whether it is justified to extrapolate as far as this, considering the vagaries of fruiting and morphological differences known to occur among basidiomycetes.

Not only the vegetative mycelium but also the fruiting of basidiomycetes is more prolific in mor sites than in mull, as illustrated by Hering's survey of decomposer species in some English woodlands (Chapter 12). Coniferous woodlands are usually the most productive. Richardson (1970), for example, recorded 16–30 kg ha^{-1} yr^{-1} dry wt of agarics and boleti in a very thorough survey of a Scottish plantation of *Pinus*. These yields are still very low in comparison with commercial mushroom farming (Hedger & Basuki: Chapter 15), but Hora (1959, 1972) and others have obtained impressive increases of production by adding fertilisers to woodland soils. Again, comparative data from the tropics are lacking, but small decomposer species such as *Marasmius, Mycena* and *Leptonia* occur in large troops in the rain forests (Dickinson & Lucas, 1979; see also Singer & de Silva Araujo, 1979).

Basidiospores are so minute that their biomass is usually ignored, but their size is counterbalanced by the vast numbers produced (Buller, 1909; Kramer: Chapter 2). Hering (1966) recovered in one season 159 fruit bodies, many of them small species, from a 100 m^2 quadrat of the relatively unproductive Meathop Wood. This represents a total spore biomass of 0.1 kg ha^{-1} dry wt assuming that the mean number of spores produced by a fruit body was 1.0×10^9 (probably a conservative estimate!) and that the mean dry weight of a spore was 0.85×10^{-11} g (Cochrane, 1958). The contribution of basidiospores to the dispersal and release of nutrients is therefore likely to be of some significance in more productive woodlands.

Concentrations of mineral elements in decomposer basidiomycetes

Evidence that basidiomycetes can form a significant proportion of the microbial biomass of a woodland has been outlined. What does this biomass represent in terms of mineral nutrients?

Most information on the chemical composition of decomposer basidiomycetes in the field comes from analysis of sheets of mycelium stripped from wood or of rhizomorphs and fruit bodies. It is extremely difficult to collect sufficient quantities of unstranded mycelium from leaf litter and soil for this purpose. In addition, the chemical composition of fungi, including saprophytic basidiomycetes, is known to depend on their substrate (Levi & Cowling, 1968), so mycelium grown on synthetic media is not a reliable guide to the field state. Analysis of the mycelium of the litter decomposer *Mycena galopus* on an almost natural substrate was achieved, however, by laboriously stripping the hyphae from *Quercus* and *Corylus* leaf litter, which had been sterilised by γ-radiation and inoculated with the fungus without addition of nutrients (J. C. Frankland, unpublished). The concentrations of major plant nutrients (N, P, K, Ca, Mg) in this mycelium are compared in Table 3 with estimates for basidiomycetes growing in litter and wood obtained by other workers. Apart from some high levels of potassium and calcium, two elements which micro-organisms can absorb in luxury amounts, the concentrations of nutrients in the simple mycelium and rhizomorphs of the litter and wood decomposers were similar. In fact the concentrations of minerals in the *Mycena* mycelium on leaf litter were all within the ranges quoted for basidiomycetes on wood (Swift, 1977*a*).

In general, the fruit bodies of the litter decomposers (Table 3) were richer in nitrogen, phosphorus and potassium, and lower in calcium than their corresponding vegetative mycelium; potassium again occurred in luxury quantities. Stark's analyses (1972) showed similar trends in the nutrient concentrations of basidiomycetes from American sites, and her fruit bodies were consistently lower than rhizomorphs in calcium in both northern coniferous and southern rain forests. Fruit bodies of nonlignicolous species also tend, it seems, to contain greater concentrations of nitrogen, phosphorus and potassium than those of lignicolous fungi (Table 3; Vogt & Edmonds, 1980).

Spores as well as sporulating tissues are usually rich in the major

Table 3. *Comparison of the concentrations of mineral nutrients in undifferentiated mycelium, rhizomorphs and fruit bodies of decomposer basidiomycetes (% dry wt)*

	Litter decomposers						Wood decomposers	
Species . . .	*Mycena galopus*		Various[a]		Various[b]		Various	
Litter/wood type . . .	Broadleaved		Broadleaved		Broadleaved and coniferous		Hardwood and coniferous	
	Mycelium	Fruit bodies	Rhizomorphs	Fruit bodies	Rhizomorphs	Fruit bodies	Mycelium & rhizomorphs[c]	Fruit bodies[d]
Mineral nutrients								
N	3.50	5.70	4.13	4.40	—	—	0.23–4.31	0.72–1.13
P	0.24	0.69	0.93	1.00	0.16	0.55	0.02–1.31	0.19
K	0.57	4.47	3.16	4.90	0.39	2.52	0.16–0.61	0.39
Ca	0.95	0.17	0.85	0.11	3.93	0.04	0.68–9.50	0.96
Mg	0.37	0.12	0.14	0.13	0.04	0.05	0.04–0.88	0.21

Sources:
[a] Ausmus & Witkamp (1974).
[b] Cromack, Todd & Monk (1975).
[c] Swift (1977*a*): review data.
[d] N: Merrill & Cowling (1966); P, K, Ca, Mg: Cromack, Todd & Monk (1975).
—, not measured.

Table 4. *Ratios of the mineral nutrient concentrations, as percentage oven dry weight, of the mycelium of two decomposer basidiomycetes to the mineral nutrient concentrations of their substrates*

Mycelium . . . Substrate . . .	*Mycena galopus*[a] *Corylus* leaf litter	*Stereum hirsutum*[b] Living *Quercus* wood
Mineral nutrients		
P	2.9	9.4
N	1.8	8.2
Mg	1.2	4.8
K	2.0	3.0
Ca	0.5	1.9

[a] J. C. Frankland (unpublished data).
[b] Swift (1977*b*).

nutrients, but information on basidiospores is limited. Merrill & Cowling (1966) found that basidiospores of some wood decomposers contained 3% N, which was equivalent to the nitrogen concentration in the current hymenia of the same fruit bodies. Dowding (1976) pointed out that the nutrient content of fungal spores is unlikely to vary as much as that of mycelium, and that for any one species the content is likely to be largely genetically fixed.

As is typical of decomposers, many elements are more concentrated in saprophytic basidiomycetes than in the plant residues on which they are growing. This is illustrated by the concentrations of macronutrients in *Stereum hirsutum* on *Quercus* wood in relation to the concentrations in its substrate (Swift, 1977*b*), and except for calcium by those in *M. galopus* on *Corylus* litter (Table 4). The concentration factors were greater for the fungus on wood, which on a dry weight basis is relatively poor in nutrients, although, as Park (1976) emphasised, the nitrogen content as parts per million of wood with a moisture content of say 50% is not particularly low for decay organisms. Phosphorus was the element most in demand by both fungi. Vogt & Edmonds (1980) have also compared the nutrient concentrations in fruit bodies, including decomposer species, with those in litter. Although the nutrient content of fruit bodies in four distinct forest types of western Washington varied widely between species (e.g. 0.66–11.27% N), nitrogen, phosphorus and potassium were generally concentrated in significantly higher amounts in the fruit bodies than in the forest floor, but the fungi were not always linked in the analyses with their specific substrates. Evidence of wide variations

Table 5. *Comparison of concentration factors for mineral nutrients in decomposer basidiomycetes and decomposer insects on wood (ratios of nutrient concentration as percentage oven dry weight in the decomposer to that in the wood).* (After Swift, 1977*a*.)

Mineral nutrients	Ratio basidiomycete mycelium:wood	Ratio insects:wood
P	14.6	27.3
Ca	13.4	0.9
N	5.8	28.3
Mg	2.8	4.4
K	1.6	3.3

in concentration factors have been obtained from field experiments using radionuclides (Witkamp, 1969). The concentration of ^{137}Cs in fruit bodies varied from <0.1 to 80 times that in their tagged litter substrate, depending on the species of basidiomycete and season. There were also indications that the ^{137}Cs concentration increased with age of the long-lived fruit body of *Clavulina cristata* over a period of several weeks.

Among the micronutrients, significantly higher concentrations of molybdenum and zinc have been found in bracket fungi than in their supporting branches (Cromack, Todd & Monk, 1975). Similarly, higher concentrations of copper, iron and zinc have been recorded in fruit bodies and rhizomorphs than in their leaf and needle substrates (Cromack *et al.*, 1975; Stark, 1972). Sodium, essential to animals but not to plants and fungi, was also accumulated. Measurement of the concentrations of trace elements in the fruit bodies of a large number of higher fungi from natural habitats has also shown that these fungi often contain much more copper and zinc (parts per million dry matter) than living higher plants (Hinneri, 1975; Mutsch, Horak & Kinzel, 1979; Tyler, 1980). Hinneri's evidence also suggests that the fruit bodies of decomposer basidiomycetes are richer in minerals than those of mycorrhizal species.

Decomposer basidiomycetes have the potential therefore to be important accumulators of plant and animal nutrients, but they should be kept in perspective with other groups. Animals can have even higher demands for mineral nutrients as shown for wood decomposers (Swift, 1977*a*; Table 5), but their relative share in nutrient uptake will depend on their biomass and rate of turnover as well as on their nutrient concentrations.

'Capture' of mineral elements by decomposer basidiomycetes

Uptake of mineral elements from plant debris by a decomposer fungus will depend *inter alia* on the balance between available energy sources and mineral supply. Basidiomycetes are particularly well fitted to acquire both of these by virtue of their enzyme systems, ability to penetrate deeply, relatively large surface area, translocating powers, and longevity. Analyses of fruit bodies and their substrates in the field have provided evidence of their ability to tap distant sources of nutrients by translocation. Merrill & Cowling (1966), for example, found that nitrogen had not been stripped from wood adjacent to the fruit bodies of various polypores, and they calculated that the nitrogen required for sporulation must have come from very large volumes of wood or from sources outside the wood itself. However, without measurement, the actual uptake of nutrients in the field usually has to be assumed.

Once a nutrient element is incorporated within a micro-organism it is temporarily unavailable to plants and can be said to be captured or immobilised. The nutrient concentration and the biomass of the organism together provide an estimate of this nutrient pool. Witkamp (1969) calculated that gross immobilisation of the readily-leached mineral ^{137}Cs by the natural microflora on *Liriodendron tulipifera* litter could be as much as 60%. It is rarely possible, however, to estimate the size of the nutrient pool in a basidiomycete in field material with current methods, but Swift (1977*b*) has achieved it for *Stereum hirsutum* on decaying branch-wood. *Quercus* branches occupied solely by this species were chosen so that the fungal biomass could be estimated by hexosamine assay. Nutrient concentrations were determined by chemical analysis of the wood and of mycelium stripped from inside the branches. As decay advanced, the fungus gradually captured the nutrients until 90–100% of the nitrogen and phosphorus, and 30–50% of the calcium, magnesium and potassium were contained in the mycelium (Fig. 2). Lower proportions of nutrients were immobilised by *Mycena galopus* growing on sterilised *Betula* and *Fraxinus* leaf litter (Table 6), which may have been due to the higher mineral content of the litter and to differences in the physiology and age of the two fungi (Frankland, Lindley & Swift, 1978). The practical difficulties are illustrated by the variation in the values obtained when two different methods were used for measuring the biomass (Table 6). The levels of immobilisation were all biologically reasonable, but the higher values, obtained by hexosamine assay, were argued to be the more accurate.

Hyphal translocation (see Jennings: Chapter 5) can lead to the accumulation of nutrients in a decomposing organic substrate in amounts exceeding those in the living material. Net accumulation or loss depends on the balance between this translocation and a variety of other import or export activities and processes, including fungal sporulation,

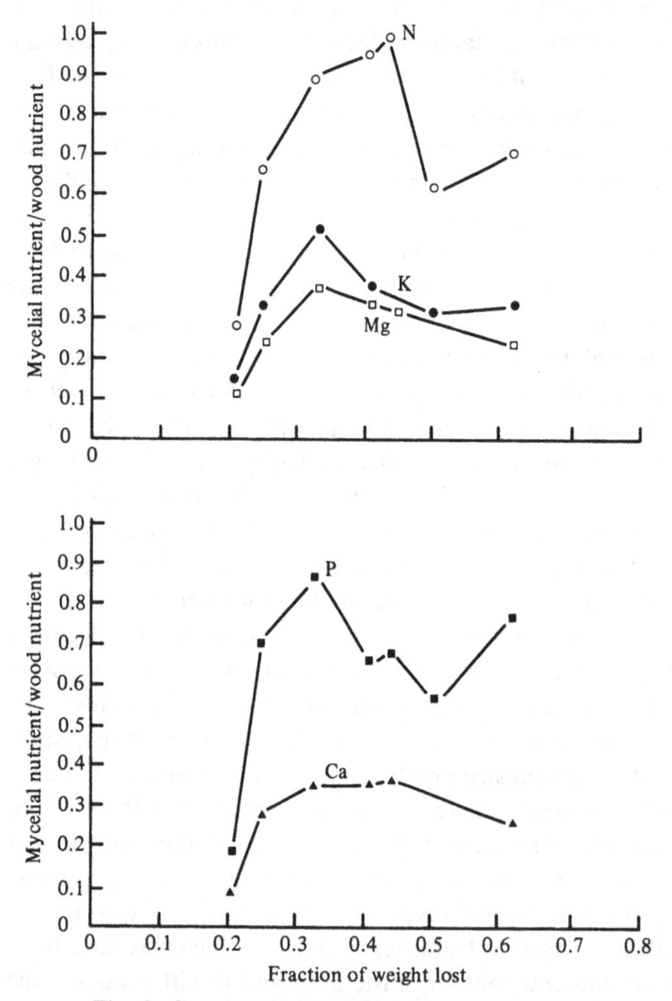

Fig. 2. Capture of mineral nutrients by *Stereum hirsutum* in *Quercus* branch-wood in relation to the state of wood decay. Capture expressed as the ratio of nutrient content (biomass × nutrient concentration) of the mycelium to the nutrient content of the wood. (From Swift, 1977*b*.)

Table 6. *Proportion of mineral nutrients immobilised in* Mycena galopus *after six months growth on sterilised leaf litter (percentage of the total amount in the substrate on an oven dry weight basis)*

		Method of biomass estimation ($n = 5$)	
Mineral nutrients	Litter	From hyphal length	By hexosamine assay
P	*Fraxinus*	16.7	43.4
	Betula	10.0	26.5
N	*Fraxinus*	6.9	17.9
	Betula	6.7	17.8
K	*Fraxinus*	6.3	16.4
	Betula	5.3	14.1
Mg	*Fraxinus*	4.1	10.3
	Betula	5.2	14.3
Ca	*Fraxinus*	2.4	6.4
	Betula	1.5	4.1

P, N, K: from Frankland, Lindley & Swift (1978).
Mg, Ca: J. C. Frankland & M. J. Swift (unpublished data).

animal migration and physical adsorption (Ausmus, 1977; Swift, 1977*b*). Swift found that the accumulation of nutrients in branch-wood decomposing on the floor of Meathop Wood depended on the type of decomposer. Significant net accumulation of nitrogen and calcium occurred in branches occupied principally by a mixed population of basidiomycetes, whereas branches at a similar stage of decay but occupied by wood-boring animals showed a net loss of nitrogen, phosphorus, calcium and magnesium (Table 7). A significant decrease in potassium in both cases was attributed to leaching of this very mobile ion. Nutrient accumulation was not evident when the branches were colonised by *S. hirsutum* alone (Fig. 2).

The jump from individual substrates to total ecosystem assessments of nutrient capture is made with a decrease in accuracy, but with this proviso in mind the exercise can give an overall picture of the relative status of particular groups of organisms. The pools of nitrogen, phosphorus and potassium in the living mycelium of the decomposer basidiomycetes of Meathop Wood, when estimated from the biomass values in Table 1, were approximately 9.5, 0.7 and 1.5 kg ha^{-1} respectively. These are small quantities in relation to the nutrient capital of a

Table 7. *Accumulation (+) or loss (−) of mineral nutrients (μg cm^{-3}) in branch-wood invaded principally by basidiomycetes or animals.* (After Swift, 1977*b*)

	N	P	K	Ca	Mg
Living wood ($n = 20$)	1057	69.5	767	1935	146
After invasion by basidio-mycetes ($n = 26$)	+247*	−15.1	−626**	+1936**	−28
After invasion by animals ($n = 25$)	−347***	−41.7**	−746**	−158***	−88***

*, **, ***: significant at $P < 0.05$, < 0.01, < 0.001, respectively.

woodland but not necessarily so in terms of the nutrient cycles or throughflow. Harrison (1978) drew up a phosphorus budget for Meathop Wood and his figures for phosphorus contents of the soil, vegetation, fauna and micro-organisms are compared in Table 8 with estimates, derived from data in Table 1, of the reservoirs of phosphorus in the basidiomycetes and other fungi. The amount of phosphorus available in the soil, about 3 kg ha^{-1}, was low, but phosphorus *uptake* by the vegetation was at least 11 kg ha^{-1} yr^{-1}, so the plants must have depended heavily on the cycling of phosphorus in organic debris by the decomposers. The living vegetation contained about 45 kg P ha^{-1} compared with 4 kg P ha^{-1} in the living fauna and micro-organisms. About one-sixth or more of the latter was in the basidiomycetes.

Release of mineral elements by decomposer basidiomycetes

Release of nutrients by micro-organisms is as crucial as nutrient capital for the perpetuation of an ecosystem. The timing of this release is likely to be one of the most critical factors. A sudden flush of nutrients could result in heavy losses by leaching. In this respect basidiomycetes may well be important, at least in temperate woodlands, through the prolonged activity and resistance to decay (see Burges, 1960) of their mycelium. They are *habitués* rather than opportunists. Thus the group includes several psychrophilic or psychrotolerant species active both in summer and winter, even under snow (Hintikka, 1964), e.g. *Mycena galopus* which can reduce the dry weight of leaf litter *in vitro* by 8% ($P < 0.001$) in 6 months even at 4 °C (J. C. Frankland, unpublished). The generation time for many micro-organisms is only a few days,

Table 8. *Portion of a phosphorus budget for a temperate deciduous woodland, comparing higher-plant requirements for phosphorus and immobilisation of phosphorus in decomposer organisms (Meathop Wood, Cumbria, UK)*

P in the living biomass of trees, shrubs and ground flora[a]		45.0	kg ha^{-1}
P uptake by vegetation[a]		11.4	kg ha^{-1} yr^{-1}
P available in the soil[a]		3.0	kg ha^{-1}
P in the living biomass of the decomposers of plant debris and soil	fauna[a]	3.3	kg ha^{-1}
	basidiomycete mycelium[b]	0.7	kg ha^{-1}
	non-basidiomycete mycelium[b]	0.3	kg ha^{-1}
	bacteria and actinomycetes[a]	0.1	kg ha^{-1}

[a] From Harrison (1978).
[b] Derived from biomass (Table 1) and percentage P in mycelium (basidiomycetes: 0.24; microfungi: 0.22).

whereas turnover and decay of *M. galopus* mycelium in Meathop Wood, for example, extended for 6 months or more (Frankland, 1975*a*). Nutrient release from the fruit bodies of basidiomycetes is likely to be more spasmodic, when they include extremely ephemeral and deliquescing forms. However, as Satchell (1974) pointed out when discussing soil invertebrates, it would be very difficult to demonstrate, in any particular situation, that retention of nutrients in biomass is more advantageous to an ecosystem in reducing leaching than it is disadvantageous in withholding plant nutrients.

Experimental evidence for nutrient release from organic matter by any type of micro-organism in the field is sparse. Like nutrient uptake it must be inferred, but fine feeder roots are often particularly well located in the organic horizons of a soil to receive any such release by decomposer basidiomycetes. Hintikka & Näykki (1967) reported a case where these roots were even more concentrated in humus containing basidiomycete mycelium than in the surrounding humus. Net release from the organic substrate will depend on certain critical levels in the concentrations of elements (Dowding, 1974; Heal, 1979). Thus, *M. galopus* utilised rather than released plant nutrients when it was grown on leaf litter for six months deprived of all external sources of mineral elements (J. C. Frankland, unpublished). Only water-soluble nitrate- and nitrite-nitrogen in *Quercus* litter increased under these conditions (Table 9). Indirect evidence of nutrient release by basidiomy-

Table 9. *Percentage gain (+) or loss (−) of water soluble mineral components of* Quercus *leaf litter after six-months incubation with* Mycena galopus *at 11 °C (n = 5)*

$(NO_3^- + NO_2^-) - N$	+ 33***
P	− 99***
$NH_4^+ - N$	− 76***
K	− 44***
Organic N	− 20***
Ca	NS
Mg	NS

*** Significantly different from controls, $P < 0.001$.
NS: not significantly different from controls.
Source: J. C. Frankland (unpublished data).

cetes in northern coniferous forests was obtained by Hintikka (1970), who found more available nitrogen and phosphorus in the so-called white-rot humus under various lignin-decomposing species of *Collybia, Marasmius* and *Mycena* than in the surrounding humus. Such remineralisation of nitrogen and phosphorus could be particularly important in forest soils, where these elements are usually less available than magnesium and potassium.

Nutrient release from the nutrient-rich sporulating tissues and substantial spore masses of basidiomycetes is again mainly conjecture from isolated observations. For example, Laursen & Miller (1977) suggested that a significantly large flush of nitrogen entered a tundra system from basidiomycete fruit bodies at snow melt. Similarly, Merrill & Cowling (1966) estimated that a single fruit body of *Fomes fomentarius* dispersed 34 g of nitrogen in its spores in 20 days, but the nutrient release from basidiospores to the soil depends on their wastage and decay rates, which are usually unknown.

Basidiomycetes may also mobilise nutrients from soil minerals by acid action. 'Weathering' of stones with disappearance of iron oxide has been observed, for example, under the mycelia of several basidiomycetes in a forest podsol by Hintikka & Näykki (1967).

Depending on its vitality, a basidiomycete can lose nutrients from its hyphae by various processes, including droplet excretion (Jennings: Chapter 5), leaching, lysis, autolysis and grazing. Healthy, living, mycelium is very resistant to leaching. When Stark (1972), for instance, exposed living rhizomorphs of basidiomycetes to leaching equivalent to a year's rainfall, 99.9% or more of the original content of nutrient elements was retained in the tissue. Whatever process of nutrient release occurs, recycling of the nutrients through animals and other micro-organisms in the habitat can follow, side-stepping uptake by plant roots. Even auto-recycling can occur. Levi, Merrill & Cowling (1968) found, for example, that some wood-destroying basidiomycetes could utilise nitrogen in their own mycelium, and they suggested that nitrogen was conserved in wood by its re-use from autolysed mycelium. Further indications of this were obtained from experiments in which white-rot fungi were found to be unique as a group in their ability to produce diffusible cellulases under conditions of nitrogen starvation (Levi & Cowling, 1969). Similarly, Merrill & Cowling (1966) suggested that several species of *Fomes* could retrieve nitrogen from the older hymenial layers of their perennial fruit bodies for re-use, since the nitrogen content of the layers decreased progressively with age. Nitrogen availability is now known to have far-reaching implications from the discovery that lignin degradation by basidiomycetes is regulated indirectly by nitrogen starvation (Kirk & Fenn: Chapter 4).

In tropical rain forests there is little available mineral in the soil, where it would be rapidly leached. This has led to the theory of direct mineral cycling (Went & Stark, 1968), which suggests that the abundant mycorrhizal fungi cycle minerals directly from dead organic matter to the living root with minimum release to the soil. In the Amazon forest area investigated by these authors, the mycorrhizas were chiefly of the non-basidiomycete endotrophic type, but Singer & da Silva Araujo (1979) found that some vegetation in Central Amazonia was rich in ectomycorrhizas. They discuss the occurrence of two basic types of tropical forest – ectotrophic and anectotrophic – one characterised by a prevalence of mycorrhizal basidiomycetes and accumulation of raw humus, and the other by saprophytic litter fungi including basidiomycetes, with the suggestion of two distinct patterns of mineral cycling. These observations draw attention to the disparity between the tropical and temperate scene, and to a need maybe for the two camps of mycologists, decomposer and mycorrhizal, to combine forces in research into nutrient cycling by basidiomycetes.

Supply of animal nutrients

The links between decomposer basidiomycetes and higher plants in nutrient cycling have been emphasised in this chapter, but basidiomycetes also provide a rich and digestible source of minerals for many animals besides the gourmet. Buller (1922) gave a particularly vivid account of Canadian squirrels hoarding stacks of fruit bodies in trees and attics, where the dry winter air delayed rotting. However, grazing of basidiomycetes by the microfauna rather than by vertebrates is probably of more relevance to the functioning of most woodland ecosystems. Grazing studies are now beginning at the microcosm level and show that complex competitive interactions and food preferences are involved (Parkinson, Visser & Whittaker, 1979; Newell, 1980). Basidiomycetes, however, may not contribute to animal nutrition via direct grazing alone. The recently formulated hypothesis on calcium cycling (Cromack *et al.*, 1977) implicates basidiomycetes, as accumulators of calcium and excretors of oxalic acid, in what could be a particularly important rôle. From field observations these authors concluded that the sparingly soluble calcium oxalate formed by fungi could be decomposed by the bacteria and actinomycetes of animal guts. The released calcium would then be utilisable by various invertebrates, such as microarthropods and earthworms, with high calcium requirements. The evidence was obtained from forest soils at pH 5–5.5; as J. P. Curry suggested in the discussion of their paper, this calcium contribution could be of even greater significance in very acid litter where calcium is severely limited.

In conclusion, therefore, the decomposer basidiomycetes are prominent in the organic horizons of many soils – the hub of biological activity. Here, they withdraw, store, accumulate and release mineral nutrients. By the very size of the hyphal production, these activities must have an impact on the ecosystem. Some of the most important facts relating to this rôle in mineral cycling have been outlined, but much research is needed to verify, quantify, and explain them.

References

Ausmus, B. S. (1977). Regulation of wood decomposition rates by arthropod and annelid populations. In *Soil Organisms as Components of Ecosystems*, ed. U. Lohm & T. Persson, pp. 180–92. Swedish Natural Science Research Council. Ecological Bulletins (Stockholm) 25.

Ausmus, B. S. & Witkamp, M. (1974). *Litter and Soil Microbial Dynamics in a Deciduous Forest Stand*. EDFB-IBP-73-10. Tennessee: Oak Ridge National Laboratory.

Bååth, E. & Söderström, B. (1977). Mycelial lengths and fungal biomasses in some Swedish coniferous forest soils, with special reference to a pine forest in central Sweden. *Technical Report*, No. 13, 1–45. Uppsala: Swedish Coniferous Forest Project.

Bååth, E. & Söderström, B. (1979). Fungal biomass and fungal immobilization of plant nutrients in Swedish coniferous forest soils. *Revue d'Ecologie et de Biologie du Sol*, **16**, 477–89.

Buller, A. H. R. (1909). Spore-deposits – the number of spores. In *Researches on Fungi*, vol. 1, pp. 79–88. London: Longmans, Green.

Buller, A. H. R. (1922). The red squirrel of North America as a mycophagist. In *Researches on Fungi*, vol. 2, pp. 195–211. London: Longmans, Green.

Burges, N. A. (1960). Dynamic equilibria in the soil. In *The Ecology of Soil Fungi*, ed. D. Parkinson & J. S. Waid, pp. 185–91. Liverpool University Press.

Cochrane, V. W. (1958). *Physiology of Fungi*. London: Chapman & Hall.

Cromack, K., Jr., Sollins, P., Todd, R. L., Fogel, R., Todd, A. W., Fender, W. M., Crossley, M. E. & Crossley, D. A., Jr. (1977). The role of oxalic acid and bicarbonate in calcium cycling by fungi and bacteria: some possible implications for soil animals. In *Soil Organisms as Components of Ecosystems*, ed. U. Lohm & T. Persson, pp. 246–52. Swedish Natural Science Research Council. Ecological Bulletins (Stockholm) 25.

Cromack, K., Jr., Todd, R. L. & Monk, C. D. (1975). Patterns of basidiomycete nutrient accumulation in conifer and deciduous forest litter. *Soil Biology and Biochemistry*, **7**, 265–8.

Dickinson, C. & Lucas, J. (1979). *The Encyclopedia of Mushrooms*. London: Orbis.

Dowding, P. (1974). Nutrient losses from litter on IBP tundra sites. In *Soil Organisms and Decomposition in Tundra*, ed. A. J. Holding, O. W. Heal, S. F. Maclean, Jr. & P. W. Flanagan, pp. 363–73. Stockholm: International Biological Programme.

Dowding, P. (1976). Allocation of resources, nutrient uptake and utilisation by decomposer organisms. In *The Role of Terrestrial and Aquatic Organisms in Decomposition Processes*, ed. J. M. Anderson & A. Macfadyen, pp. 169–83. Oxford: Blackwell Scientific Publications.

Flanagan, P. W. & Scarborough, A. M. (1974). Physiological groups of decomposer fungi on tundra plant remains. In *Soil Organisms and Decomposition in Tundra*, ed. A. J. Holding, O. W. Heal, S. F. Maclean, Jr. & P. W. Flanagan, pp. 159–81. Stockholm: International Biological Programme.

Frankland, J. C. (1975*a*). Fungal decomposition of leaf litter in a deciduous woodland. In *Biodegradation et Humification*, ed. G. Kilbertus, O. Reisinger, A. Mourey & J. A. Cancela da Fonseca, pp. 33–40. Rapport 1er Colloque International, 1974, Nancy. Sarreguemines: Pierron.

Frankland, J. C. (1975*b*). Estimation of live fungal biomass. *Soil Biology and Biochemistry*, **7**, 339–40.

Frankland, J. C., Lindley. D. K. & Swift, M. J. (1978). A comparison of two methods for the estimation of mycelial biomass in leaf litter. *Soil Biology and Biochemistry*, **10**, 323–33.

Gray, T. R. G., Hissett, R. & Duxbury, T. (1973). Bacterial populations of litter and soil in a deciduous woodland. II. Numbers, biomass and growth rates. *Revue d'Ecologie et Biologie du Sol*, **11**, 15–26.

Harrison, A. F. (1978). Phosphorus cycles of forest and upland grassland ecosystems and some effects of land management practices. In *Phosphorus in the Environment: its Chemistry and Biochemistry*, ed. R. Porter & D. W. Fitzsimmons, pp. 175–99. Amsterdam: Elsevier.

Heal, O. W. (1979). Decomposition and nutrient release in even-aged plantations. In *The Ecology of Even-aged Forest Plantations*, ed. E. D. Ford, D. C. Malcolm & J. Atterson, pp. 257–91. Cambridge: Institute of Terrestrial Ecology.

Hering, T. F. (1966). The terricolous higher fungi of four Lake District woodlands. *Transactions of the British Mycological Society*, **49**, 369–83.

Hinneri, S. (1975). Mineral elements of macrofungi in oak-rich forests on Lenholm Island, inner archipelago of SW Finland. *Annales Botanici Fennici*, **12,** 135–40.
Hintikka, V. (1964). Psychrophilic basidiomycetes decomposing forest litter under winter conditions. *Communicationes Instituti Forestalis Fenniae*, **59:2,** 1–20.
Hintikka, V. (1970). Studies on white-rot humus formed by higher fungi in forest soils. *Communicationes Instituti Forestalis Fenniae*, **69:2,** 1–68.
Hintikka, V. & Naykki, O. (1967). Notes on the effects of the fungus *Hydnellum ferrugineum* (Fr.) Karst. on forest soil and vegetation. *Communicationes Instituti Forestalis Fenniae*, **62:2,** 1–22.
Hora, F. B. (1959). Quantitative experiments on toadstool production in woods. *Transactions of the British Mycological Society*, **42,** 1–14.
Hora, F. B. (1972). Productivity of toadstools in coniferous plantations – natural and experimental. *Mycopathologia et Mycologia Applicata*, **48,** 35–42.
Jones, P. C. T. & Mollison, J. E. (1948). A technique for the quantitative estimation of soil micro-organisms. *Journal of General Microbiology*, **2,** 54–69.
Laursen, G. A. & Miller, O. K., Jr. (1977). The distribution of fungal hyphae in arctic soil in the tundra biome site, Barrow, Alaska. In *The Belowground Ecosystem: a Synthesis of Plant-Associated Processes*, ed. J. K. Marshall, pp. 273–8. Range Science Department, Science Series, No. 26. Fort Collins: Colorado State University.
Levi, M. P. & Cowling, E. B. (1968). The effect of C to N ratio of substrate on the growth, composition, cellulase production, and wood-destroying capacity of *Polystictus versicolor*. In *Biodeterioration of Materials*, ed. A. H. Walters & J. J. Elphick, pp. 575–83. Amsterdam: Elsevier.
Levi, M. P. & Cowling, E. B. (1969). Rôle of nitrogen in wood deterioration. VII. Physiological adaptation of wood-destroying and other fungi to substrates deficient in nitrogen. *Phytopathology*, **59,** 460–8.
Levi, M. P., Merrill, W. & Cowling, E. B. (1968). Rôle of nitrogen in wood deterioration. VI. Mycelial fractions and model nitrogen compounds as substrates for growth of *Polyporus versicolor* and other wood-destroying and wood-inhabiting fungi. *Phytopathology*, **58,** 626–34.
Merrill, W. & Cowling, E. B. (1966). Rôle of nitrogen in wood deterioration: amount and distribution of nitrogen in fungi. *Phytopathology*, **56,** 1083–90.
Miller, O. K., Jr., & Laursen, G. A. (1974). Belowground fungal biomass on US Tundra Biome sites at Barrow, Alaska. In *Soil Organisms and Decomposition in Tundra*, ed. A. J. Holding, O. W. Heal, S. F. Maclean, Jr. & P. W. Flanagan, pp. 151–8. Stockholm: International Biological Programme.
Mutsch, F., Horak, O. & Kinzel, H. (1979). Spurenelemente in hohoren Pilzen. *Zeitschrift fur Pflanzenphysiologie*, **94,** 1–10.
Newell, K. (1980). The effect of grazing by litter arthropods on the fungal colonization of leaf litter. PhD thesis, University of Lancaster.
Park, D. (1976). Carbon and nitrogen levels as factors influencing fungal decomposers. In *The Role of Terrestrial and Aquatic Organisms in Decomposition Processes*, ed. J. M. Anderson & A. Macfadyen, pp. 41–59. Oxford: Blackwell.
Parkinson, D., Visser, S. & Whittaker, J. (1979). The effects of collembolan grazing on fungal colonization of leaf litter. *Soil Biology and Biochemistry*, **11,** 529–35.
Reichle, D. E. (1977). The rôle of soil invertebrates in nutrient cycling. In *Soil Organisms as Components of Ecosystems*, ed. U. Lohm & T. Persson, pp. 145–56. Ecological Bulletins (Stockholm) 25. Swedish Natural Science Research Council.
Richardson, M. J. (1970). Studies on *Russula emetica* and other agarics in a Scots Pine plantation. *Transactions of the British Mycological Society*, **55,** 217–29.

Satchell, J. E. (1974). Litter-interface of animate/inanimate matter. In *Biology of Plant Litter Decomposition*, vol. 1, pp. xiii–xliv. London, New York and San Francisco: Academic Press.

Singer, R. & da Silva Araujo, I. de J. (1979). Litter decomposition and Ectomycorrhiza in Amazonian forests. *Acta Amazonica*, **9**, 25–41.

Stark, N. (1972). Nutrient cycling pathways and litter fungi. *Bioscience*, **22**, 355–60.

Swift, M. J. (1977*a*). The ecology of wood decomposition. *Science Progress, Oxford*, **64**, 175–99.

Swift, M. J. (1977*b*). The rôles of fungi and animals in the immobilisation and release of nutrient elements from decomposing branch-wood. In *Soil Organisms as Components of Ecosystems*, ed. U. Lohm & T. Persson, pp. 193–202. Ecological Bulletins (Stockholm) 25. Swedish Natural Science Research Council.

Tyler, G. (1980). Metals in sporophores of basidiomycetes. *Transactions of the British Mycological Society*, **74**, 41–9.

Vogt, K. A. & Edmonds, R. L. (1980). Patterns of nutrient concentration in basidiocarps in Western Washington. *Canadian Journal of Botany*, **58**, 694–8.

Went, F. W. & Stark, N. (1968). Mycorrhiza. *Bioscience*, **18**, 1035–9.

Witkamp, M. (1960). Seasonal fluctuations of the fungusflora in mull and mor of an oak forest. *Instituut voor Toegepast Biologisch Onderzoek in de Natuur, Arnhem*, Publication No. 46, 1–51.

Witkamp, M. (1969). Environmental effects on microbial turnover of some mineral elements. II. Biotic factors. *Soil Biology and Biochemistry*, **1**, 177–84.

15 The rôle of basidiomycetes in composts: a model system for decomposition studies

J. N. HEDGER and T. BASUKI
Department of Botany and Microbiology, University College of Wales, Aberystwyth, UK

Introduction

A number of other chapters in this book draw attention to the difficulty of accurately estimating the importance of basidiomycetes as primary decomposers, although all conclude that they are of major significance (Hering: Chapter 12; Frankland: Chapter 14; Swift: Chapter 16). This difficulty is due, amongst other things, to the heterogeneity of the substrates available to the decomposers, and to the diversity of species and substrate specificity within the fungal decomposer community, making it difficult to study the basidiomycetes *per se*. In addition the decomposition process may extend over a long period of time. Calculation of turn-over rates (k values) for plant litters (Olson, 1963) shows that, in some cases, complete decomposition may take many years, during which time a number of decomposer communities may appear and disappear. For example, decomposition of the massive resource presented by timber may take decades to complete (Mangenot, 1952). Such slow decomposition causes problems in studying the relationship between the basidiomycetes and other saprophytes that have previously colonised the substrate, or that co-exist with the basidiomycetes.

At high temperatures the whole decomposition process is accelerated and, in common with other extreme environments, the species diversity is lower than would occur on a substrate at 'normal' temperatures. It is easier, therefore, to study the populations of individual decomposer basidiomycetes and their interactions with other micro-organisms in these situations. Composts, the high-temperature habitats that are the subject of much of this chapter, have the additional advantage of a degree of homogeneity, especially when made from one type of plant material. Swift (1976 and Chapter 16), points out that the food resources available to most decomposers are heterogeneous, consisting of 'units' of variable size and origin, such as the twigs, leaves and branches in litter. Generalisations about fungal successions and substrate changes may therefore be misleading, since it is only at the 'unit' level that such parameters can be measured accurately, the decomposer community at this level being defined as the 'unit community'. Composts consist of such large areas of uniform environment and substrate composition, that most of the 'unit communities' are likely to be similar, unlike those of litter and soil (Swift, 1976), making the study of their composition in compost much easier.

The purpose of this chapter is to illustrate how composts may be used as model systems demonstrating the rôle of decomposer basidiomycetes. The position of these fungi in the various microbial communities of compost provides a useful comparison with the findings at 'normal' temperatures. First, the basidiomycetes associated with high-temperature habitats, especially composts, will be briefly described, and then the ecology of basidiomycetes in composts will be considered in more detail.

Basidiomycete decomposers in composts and other high-temperature habitats

What constitutes a high-temperature versus a 'normal' – or meso-temperature – habitat? The temperature growth range of fungi varies quite widely (see Crisan, 1973; Tansey & Brock, 1978), but decomposition normally occurs at temperatures ranging from −7.5 °C in the tundra biome to 35 °C or more in the savannah biome (Swift, Heal & Anderson, 1979). A warm habitat, when defined as one that is at 30–35 °C or above, would be accepted by most mycologists as being outside the normal range of habitats in which decomposers operate. Even in a tropical rain-forest, temperatures are usually well below this figure, which is reached only in hot, dry, conditions.

Tansey & Brock (1978) reviewed a variety of high-temperature habitats and their fungal populations. They listed sixty-seven taxa which were able to grow at 50 °C or above (i.e. *thermophilic* or *thermotolerant* fungi, according to the definition of temperature–growth relationships proposed by Cooney & Emerson, 1964). Only two of the taxa were basidiomycetes: *Coprinus* sp. and *Phanerochaete chrysosporium.* These are listed in Table 1, and we have added a number of other basidiomycetes, which, although not truly thermophilic, are capable of growth

Table 1. *Some decomposer basidiomycetes characteristic of high-temperature habitats*

Basidiomycete	Cardinal temperatures for growth (°C)			Substrate	Source
	Optimum	Maximum	Minimum		
Agaricus bisporus	24	32	1	Mushroom compost	Treschow, 1944
Coprinus cinereus	35–37	40–45	10–15	Wheat- and rice-straw compost	Chang, 1967; Basuki, 1981
Coprinus congregatus	32	41	?20	Mushroom compost	Fergus, 1978
Coprinus delicatulus	?	?40	?25	*Agropyron* stems	Apinis, 1965
Coprinus ephemeroides	37	40–45	10–15	Rice-straw compost	Basuki, 1981
Coprinus patouillardii	37	40–45	10–15	Rice-straw compost	Basuki, 1981
Coprinus species	40	*c.* 55	< 20	Cellulose	Cooney & Emerson, 1964
Phanerochaete chrysosporium	40	50	10	Wood-chip piles	Bergman & Nilsson, 1971
Volvariella volvacea	37	43	16	Rice-straw compost	Basuki, 1981

at 40 °C or above and may be termed *thermophilous* (Evans, 1971). *Agaricus bisporus* is also included, since, although not thermophilic or thermophilous, it is characteristic of a low-temperature phase of a high-temperature habitat. The list is somewhat arbitrary, but it brings together examples of the basidiomycetes characteristic of self-heating materials such as compost.

Even the two basidiomycetes listed by Tansey & Brock (1978) are not thermophilic *sensu* Cooney & Emerson (1964) but correspond to their category of thermotolerant fungi. Of these, only *P. chrysosporium* (Corticiaceae) has been well characterised. This fungus has been isolated repeatedly from self-heating piles of industrial wood chips. It appears to grow in aerobic regions at temperatures of 30–50 °C and causes a rapid white-rot degradation of the chips (Bergman & Nilsson, 1971; Tansey, 1971; Ofosu-Asiedu & Smith, 1973). Most reports of this fungus have been of possible anamorphs, variously described as *Sporotrichum* sp., *S. pulverulentum, S. pruinosum* or *Chrysosporium lignorum*. Whether all these belong to *P. chrysosporium* is not clear, and the reader is referred to Burdsall & Eslyn (1974). Currently, a great deal of interest is being shown in the biotechnological applications of this basidiomycete in upgrading low quality foods (von Hofsten, 1976) and as a selective degrader of lignin in paper-pulp manufacture (Ander & Eriksson, 1975, cited by von Hofsten, 1976). The rapidity of its growth has proved useful in the study of the mechanism of white-rot degradation of lignins (Ander & Eriksson, 1976) and these aspects are considered in detail by Kirk & Fenn (Chapter 4).

The other thermotolerant basidiomycete listed by Tansey & Brock (1978), '*Coprinus* species', is a poorly characterised fungus, mentioned by Cooney & Emerson (1964) as having been isolated from cellulose at 40 °C, and above this temperature by Reese (1946, cited by Cooney & Emerson, 1964). There seems to be considerable vagueness about the identity of the many *Coprinus* species that have been found on self-heating materials, probably reflecting the difficult taxonomy of this genus. The majority of reports have been of an agaric, producing large (over 10 cm high) fruit bodies on straw, dung and compost. Some of these fungi have been reported to grow or survive at temperatures as high as 55–57 °C (Perrier, 1929), but there is considerable variation in reported cardinal temperatures and the conditions under which they were measured. These thermotolerant fungi have been variously described as *C. fimetarius* (Rege, 1927; Kurtzman, 1978), *C. lagopus* (Cooney & Emerson, 1964), *C. cinereus* (Chang & Hudson, 1967), and

C. macrocephalus (Chang & Hudson, 1967) cited erroneously as *C. megacephalus*; Perrier (1929) has also recorded a species similar to *C. stercorarius*. All but the last taxon are referable to Section Lanatuli (Kühner & Romagnesi, 1953) or Stirps Lagopus (Orton & Watling, 1979) and are similar in appearance. Even Perrier's isolate may well have been referable here also. Using the taxonomic approach of Orton (1957), Pinto-Lopes & Almeida (1971) and Orton & Watling (1979), most of the isolates would be placed in *C. cinereus* although there has been confusion with *C. radiatus*; unfortunately *C. fimetarius* has been used in the past for both species. *C. cinereus* is a species characterised by large (up to 10–12 cm high) fruit bodies with a white, densely scaly cap covered in relatively broad (up to 40 μm diameter) veil cells which are constricted at the septa. The stipe often has a long rooting base. The elliptical basidiospores measure 10–11 × 6–7 μm. In culture at 30–36 °C, a white, rapidly extending, colony is formed which soon produces numerous brown microsclerotia. *C. macrocephalus* (Orton, 1957) is very similar, but lacks the rooting base and has larger basidiospores (11–16 × 7–9 μm).

The temperature relations of these *Coprinus* species, which will be referred to from now on under the umbrella of *C. cinereus*, vary considerably but most fall into the range given in Table 1. Clearly they represent a species complex, and R. F. O. Kemp (personal communication) has found a number of distinct taxa, based on compatibility tests. R. Watling (personal communication) believes that *C. cinereus* represents a species complex in the process of rapid evolution, in response to the increasing availability of suitable high-temperature habitats and wastes. *C. cinereus* has received considerable attention as an agent for the biodegradation and upgrading of waste materials such as straw and paper. These aspects are reviewed by Burrows, Seal & Eggins (1979) and Penn (1977).

Coprinus delicatulus was described from *Agropyron pungens* stems on the Lincolnshire coast by Apinis (1965) who characterised it by its very small grey fruit bodies (2–6 μm high), which were produced at 38 °C. Recently, Kuthubutheen & Pugh (1977) and Kuthubutheen (1980) reported this fungus as present on leaf surfaces in Britain and Malaysia. They consider it to be distinct from *C. cinereus*, but both Orton & Watling (1979) and R. F. O. Kemp (personal communication) found it to be generally compatible with strains of *C. cinereus*, so the status of this species is uncertain. A similar thermophilous *Coprinus* which has been isolated from rice (*Oryza sativa*) straw by T. Basuki (unpublished data) seems to be *C. bilanatus* Kemp, *nomen provisorium* (Kemp, 1974).

The other thermophilous *Coprinus* species listed in Table 1 are less regularly reported from self-heating materials than *C. cinereus*. Fergus (1978) found the mildly thermophilous *Coprinus congregatus* (Stirps Ephemerus) on *Agaricus bisporus* compost. *Coprinus ephemeroides* and *Coprinus patouillardii* (Stirps Niveus) were found by the authors (unpublished data) to be very common on rice-straw composts used in the cultivation of *Volvariella volvacea* in Indonesia (see also Hedger, 1975). They both produced small white fruit bodies (5–10 cm high) with a conical, later flattened and involuted, cap which was covered with a brown or white powdery veil made up of round cells. Isolates showed a similar temperature growth range to *C. cinereus* (Table 1). It may be that these species of coprophilous fungi are often mistaken for a form of *C. cinereus*, since a number of cultures isolated by other workers from composts and identified as this species have often been found to be referable to *C. patouillardii* (J. N. Hedger, unpublished data).

V. volvacea, the 'padi-straw' or 'rice-straw mushroom', is an agaric of considerable commercial importance and is widely cultivated throughout the South-east Asian tropics, as well as in Hong Kong, Taiwan and China. World production amounted to 42 000 tons in 1975 (Delcaire, 1978). The fruit bodies are completely enclosed by a volva in the early stages of development, the mushroom being harvested at the 'egg' stage, before stipe elongation has begun. The mycelium, which appears to be homothallic and lacks clamp connections, has a similar temperature growth range to *C. cinereus* (Table 1), and the fruit bodies, which require light for development, are formed optimally between 30 and 35 °C. The biology and cultivation of this mushroom have been reviewed by Chang (1972, 1978). It is clear that it is able to colonise a wide variety of substrates. In Indonesia it occurs as a saprophyte on a number of plant species (M. Rifai, personal communication). Chang (1979) and Saono & Sastrapradja (1979) emphasised the potential of this basidiomycete for the utilisation of agricultural wastes in tropical areas, and it has been successfully cultivated on many materials, including rice straw and cotton waste (Chang, 1974), sugar-cane waste (Hu, Song, Liu & Peng, 1974), oil-palm pericarps (Yong & Graham, 1973), water hyacinth (Cheng & Mok, 1971; de Vries, 1973) and banana leaves (Chua & Ho, 1973).

Agaricus bisporus (syn. *A. brunnescens*) with a related species, *A. bitorquis*, ranks first in the world order of mushroom cultivation, with a total production figure of over 670 000 tons for 1975 (Delcaire, 1978). Its biology and cultivation have been reviewed by Hayes (1978*a*,*b*). This agaric is probably one of the best understood of all the decomposer basidiomycetes and a number of investigations have shed light on the physiological ecology of this fungus; these will be reviewed later in the chapter. *A. bisporus* is not in any sense thermophilic, as it will not grow above 32 °C (Treschow, 1944). However, the cultivation of this mushroom involves a composting phase, followed by 'pasteurisation', cooling and addition of the mushroom spawn. Subsequently, growth of *A. bisporus* mycelium must be considered as an interaction between a mesophilic fungus and a pre-established thermophilic population of micro-organisms. The significance of this confrontation will be emphasised later in the chapter.

It appears, then, that only a limited number of basidiomycetes are able to act as decomposers in high-temperature habitats. Even these show only a moderate degree of thermophilism. In contrast, there are many species of ascomycete and phycomycete which are thermophilic, growing and sporulating at temperatures in excess of 50 °C. Why there should be this eco-physiological difference is a matter for speculation. A working hypothesis might be that the development of fruit bodies is aberrant at high temperatures. *Phanerochaete chrysosporium* does not form fruit bodies above 35 °C and reproduces only by the anamorph at higher temperatures (R. A. Samson, personal communication). Fruit bodies of *C. cinereus* are sterile above 35 °C (J. N. Hedger, unpublished) but R. A. Samson (personal communication) has recently isolated from corn-cob waste a thermophilic *Coprinus*, resembling *C. cinereus*, which produces fruit bodies at 40–45 °C.

It may be that more thermophilous and even thermophilic basidiomycetes await discovery, perhaps in tropical areas. In addition, some thermophilic Hyphomycetes and mycelia sterilia may, like *Sporotrichum pruinosum*, eventually, prove to have basidiomycete teleomorphs. *Burgoa*, close to *Papulaspora* and described by Tansey (1973), and *Papulaspora thermophila*, described by Fergus (1971), which have an optimum growth temperature of 45 °C, are likely candidates.

General principles of composting

Thermogenesis raises the temperature of a self-insulating mass of organic material, such as a compost, to a level at which surface heat loss balances heat production, and the temperature may reach a level which inhibits microbial activity. Pioneer studies were carried out on the process by Miehe (1907), but the investigations by Waksman and his co-workers in the 1930s were the first relevant to decomposer basidiomycetes, since they concentrated on mushroom compost and its mode of preparation (Waksman & McGrath, 1931; Waksman & Nissen, 1932; Waksman & Cordon, 1939). These fundamental studies confirmed the rôle of thermophilic fungi, bacteria and actinomycetes in the self-heating of the mushroom compost and identified the principal substrates utilised. Fergus (1964, 1978), Fordyce (1970), Eicker (1977, 1980), Hayes (1968) and Fermor, Smith & Spencer (1979) have since provided data on the changes in microbial populations in mushroom composts, so that we now know something of the microbial background colonised by *Agaricus bisporus* mycelium. Microbial populations in other types of compost which have been evaluated include those of wheat-straw compost (Chang & Hudson, 1967), municipal composts (Gray, Sherman & Biddlestone, 1971*a,b*; Anonymous, 1978), and the composts used by alligators and oven birds in which to incubate their eggs (Tansey, 1973; Tansey & Jack, 1975). In all composts, the mesophilic decomposer community is replaced by a thermophilic community and the decomposition process follows a different pathway.

The initiation of thermogenesis in compost is due to the combination of a number of circumstances. The physical structure of composts promotes rapid aerobic thermogenic decomposition, usually aided by convective movement of air through the mass of the compost from the base to the surface, as well as by the practice of regular turning and mixing of the material. To promote optimum microbial activity, water contents are usually adjusted to between 70–80% on a dry-weight basis, and the level of inorganic nutrients, especially nitrogen and phosphorus, may be raised. Gray & Biddlestone (1974) considered that C:N ratios of compost should usually be adjusted between 50:1 and 25:1 initially, reaching about 10:1 after the end of thermogenic decomposition, which may last days or weeks. In the 'normal' decomposer system, such C:N ratios are also reached in the lower layers of woodland leaf litter for example, but often after months or years of decomposer activity. Low nutrient status in the 'normal' decomposer system must limit the rate of decomposition when compared to that of the compost, especially in woody substrates where the initial C:N ratio is usually greater than 100:1.

Apart from differences in nutrient status, composts differ in some other respects from many natural decomposer systems. They operate at high pH values, usually well above neutrality, and often as high as pH

8.0–9.0. This alkalinity is often associated with the presence of ammonia during high-temperature phases (Gerrits, 1978). As decomposition proceeds, the pH usually falls, but even so it is comparatively rare to find values much below pH 6.0–7.0, even in old material.

Succession of fungi in composts

Unfortunately, many of the studies of composts, as well as of other self-heating materials such as hay, grain and wood-chip piles, have been rather generalised, usually recording only total populations of fungi and prokaryotes, so that it is not possible to estimate the populations of basidiomycetes. However, Chang & Hudson (1967) demonstrated a succession of fungi in small (1 m^3) composts of chopped wheat straw amended with ammonium nitrate (0.75 parts of nitrogen to 100 parts of straw). This system serves as a useful model for studying compost microbiology. The microbial succession in composts used for cultivating *Agaricus bisporus* and *Volvariella volvacea* may be interrupted by a pasteurisation phase, and will be discussed separately.

Figure 1 shows the changes in temperature, pH and total populations of mesophilic and thermophilic fungi in the centre of a wheat-straw compost of the same design as that of Chang & Hudson (J. N. Hedger, unpublished). The temperature rose to 70 °C within five days, probably due to the activity of thermophilic bacteria and actinomycetes which utilise simple sugars, hemicellulose and cellulose (Chang & Hudson, 1967; Chang, 1967). After this first peak, the temperature fell below 50 °C but was maintained at or above 30 °C (ambient: *c.* 15 °C) for up to 35 days after the beginning of composting, during which time thermophilic fungi dominated the straw.

The fungal succession in this wheat-straw compost consisted of a short colonisation phase, characterised by *Rhizomucor pusillus* and other members of the Mucorales, followed by an extended period of decomposition by thermophilic Hyphomycetes and ascomycetes. Most activity was shown by the Hyphomycetes *Humicola insolens* and *Thermomyces lanuginosus* and by the ascomycete *Chaetomium thermophilum*. The final stage of the succession was dominated by a basidiomycete, *Coprinus cinereus*, associated apparently with *Mortierella wolfii* (Fig. 1).

Mycelium of *C. cinereus* began to colonise the centre of the compost 20 days after the start of composting and became dominant by day 30 (Fig. 1). Its fruit bodies had already appeared on the surface of the compost by day 20. By examining the distribution of the mycelium in a number of beds, it was found that it invaded the base and corners of the compost from day 5 onwards, and colonised inwards to the centre following the line of the 35–40 °C isotherm as the material cooled. Its colonisation of the centre was often associated with an increase in thermogenesis, as shown by the temperature/time course in Fig. 1. Growth of its mycelium began to decline after day 40 and few fruit bodies occurred after this time. J. N. Hedger found that colonisation of areas of the compost by this basidiomycete was accompanied by a rapid decline in the activity of the thermophilic decomposers other than *M. wolfii*, even though the temperatures were still high enough for their growth. This is illustrated for the central region of a compost by Fig. 1, in which a decline in the activity of *H. insolens* mirrors the increase in activity of *C. cinereus*. Chang & Hudson (1967) found a second basidiomycete, *Clitopilus pinsitus*, at this stage of the succession.

How does the above succession differ in composts designed for 'mushroom' cultivation? Two such composts will be briefly compared.

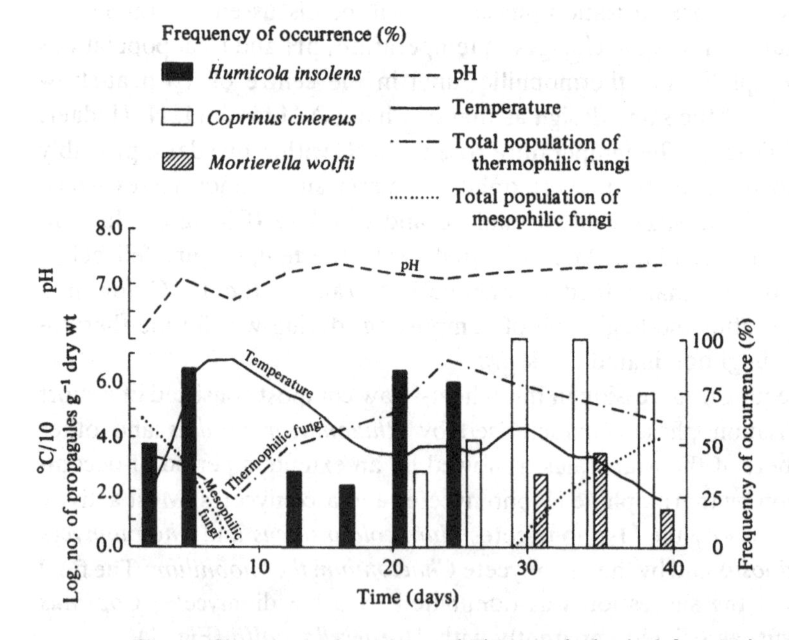

Fig. 1. Changes in fungal populations and physical factors in wheat-straw compost. All measurements were taken in the centre of a 1 m³ compost amended with ammonium nitrate (0.75%). Samples were withdrawn at regular intervals for assessment of the fungal populations by dilution plating at 22.5 °C and 45 °C, and by direct observation for fungal activity at the same temperatures. Temperature was recorded twice daily. For media and methods see Chang & Hudson (1967). J. N. Hedger (unpublished).

The authors (unpublished) studied rice (padi)-straw/cotton-waste compost used for the cultivation of *V. volvacea* in Indonesia, and measured changes in the populations of mesophilic and thermophilic micro-organisms. The cultivation procedure used in Indonesia today is an adaptation of that developed in Taiwan (Chang, 1978). Rice straw and cotton waste are composted for 9–10 days, turning at three-day intervals, and are amended with only rice bran and lime. The compost is then laid in 20 cm deep beds in mushroom houses, covered with a 2–3 cm layer of composted cotton waste, and steam 'pasteurised' at 80 °C. When the temperature has fallen below 40 °C, the surface of the beds is inoculated with spawn of *V. volvacea.* Fig. 2(*a*) summarises changes in pH and temperature, and Fig. 2(*b–d*) illustrates the microbial succession. Following inoculation, thermogenesis maintained temperatures at 30–35 °C for the first 2–3 weeks, and relative humidity was 80% or above.

The successional pattern in rice-straw mushroom compost was similar to that described for wheat-straw compost, although there were some differences in species composition. The later stages of the decomposition were dominated again by basidiomycetes, but the introduced basidiomycete may be replaced by species of *Coprinus*. The inoculum of these *Coprinus* species probably comes from incompletely pasteurised areas of the bedding, the housing and the soil floor. Composting and pasteurisation are not, however, essential for colonisation by *V. volvacea.* In the traditional Javan system, it is inoculated directly into heaps of rice straw or into composts prepared by placing rice-husk ash between alternate layers of soaked rice straw which is tied in bundles. Hedger (1975) found that *V. volvacea* was able to grow into these self-heating masses (centre temperature: 40–50 °C) in which *Papulaspora thermophila, Humicola grisea* var. *thermoidea* and *Scytalidium thermophilum* were dominant thermophilic fungi. Fruit bodies were produced, together with those of *C. patouillardii*, within 1–2 weeks. However, the productivity of this method is much lower (3–6% conversion, dry weight of straw to fresh weight of fruit bodies) than in the modern pasteurised composts of Indonesia. In the latter, fruit-body production begins after 7–10 days, continuing as weekly flushes, and conversions of 10–15% can be obtained. In Hong Kong and Taiwan even higher conversions, 28% or more, have been recorded on rice straw and cotton waste by Chang (1978). The uncontrolled temperature and low relative humidity of traditional systems play a rôle in reducing productivity, but basidiomycete–basidiomycete competition may also be important, as will be discussed later.

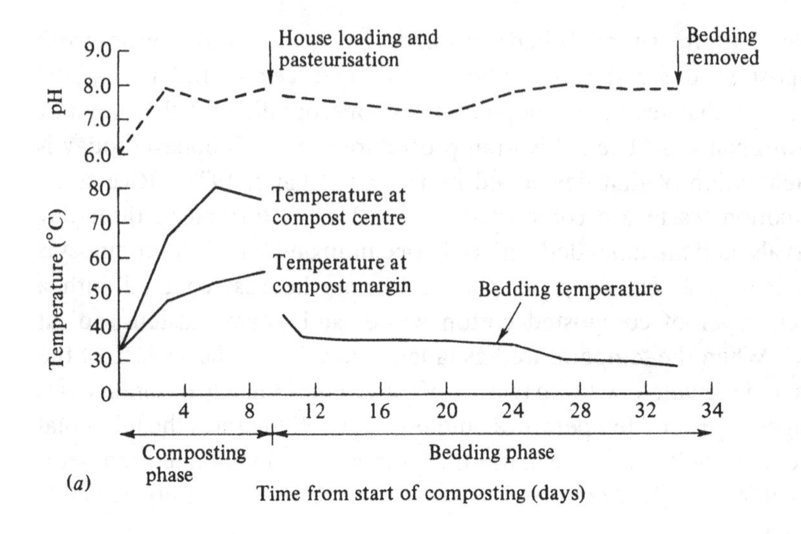

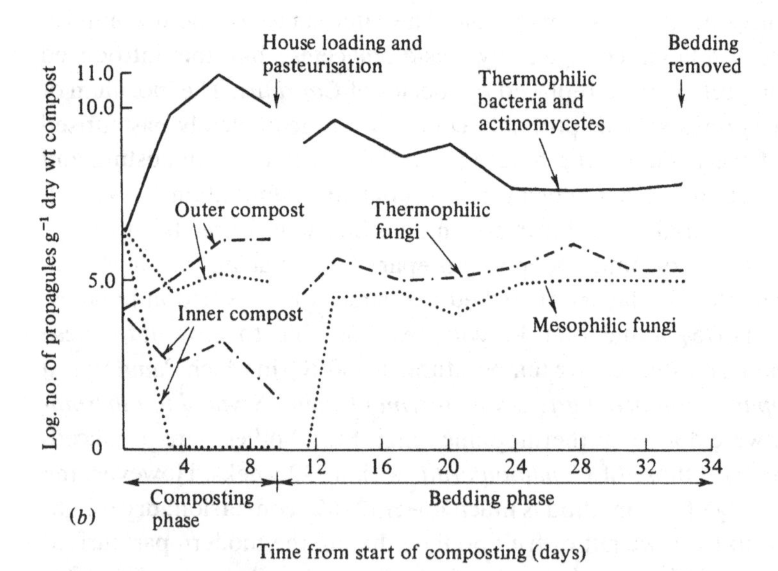

Fig. 2. Changes in microbial populations and physical factors in rice-straw mushroom composts. (*a*) Changes in pH and temperature in compost and bedding measured directly. (*b*) Changes in total populations of fungi and prokaryotes g^{-1} dry weight of material in withdrawn samples, measured by dilution plating on a yeast starch medium (Cooney & Emerson, 1964) plus bacteriostats at 30 °C and 45 °C for mesophilic and thermophilic fungi, and on nutrient agar plus cycloheximide for bacteria at 45 °C; populations at 10 cm (outer) and 40 cm

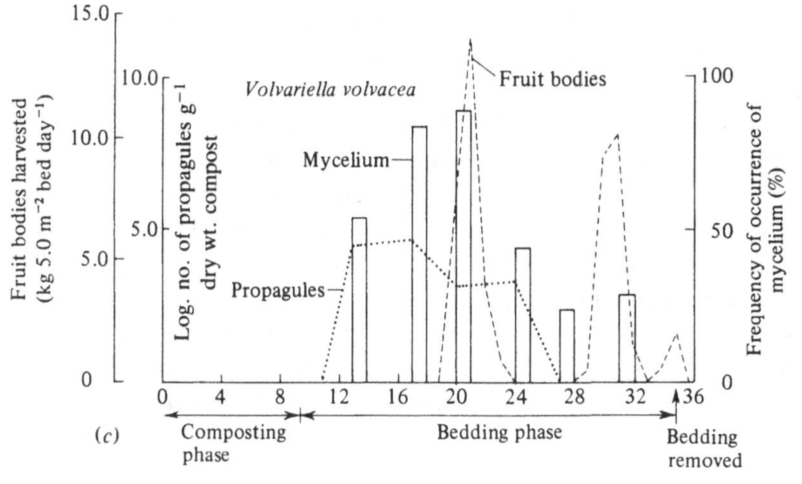

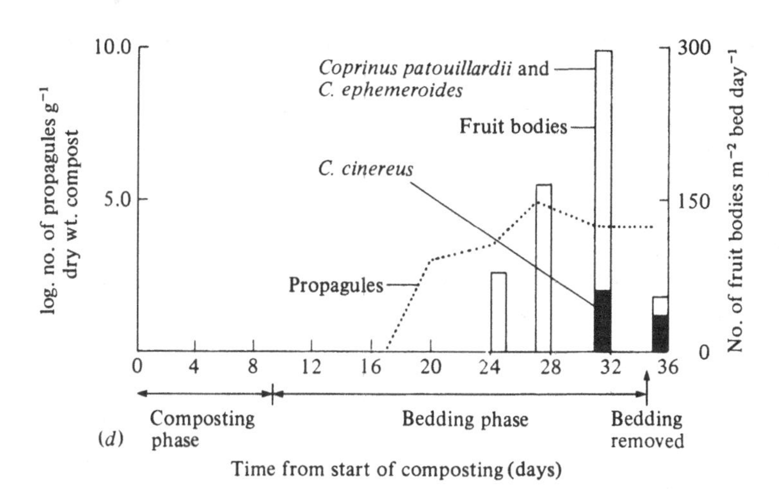

(inner) depths are shown. (*c, d*) Changes in the populations of *Volvariella volvacea* and *Coprinus* species; 'activity' of *V. volvacea* estimated by direct observation of mycelium (percentage frequency of occurrence on incubated 1.0 cm straw samples, $n = 20$, Chang & Hudson, 1967); the total population (log. no. of propagules g^{-1} dry wt compost) measured by dilution plating at 30 °C (medium as above), and fruit-body production as kg fresh weight 5.0 m^{-2} of bed day^{-1} for *V. volvacea* and as numbers of fruit bodies m^{-2} of bed day^{-1} for the *Coprinus* species. Data from Basuki (1981).

The populations of thermophilic fungi and prokaryotes were unaffected by the brief pasteurisation phase of the rice straw/cotton waste composts, although the mesophilic fungi were eliminated (Fig. 2*b*), but the coarse grey mycelium of *V. volvacea* had penetrated all regions of the bedding within 2–3 days of spawning. Fig. 2*c* shows mycelial activity, numbers of propagules and fruit-body production during the production run. Mycelial activity appeared to increase steadily up to the appearance of the first flush of fruit bodies, after which it declined steadily; the second flush (18–19 days after spawning) produced fewer fruit bodies than the first, and, after this, mycelial activity could not be detected.

Apart from *V. volvacea*, the only fungi to show activity in the compost bedding within 2–3 days of pasteurisation were *H. grisea* var. *thermoidea* and *S. thermophilum*, but populations of these fungi, together with those of the thermophilic prokaryotes, fell steadily during the main period of active growth of *V. volvacea*. However, mycelium bearing clamp connections (*V. volvacea* has simple septa) was detected in the compost 6–7 days after spawning, and *Coprinus ephemeroides* and *C. patouillardii* were isolated on dilution plates. Numerous fruit bodies of these two species appeared on the surface of the beds two weeks after pasteurisation and were joined by the larger fruit bodies of *C. cinereus* about one week later. The activity of these *Coprinus* species was strongly correlated with a decline in the activity and productivity of *V. volvacea* as can be seen by comparing Fig. 2*c* and *d*.

In contrast to *V. volvacea, Agaricus bisporus* does seem to have an absolute requirement for both composting and pasteurisation, if successful mycelial colonisation and fruit-body production is to be ensured. The detailed microbiology of *A. bisporus* cultivation is still not understood, but studies by Hayes (1968), Eicker (1977, 1980), Fermor, Smith & Spencer (1979) and Fermor & Wood (1979) have described the populations of micro-organisms which occur during the different stages of cultivation. The substrate which is employed usually consists of compost mixtures of a nitrogen source (usually animal manure) and straw, although a number of other compost materials such as corn-cob waste, sawdust (Rao & Block, 1962) and even manureless mixtures ('rapidly prepared compost' or RPC, Smith & Spencer, 1976) have been successfully employed (Hayes, 1978*b*). In conventional *A. bisporus* cultivation systems based on the processes developed by Lambert (1941) and Sinden & Hauser (1950), cultivation may be divided into three phases, although shortened production sequences have since been developed (see Fermor & Wood, 1979; Hayes, 1978*b*). In phase I, which may last up to two weeks, the material is composted in long heaps or windrows, which may be turned mechanically every 2–3 days. Temperatures may be maintained at 60–70 °C in the centre of these windrows for several weeks and the microflora is dominated by thermophilic bacteria and actinomycetes. The fungi occur in lower numbers, but a successional pattern similar to that of wheat-straw compost occurs. Hayes (1968), Fermor & Wood (1979) and Eicker (1977) found that there was an initial mucoraceous phase (*Rhizomucor pusillus*) with subsequent invasion after 2–3 days by *Humicola* species and *Chaetomium thermophilum. Coprinus* species (?*C. cinereus*) are often recorded during phase I, but detailed information as to their importance is not available.

Following phase I, there is usually a 'peak heating' or 'pasteurisation' phase (phase II). During this time, considerable growth of the actinomycetes and thermophilic fungi may occur. The material, after being placed in the mushroom house, is then allowed to cool to 25 °C, and it is inoculated with spawn of *A. bisporus*. Mycelium of *A. bisporus* grows through the compost, and after about 14 days a 'casing' layer, usually of peat, chalk or soil, is added to the surface. The temperature is maintained at 25 °C for about a week after casing, but then it is lowered by several degrees (sometimes to as low as 10 °C); this fluctuation is apparently necessary to induce maximum initiation of fruit-body primordia, but correct carbon dioxide levels (less than 0.15%) are also important (Hayes, 1978*a*). Mature fruit bodies appear about three weeks after casing, and flushes then occur at 7–10 day intervals; unlike *V. volvacea*, they show no photoresponse and will develop in the complete absence of light.

During the growth of *A. bisporus* in the compost, there is evidence that the previously established populations of thermophilic prokaryotes and fungi such as *Humicola* species, which survive the pasteurisation phase, gradually decline (Fermor *et al.*, 1979; Fordyce, 1970), but progressive colonisation by mesophilic fungi takes place. Eicker (1980) found that these fungi included species of *Aspergillus* and *Trichoderma; Coprinus* species do not seem to be of much importance in the mushroom beds unless the material is insufficiently composted or free ammonia is present.

In these three examples of composts, the mycelium of the decomposer basidiomycetes represents the later stages of a succession and it colonises an environment already occupied by large populations of other micro-organisms. It could be argued that the *Coprinus* species represent a 'natural' colonisation stage of composts and they appear to compete effectively with the established microflora, whereas *V. volvacea* and *A. bisporus* represent a stage in a succession deflected by pasteurisation and artificial inoculation. However, *V. volvacea* can successfully and naturally colonise unpasteurised rice straw and other materials in tropical environments, and even *A. bisporus*, or at least the closely related wild species, *A. campestris*, is capable presumably of colonising straw and manure at an advanced stage of decomposition in soils. A common feature of all three basidiomycetes is that they establish their mycelial system in spite of a large background population of other micro-organisms, and their growth is accompanied by a decline in this pre-established thermophilic population.

Substrate availability in composts

Decomposer basidiomycetes growing into composts are colonising material which has already undergone some decomposition, particularly of the carbohydrate fractions of the material. The actual composition of composts is variable (Anonymous, 1978; Gray *et al.*, 1971*a*), but data on the carbon and nitrogen fractions are available for some systems. Wheat-straw compost was studied by Chang (1967) using fractional hydrolysis and was found to consist originally of 45.3% cellulose, 35.6% hemicellulose, 9.5% 'lignin' and 9.3% starch and sugars. By the end of composting (60 days), about 70% of the cellulose and 50% of the hemicellulose had been degraded, with a total loss in dry weight of 50.1%. After 165 days, a further loss of only 3.7% had occurred. The 'lignin' fraction remained unattacked. Hemicellulose was degraded at a uniform rate, but maximum cellulose decomposition took place in two phases: during the first temperature peak (day 0–5 of composting) and between days 16–34. Reference to Fig. 1 will show that this latter period corresponds to active growth of the basidiomycete *C. cinereus* in the compost.

T. Basuki (unpublished data) also employed sequential digestion of the material (Chang, 1967) to study biochemical changes in the cotton-waste and rice-straw compost during cultivation of *V. volvacea* in Indonesia. The results are shown in Fig. 3. The rice straw was found to contain 43.1% hemicellulose and 33.8% cellulose. By the end of the production run (32 days after the start of composting) some 50% of the hemicellulose but only 24% of the cellulose had been degraded. The 'lignin' fraction remained little altered. Two major periods of decomposition occurred: during the early composting phase (days 0–6) when the total loss in dry weight averaged 1.3% day^{-1}, and during the first 15 days of the bedding phase when a similar rate of dry-weight loss occurred. Reference to Fig. 2*c* shows that the second period of decomposition corresponded with the period of active growth and fruit-body production of *V. volvacea*, although colonisation by *Coprinus* species and disappearance of *V. volvacea* were followed by slowing of the rate to 0.6–0.7% day^{-1} towards the end of the production run.

Biochemical changes in *A. bisporus* compost were first studied in detail by Waksman and his co-workers in the 1930s. They found that losses of cellulose and hemicellulose took place in phase I but there was little decomposition of lignins (Waksman & McGrath, 1931; Waksman & Nissen, 1932). Their figures are similar to those of Chang (1967) for

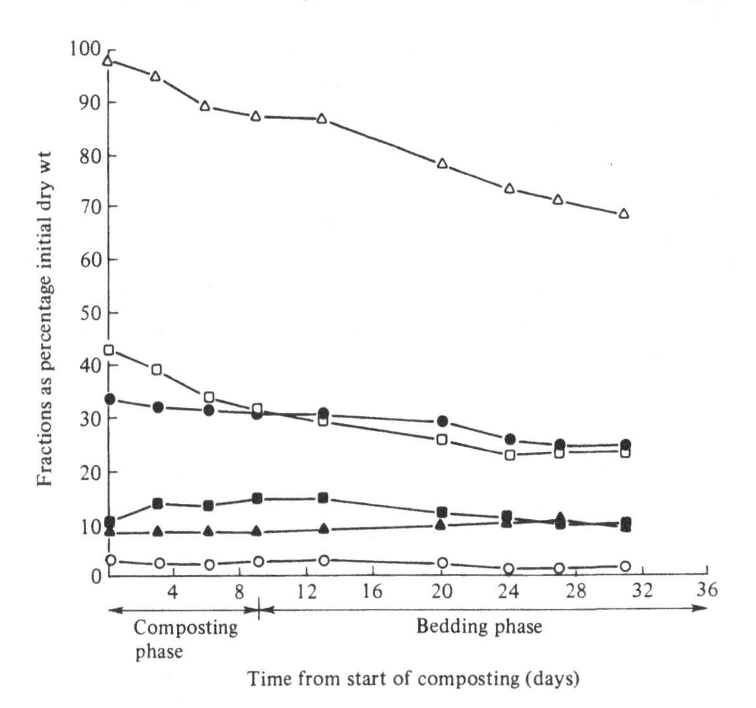

Fig. 3. Loss in dry weight and changes in biochemical composition in rice-straw compost used for the cultivation of *Volvariella volvacea*. Nylon mesh bags containing weighed rice straw were incorporated into the compost and samples (n = 6) withdrawn at regular intervals using the method of Chang (1967) for measurement of total loss of dry weight (△), and, by sequential analysis, material soluble in: ethanol (= sugars, ○), diastase (≃ starch, ■), 24% potassium hydroxide (≃ hemicelluloses, □) and 72% sulphuric acid (≃ cellulose, ●); the insoluble remainder was considered to represent lignin (▲). Data from Basuki, 1981.

wheat-straw compost. However, during the colonisation of the prepared compost by *A. bisporus* mycelium, not only does further decomposition occur, but there is also a considerable reduction in the lignin fraction. Gerrits, Bels-Koning & Muller (1967) found that lignin degradation was associated with the mycelial colonisation phase, but cellulose and hemicellulose appeared to be broken down at a faster rate during periods of fruit-body production.

Apart from carbon, nitrogen is a key nutrient in composts (Gray & Biddlestone, 1974), although in mushroom cultivation (both of *A. bisporus* and *V. volvacea*) other inorganic nutrients such as phosphorus and calcium may be added to the compost mixture (Watson, 1973; Chang, 1978; Hayes, 1978*b*). Although the nitrogen may be initially in an inorganic form such as ammonia or nitrate, it is rapidly converted to an organic form, probably largely by incorporation into microbial biomass during the initial high-temperature phase of composting, although nitrogen may be volatilised as ammonia during this phase (Gray & Biddlestone, 1974; Gerrits, 1978). Basidiomycetes that colonise compost may therefore grow into an environment with an elevated level of organic nitrogen. Chang (1967) found that ammonium and nitrate fell by 46% in the first five days of composting of wheat straw amended with ammonium nitrate. Inorganic nitrogen had reached a very reduced level by day 16 of composting, but total nitrogen showed comparatively little change over this period.

A. bisporus compost may be prepared with a variety of nitrogen amendments, usually animal manures, although non-manure formulations have been developed (Smith & Spencer, 1976). During phase I composting, there is conversion of nitrogen in the manure to microbial nitrogen and loss of free ammonia. It is generally recognised that the presence of free ammonia in compost is inhibitory to growth of *A. bisporus*, so that composting and peak heating must condition the substrate by fixation or volatilisation of ammonia to non-toxic levels (Gerrits, 1977; van Faasen & Van Dijk, 1979). During phase I composting, there is also adjustment of the C:N and C:P ratios, the C:N falling from 30:1 to as low as 15:1 (Gerrits, Bels-Koning & Muller, 1967), and the literature suggests that ratios of this order are optimal for fruit-body production (Hayes, 1980).

The type of compost required by *V. volvacea* is less well understood. T. Basuki (unpublished) found that the initial C:N ratios of cotton-waste and rice-straw composts were 45–50:1, falling to 30:1 by the time

V. volvacea was spawned. By the end of the production run, the ratio had reached 19–21:1. These figures are probably typical of most *V. volvacea* cultivation, and nitrogen amendment of the compost does not seem necessary (although such amendments as rice bran must increase the level slightly). Chang-Ho & Ho (1978) studied the nitrogen composition of waste-cotton composts and concluded that *V. volvacea* could grow over a very wide range of C:N ratios, from 32:1 to 150:1, with optimum growth on composts with organic nitrogen amendments. Furthermore, Yong & Graham (1974) added 0.1% urea to palm-pericarp composts and obtained satisfactory fruit-body yields. When *V. volvacea* is grown by the traditional method, the nitrogen status must be even lower, probably at a C:N ratio of 40–50:1.

Chang Ho & Yee (1977) carried out a comparative study on the nitrogen nutrition of *V. volvacea* and *C. cinereus*. They showed that the nitrogen source had a profound effect on the rates of cellulolysis and utilisation of xylan by these two fungi; *V. volvacea* showed greatest activity when asparagine was the sole nitrogen source, as compared to nitrate nitrogen; *C. cinereus* showed a reverse trend. In both fungi, cellulolysis was optimal at a C:N ratio of 24–36:1, but it was found that levels of nitrate nitrogen above 1650 mg ml^{-1} in the nutrient medium depressed the rate of cellulolysis by *V. volvacea*, whereas levels as high as 3300 mg ml^{-1} stimulated optimum cellulolysis by *C. cinereus*. The concept of *C. cinereus* as a fungus which responds to high levels of inorganic nitrogen was developed by Penn (1977) and Burrows *et al.* (1979), who found that addition of ammonia to cereal straw promoted rapid, almost exclusive, colonisation by this fungus. However, they also found that it had an alkaline pH requirement, and that colonisation was poor in acid, high nitrogen, straw. A similar picture was established by T. Basuki (unpublished) for the *Coprinus* species characteristic of rice-straw compost.

All three decomposer basidiomycetes discussed show differences therefore in their requirements for a correct growth medium, in particular the nitrogen level. The *Coprinus* species seem to be the most flexible, and can colonise environments with both high and low levels of nitrogen, although they may respond best to inorganic nitrogen sources, especially ammonia. In contrast, *V. volvacea* is inhibited by high nitrogen levels and requires a high C:N ratio with an organic nitrogen source. *A. bisporus* apparently responds best to organic nitrogen and to low C:N ratios; it is inhibited by free ammonia.

Utilisation of compost polymers by decomposer basidiomycetes

We have seen that wheat-straw compost, rice-straw compost and mushroom compost present very different backgrounds for the growth of the basidiomycetes. Are these differences related to the nutritional requirements of decomposer basidiomycetes which colonise these materials?

Chang (1967) carried out a comparative study on the utilisation of polysaccharides by fungi isolated from wheat-straw compost. The basidiomycetes (*Coprinus cinereus, C. macrocephalus* and *Clitopilus pinsitus*) were moderately cellulolytic, the respiratory loss in dry weight of filter paper (Garrett, 1963) being about half that caused by many of the cellulolytic thermophilic Hyphomycetes and ascomycetes. However, the basidiomycetes were able to utilise xylan (hemicellulose) more effectively than these fungi. T. Basuki (unpublished) carried out a similar comparative study of the utilisation of polysaccharides and lignins by 15 species of fungi isolated from rice-straw composts, including the basidiomycetes *Volvariella volvacea*, *C. cinereus*, *C. patouillardii* and *C. ephemeroides*, and the data are summarised in Fig. 4 and Table 2. Again, some Hyphomycetes and ascomycetes proved to be more cellulolytic and also utilised rice-straw at a faster rate than the basidiomycetes (Fig. 4). The cultivated basidiomycete, *V. volvacea*, had the lowest cellulase activity of all the cellulolytic species. However, in an evaluation of fractionated culture filtrates from fungi grown on rice straw, it was found that those from *V. volvacea* cultures had the highest activity against hemicelluloses (araban, xylan, galactan and mannan), whilst the activity of culture filtrates of some of the highly cellulolytic Hyphomycetes, such as *Aspergillus fumigatus*, was weak (Table 2). Lignin degradation by cultures of the fungi was assessed by the oxidase test (Lindeberg, 1948), and by ability to dephenolise lignin agar (Sundman & Näse, 1971). Allowing for the limitations of these procedures (Kirk & Fenn: Chapter 4), it is possible to suggest that the basidiomycetes were the only fungi able to modify lignins in the rice straw (Fig. 4). Burrows *et al.* (1979) also found that isolates of *C. cinereus* from wheat straw were able to clear lignin preparations, and their isolates were also able to utilise hemicellulose and cellulose at almost equal rates. However, decomposition of wheat straw by this basidiomycete did not cause any significant decrease in the lignin fraction.

Turner (1974), Turner *et al.* (1975), and Wood & Goodenough (1977) studied the production of extracellular enzymes by mycelium of

Table 2. *Hydrolysis of hemicelluloses by protein fractions of culture filtrates of selected fungi.* (Data from Basuki, 1981)

Species	Units of enzyme activity against hemicelluloses[c]			
	Araban	Xylan	Mannan	Galactan
Coprinus patouillardii[a]	1.24	0.54	0.93	0.55
Volvariella volvacea[a]	7.35	7.14	4.33	2.56
Chaetomium thermophilum[b]	3.26	3.08	2.91	1.17
Thermomyces lanuginosus[b]	3.24	6.51	0.81	0.64
Aspergillus fumigatus[b]	0.52	0.34	0.22	0.00

Cultures were grown on 3.0 g rice straw for 3 weeks at 30 °C[a] and 45 °C[b]. The straw was then extracted with 50.0 ml of 0.55 M phosphate buffer, followed by filtration. Ammonium sulphate precipitation was employed to isolate the protein fraction, and release of sugars by this fraction from 1% w/v solutions of hemicellulose was estimated by method of Albersheim, Nevins, English & Karr (1967). [c]Units are μg sugar released from 4.0 mg of hemicellulose by 1.0 μg protein in 6 h at 37 °C.

A. bisporus as it grew through compost, as well as measuring their activity in culture filtrates. Wood (1979) listed eight extracellular enzymes released into wheat-straw compost during the growth of *A. bisporus* mycelium, including proteases, phosphatases, xylanase, laccase and cellulase. The cellulase is at a low level of activity in the compost during the mycelial colonisation phase (Turner, 1974; Wood & Goodenough, 1977) but shows a rapid increase during the formation of the first flush of fruit bodies, after which it declines again (Fig. 5*a*). Xylanase activity is not so correlated and increases steadily after casing (Fig. 5*b*), but laccase activity in the compost (Fig. 5*c*) is negatively correlated with flushing, steadily increasing during mycelium colonisation but declining sharply during the flushing (Turner, 1974; Turner *et al.*, 1975; Wood & Goodenough, 1977). Detailed information on the rates of degradation of cellulose, hemicellulose and lignin in the compost during mycelial growth and flushing is not available, but Hayes (1980) and others all concluded that the laccase was responsible for the decrease in the lignin fraction, which occurs most rapidly during the mycelial colonisation phase (Gerrits *et al.*, 1967). However, an actual demonstration of lignin metabolism by *A. bisporus* has yet to be made, in spite of the fact that lignin is often assumed to be an energy source for this basidiomycete (Gerrits, 1969; Hayes, 1980). Kirk & Fenn (Chapter 4) discuss the question as to whether lignin degradation is in fact an

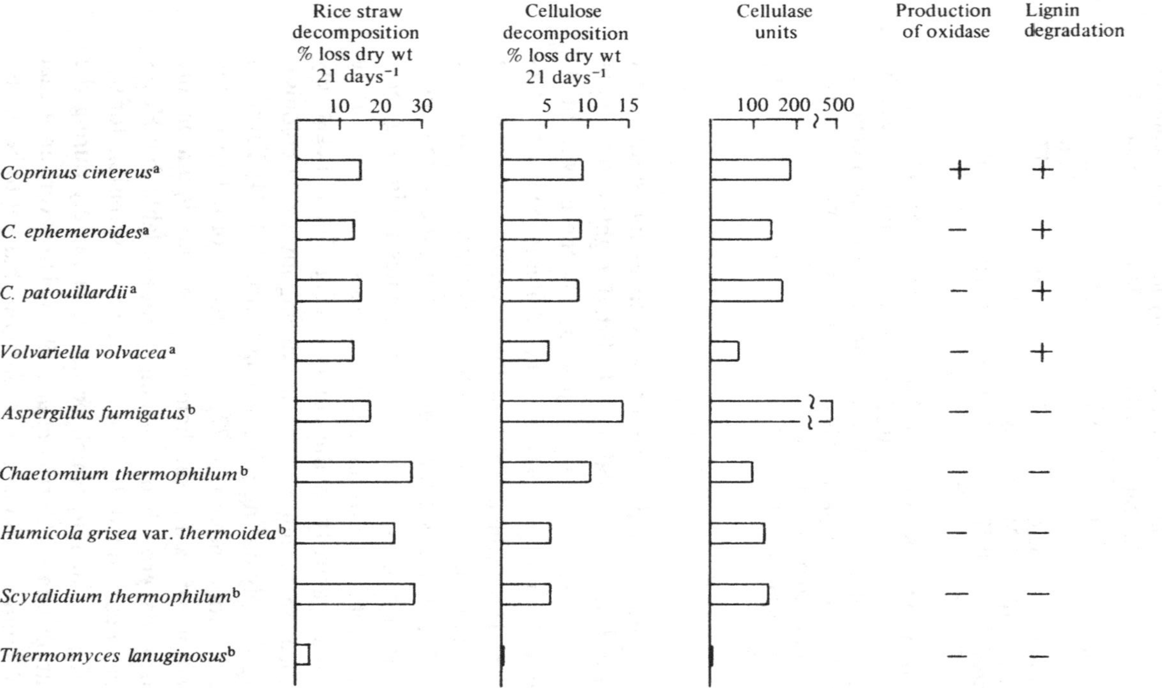

Fig. 4. Degradation of rice-straw, cellulose and lignin by fungi isolated from rice-straw compost. Rice-straw utilisation expressed as percentage loss in dry weight of 3.0 g samples of straw following 21 days growth of the fungi ($n = 5$); cellulose utilisation expressed as percentage loss in dry weight of 5 Whatman No. 3 7.0 cm filter papers amended with 15.0 ml cellulolysis medium (Garrett, 1962) after 3 weeks growth ($n = 5$); cellulase activity assessed by extracting, with 30.0 ml 0.55 M phosphate buffer, cultures pre-grown on sterilised 3.0 g Whatman cellulose powder plus cellulolysis medium for 3 weeks (units are μg reducing sugar released by 0.5 ml culture filtrate from 5.5 mg carboxymethyl cellulose (as 0.5% w/v solution) in 0.5 h); lignin degradation assessed by dephenolisation of lignin media by the cultures (Sundman & Näse, 1971) and by oxidase production (Lindeberg, 1948) on 0.5% gallic acid/0.5% malt extract agar. [a]Incubation of culture at 30 °C; [b] incubation at 45 °C; –, negative result. Data from Basuki, 1981.

energy-yielding process for the basidiomycetes and demonstrate that it is linked strongly to simultaneous polysaccharide utilisation. Clearly, the importance of cellulase and laccase levels in both growth and sporulation by *A. bisporus* has yet to be elucidated, and Wood (1979) has recently made the interesting suggestion that the decline in laccase levels during flushing is connected to its breakdown by proteases, facilitating recovery by the mycelium of amino acids required for fruit-body production.

If a link between laccase levels and lignin degradation is postulated, it is by analogy with the white-rot basidiomycetes (Kirk & Fenn: Chapter 4). Wood (1979), in discussing this analogy, points out that the cellulase produced by *A. bisporus* seems to have only an endocellulase component, and an exocellulase (and hence C_1 action) may be absent (Manning & Wood 1978). D. A. Wood (personal communication) suggests that *A. bisporus* may in fact attack crystalline cellulose initially by non-enzymic oxidation of the type found in brown-rot fungi (Montgomery: Chapter 3). He feels that additional evidence of brown-rot affinities is also afforded by the well-known production of oxalate crystals on the surface of the mycelium of *A. bisporus* (de Bary, 1887), a characteristic also shown by many brown-rot basidiomycetes.

Utilisation of microbial tissue in composts by basidiomycetes

The *Coprinus* species and *Volvariella volvacea* will colonise compost but can also successfully colonise uncomposted materials; *Agaricus bisporus* apparently requires conditioning of the material by composting before it will successfully colonise and sporulate. It has now been established that, apart from conditioning of the compost, including reduction in levels of toxic materials such as ammonia, the microbial population established during composting has a direct effect on growth of the mycelium of *A. bisporus*. Fermor & Wood (1979) have demonstrated that 3% or more of the compost may consist of microbial tissue prior to spawning. Stanek (1972) and Eddy & Jacobs (1976) showed that growth factors and polysaccharide capsular material produced by bacteria were probably utilised by *A. bisporus* mycelium as it grew through the compost. Recently, Fermor & Wood (1979) have demonstrated that *A. bisporus* is able to utilise dead cells of *Bacillus subtilis* as a carbon and nitrogen source, and Wood (1979) has emphasised the rôle of proteases in this process. Fermor & Wood (1981) have now extended these studies to a range of gram-negative and gram-positive bacteria. They found that not only could *A. bisporus* utilise dead cells as a sole

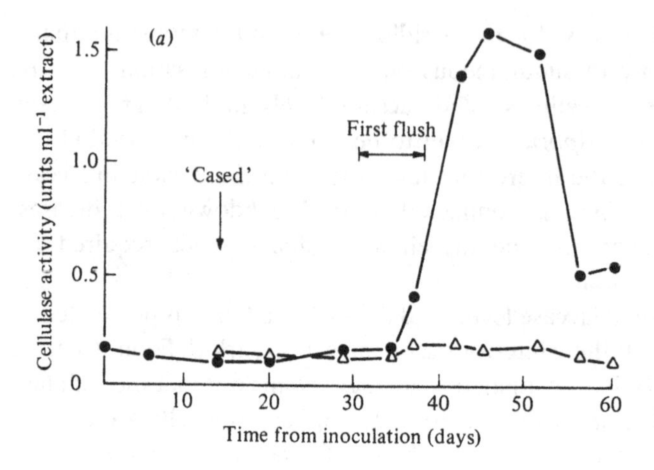

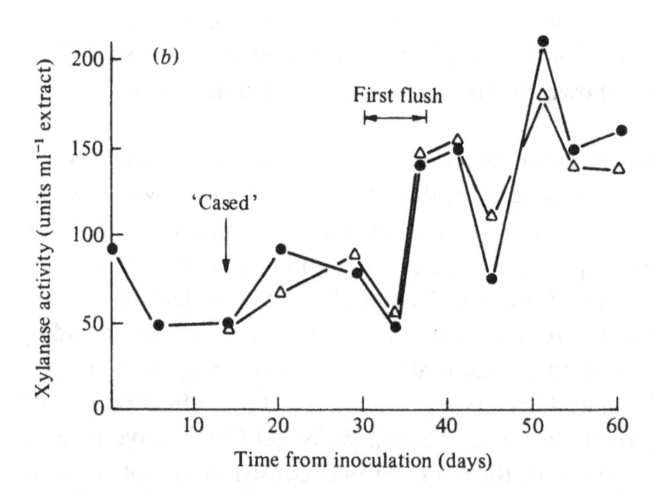

Fig. 5. Enzyme activity during growth and fruit-body production by *Agaricus bisporus* in compost: (*a*) cellulase; (*b*) xylanase; (*c*) laccase. Samples of compost were extracted with 0.01 M phosphate buffer. After purification of the filtrate, laccase activity was measured by use of an oxygen electrode (unit = amount of enzyme which consumed 1 μmol O_2 min^{-1}); cellulase activity measured viscometrically using 2.5% w/v carboxymethyl cellulose solution, and xylanase activity by determining reducing groups liberated from 1% w/v solutions of xylan in buffer. Growth *with* fruit-body production (•) and growth *without* fruit-body production (△) were examined in this way. Data from Wood & Goodenough, 1977.

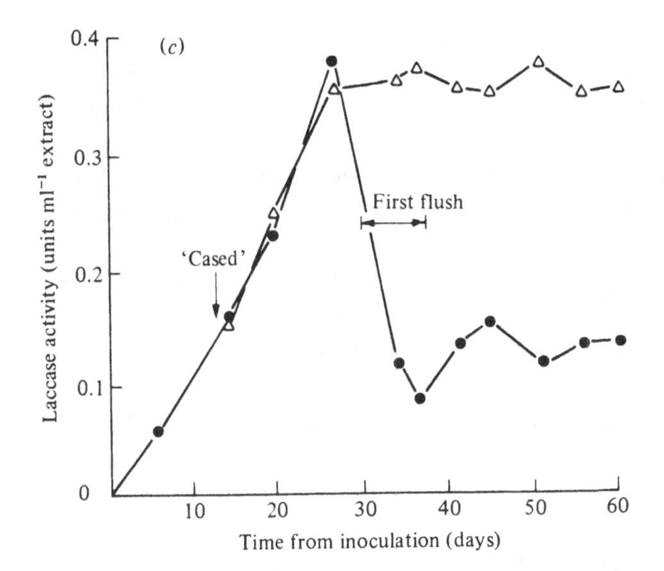

carbon and nitrogen source, but it could also lyse and degrade living bacteria, using a battery of extracellular enzymes to do so. D. A. Wood (personal communication) has preliminary evidence that *A. bisporus* will also utilise the mycelium of thermophilic fungi such as *Humicola grisea* var. *thermoidea* in a similar way.

Although *V. volvacea* will colonise and fruit on uncomposted rice straw, Hayes & Lim (1979) underline the increased productivity which results from cultivation employing a compost phase. How this effect is mediated is not known and may well include a response to conditioning of the compost, but at present there is no evidence that this basidiomycete can utilise the thermophilic microbial population in the same way as *A. bisporus*.

Interactions involving decomposer basidiomycetes in composts

When a basidiomycete grows into compost, it is usually colonising material that has a pre-established microbial population, which, as we have seen, will have already begun to exploit available substrates. Interactions between the basidiomycetes and these populations are therefore likely, and they could be neutral, negative or positive, and could be mediated by substrates (nutrient competition or synergism), metabolites (inhibitory or stimulatory) or even by direct physical contact. A good example of such an interaction is the utilisation of thermophilic micro-organisms by *A. bisporus*.

In the successions in wheat-straw and rice-straw composts, evidence has already been presented for replacement colonisation by basidiomycetes. The activity of *Coprinus cinereus* in wheat-straw compost coincides with a decline in the activity of thermophilic fungi such as *Humicola insolens*. It is difficult to view this replacement as due simply to nutritional competition in view of the rather poorer performance of *C. cinereus* in utilising polysaccharides when compared to fungi such as *H. insolens*. The slight ligninolytic ability of *C. cinereus* may, however, convey an advantage in enabling it to utilise remaining cellulose and hemicellulose which may be masked by lignins at this stage of the decomposition process. J. N. Hedger (unpublished) explored the possibility that replacement might be by antagonism. *C. cinereus* was inoculated opposite other fungi on 2.0 cm sterile cellophane strips mounted on a mineral agar medium (Tribe, 1966; Hedger & Hudson, 1974). Observation of the contact zone between the colonies showed that hyphae of most fungi likely to interact with *C. cinereus* in the compost, such as *H. insolens*, were adversely affected. *C. cinereus* was one of the fungi examined by Ikediugwu & Webster (1970), who also found that it was a moderately aggressive antagonist.

T. Basuki (unpublished) used a similar method to study antagonism by the *Coprinus* species characteristic of rice-straw compost (*C. cinereus, C. patouillardii* and *C. ephemeroides*). These were all able to suppress and overgrow colonies of fungi already established in the succession, such as *Humicola grisea* var. *thermoidea, Scytalidium thermophilum* and *Thermomyces lanuginosus*. A striking feature was that the growth of the *Coprinus* species was greatly stimulated in the dual cultures and, in the contact zone between the colonies, fruit bodies were initiated in larger numbers and at an earlier time than in single cultures. Basuki found two exceptions to this behaviour: fungi which occurred very late in the succession on the straw, such as *Trichoderma polysporum*, were unaffected, and *Mortierella wolfii*, a mucoraceous fungus which appeared in the compost coincidentally with the *Coprinus* species, formed an intermingling growth with the *Coprinus* mycelium on sterile cellophane.

T. Basuki (unpublished) also examined the interactions of *Volvariella volvacea* and 20 fungi prominent in the rice-straw compost succession. As well as studying confrontation between colonies on agar plates and cellophane, he evaluated the effect on *V. volvacea* of volatiles produced by these fungi by use of the double-dish technique of Gibbs (1967). Direct confrontation was assessed by the criteria of Skidmore &

Dickinson (1976). *V. volvacea*, like the *Coprinus* species, was able to suppress and overgrow the colonies of the major colonisers from the composting phase, such as *H. grisea* var. *thermoidea*, but was itself antagonised by *Trichoderma* species, *Aspergillus niger* and *Gliocladium virens*. Surprisingly, many of the Mucorales were antagonistic, especially *Rhizopus oligosporus*. Yee & Chang Ho (1980) have recently described antagonism of *V. volvacea* by *A. niger* as being due to production of organic acids. Volatiles released by growing colonies of nearly all the isolates screened by Basuki stimulated growth of *V. volvacea*, up to 20% increases in growth being recorded as compared to the control (*V. volvacea* paired with itself), the only exceptions being again the Mucorales.

Colonisation of the rice-straw compost by *V. volvacea* and the *Coprinus* species may therefore involve antagonistic suppression of other fungi and possibly stimulation by volatiles or other materials released into the compost by these fungi. Basuki also found some indications of an antagonistic interaction between the *Coprinus* species and *V. volvacea* on agar plates at 30 °C. Branch hyphae of *Coprinus* dichotomised on tip contact with a hypha of *V. volvacea* and both branches then grew in a coiling fashion around it. Vital staining with fluorescein diacetate revealed that the hyphae of *V. volvacea* remained viable, even when covered in dense coils of *Coprinus* mycelium. This coiling appeared similar to that described by Dennis & Webster (1971) when hyphae were parasitised by *Trichoderma*, but no penetration was observed. In experiments with sterile lengths of rice straw, Basuki found that *V. volvacea* fully penetrated straw occupied by the *Coprinus* species, but *C. cinereus* showed a limited ability to penetrate sections occupied by mycelium of *V. volvacea.* However, when this experiment was scaled up by inoculating pots of *c.* 1.0 kg of sterilised rice-straw compost at 30 °C with combinations of *V. volvacea* and *C. cinereus*, the mixed cultures all showed a significant reduction in the yield of *V. volvacea* fruit bodies when compared to single cultures of *V. volvacea*, and *C. cinereus* had the most profound effect, causing a 75% reduction in the number of *V. volvacea* fruit bodies.

Yee & Chang-Ho (1980) also examined confrontation between *C. cinereus* and *V. volvacea*, and considered it to be a mutual inhibition. They concluded that nutrient competition between these species in the compost, rather than hyphal antagonism, was of the greater importance. However, the hyphal coiling may also be of significance in causing a drain of nutrients from *V. volvacea* to the *Coprinus* species. A balance appears to exist initially in the compost between the species of basidiomycete, but it is gradually eroded in favour of the *Coprinus* species as the bedding compost ages. A particularly striking demonstration of this balance is the effect of adding nitrogen; T. Basuki & J. N. Hedger (unpublished) found that remarkable stimulation of the growth of *C. cinereus* occurred when beds were watered with 0.15% urea (Yong & Graham, 1974), whilst the fruit-body yield of *V. volvacea* was reduced to zero. Possibly, the shift in the C:N ratio which occurs as the bedding material is decomposed may assist the *Coprinus* species to compete more effectively in the later stages of production.

Interactions between *Agaricus bisporus* and the microbial population of composts and casing materials have long been considered to have an effect on the yield of the fruit bodies, and one of the objects of the composting and pasteurisation phase is to reduce the level of competitors, antagonists and parasites of *A. bisporus* (Sinden, 1971). Mycoparasitism of mycelium and fruit bodies by fungi, such as *Mycogone perniciosa* and *Verticillium* species, as well as bacterial pathogens, such as *Pseudomonas tolaasii*, necessitate the use of fumigants such as methyl bromide, or selective fungicides such as benomyl, to prevent build-up of infection in mushroom houses (Hayes & Nair, 1975).

Fermor & Wood (1979) and others have shown that the thermophilic population of the compost is stimulatory to *A. bisporus*, providing a carbon and nitrogen source for the basidiomycete. This population must be regarded as being relatively inactive. What of the more active mesophilic populations of bacteria and fungi? Some may act as nutrient competitors or even antagonists. Eicker (1980) found that there was a steady build-up of mesophilic colonisers in the compost, both before and after casing; the fungi were principally *Trichoderma viride* and *Aspergillus* species, which he considered may act as direct antagonists of *A. bisporus*. It is also possible to show that some of this mesophilic population stimulates growth and fruit-body production in *A. bisporus*. Wood & Matcham (1980) have recently investigated the growth of *A. bisporus* mycelium into sterile and unsterile composts. They found that the linear growth rate in the composts, sterilised either by autoclaving or irradiation, was about half that in the unsterilised compost. Addition of growth factors, such as thiamine and biotin, to the sterilised materials did not affect the result. They suggested that this effect could be mediated by removal of a toxic metabolite of *A. bisporus*, by the compost microflora or by a protoco-operative type of interaction, the latter being the most likely.

A. bisporus also interacts with the microbial population of the casing layer. Hayes, Randle & Last (1969) considered that pseudomonads from this layer stimulated formation of fruit-body primordia. One species in particular, *Pseudomonas putida*, seemed to be particularly effective, and would induce fruiting when introduced to mycelium grown on sterilised compost and casing material. The nature of this relationship was reviewed by Hume & Hayes (1972) and Hayes & Nair (1975), who found that cultures, or even culture filtrates, of *P. putida*, induced fruiting of *A. bisporus* on agar media. The fact that a range of soil bacteria and culture filtrates will all promote this effect when added to agar (Park & Agnihotri, 1969) indicates that it may be non-specific. Hayes (1974) suggested that compounds released by bacteria act as chelating agents, making iron available in the alkaline casing, but an investigation by Wood (1979) failed to confirm this hypothesis. He found that chelating agents and iron compounds, such as haemoglobin or ferrous sulphate, had no effect on the formation of primordia, nor did iron-binding compounds of microbial origin, such as coprogen and ferritin. He concluded that the stimulation of primordia by bacteria may be due to the removal of compounds which are produced by the *A. bisporus* mycelium and act as self or auto-inhibitors. He cites as evidence the fact that fruit bodies can be initiated in axenic culture by the use of a casing layer of sterile autoclaved charcoal (Long & Jacobs, 1974; Wood & Hammond, 1977).

Whatever the mechanism of this interaction, Hayes (1974) makes the important point that addition of the casing layer functions by switching *A. bisporus* growth into a reproductive phase. This could be by the action of the casing bacteria and by the trapping of volatiles, such as acetaldehydes, produced both by *A. bisporus* itself and by other fungi, which previously had been lost from the compost. Hayes & Lim (1979) reported that casing of compost also increased fruit-body formation by *V. volvacea*, but *V. volvacea* and the *Coprinus* species will form fruit bodies in axenic culture or on simple media such as malt extract, so that they obviously differ considerably from *A. bisporus* in this respect.

Discussion: composts as a model for decomposition by basidiomycetes

The successions of fungi which occur as plant material is decomposed can be viewed quite simply as the replacement of one species assemblage of fungi by another. Garrett (1951, 1963) defined these assemblages as nutritional groupings which he related to stages of decomposition of the substrate. Hudson (1968) refined this concept to emphasise the place of individual species of fungi in successions on plant remains above soil, communities being continually re-adjusted by loss and gain of species over the period of decomposition, so that mycelium of initial colonisers had disappeared by the time later stages of decomposition had been reached. He related these changes in fungal communities to the physiological adaptations of species of fungi to the changing environment, as well as to interactions between species. The communities which developed were often characteristic of particular plant debris.

Swift (1976) and Swift, Heal & Anderson (1979) have come to question the reality of this approach, and reference has already been made to the 'unit community' concept of Swift (1976) in the introduction to this chapter. Swift *et al.* (1979) pointed out that the major resource available to decomposers, plant material, has a relatively uniform composition, consisting largely of celluloses, hemicelluloses and lignins; the fungal community which develops may be as much a result of random arrival of species from a background of spore inoculum as of any substrate specificity, although Swift (1976) did divide colonisers into two groupings: resource specific (RS) and resource non-specific (RN) (see Chapter 16). These approaches imply that the attempt to resolve fungal colonisation of material as a successional sequence of communities may be fruitless. In fact we know, as yet, too little about the detailed biochemistry of the decomposition of plant remains, or about the physiology of decomposer fungi and their interactions, to decide how much the species composition of the decomposer communities is a result of random assemblage, and how much is determined by the substrate or by biotic factors. However, even if only major changes in species composition are accepted as being of significance during decomposition, the decomposer basidiomycetes represent one of the most easily identifiable components of any successional sequence.

Throughout this book continual reference can be found to the relatively late colonisation of substrates by these fungi, although in some materials, such as wood, the prior period of colonisation by other organisms may be brief (Käärik, 1975). Hudson (1968) emphasised this point and, in his general scheme for succession of fungi on plant litter above soil, considered that basidiomycetes may not appear on some materials until as much as two to three years decomposition has taken place. To take a specific example, Frankland (1966) found that basidiomycetes only became prominent as decomposers of *Pteridium* petioles after two years. Usually, mention of the late successional position of basidiomycete decomposers is accompanied by a generalisation about their physiology: that they grow slowly but are able to utilise the most recalcitrant molecules in the substrate, the lignins and celluloses, although they may be accompanied by fast-growing, non-basidiomycete, fungi such as *Mortierella* or *Mucor*, which lack this attribute but grow in a protoco-operative relationship (Hedger & Hudson, 1974; Hudson, 1975).

Swift (1976, and Chapter 16) considered that the basidiomycete life cycle is similar to the *K*-selection life strategy of higher plants (MacArthur & Wilson, 1967). In contrast, other fungi may show *r*-selection life strategies – rapid growth followed by sporulation. Such fungi can be found in the initial stages of fungal successions, in which they exist in a rapidly changing environment from which they may be displaced quickly, for example, in early stages of litter decomposition.

We have seen that the succession of fungi in self-heating composts culminates in a basidiomycete phase, with the exception of highly acidic composts in which it may be delayed or absent (Burrows *et al.* 1979). The initial phase of compost colonisation by actinomycetes and bacteria is accompanied by a brief growth of thermophilic Mucorales which do not recur. These fungi, corresponding to Garrett's (1963) 'primary saprophytic sugar fungi', represent good examples of the *r*-selection life strategy, and they are soon replaced by a community of cellulose-decomposing Hyphomycetes and ascomycetes. In composts that are not interrupted by pasteurisation, the next community is dominated by the basidiomycete *Coprinus cinereus*, together with an associated 'secondary saprophytic sugar fungus' (Garrett, 1963), *Mortierella wolfii*. In mushroom composts, the cultivated basidiomycete is added after a pasteurisation phase. In the case of *Volvariella volvacea*, this inoculation is done after 6–10 days composting, at about the same time as the *Coprinus* species would probably develop naturally, whilst *A. bisporus* inoculation may be performed after a considerably longer period of composting, when *Coprinus* might have already colonised the material.

A. bisporus represents what would be regarded by most mycologists as a 'typical' decomposer basidiomycete. We have seen that it is a slow-growing fungus (the hyphal extension rate is 1 mm day^{-1} on malt agar at 25 °C) and it is a late coloniser of the compost material. Obviously in commercial cultivation this colonisation is artificially initiated, but we can guess that the mycelium of related wild species, such as *Agaricus campestris*, also colonises rather late in the decomposition sequence of herbivore dung and plant litter in soil. The prior period of decomposition is represented by the composting phase in cultivation systems. An extended period of mycelial growth is required before the fruit bodies are initiated (4–5 weeks), and, during this period, cellulose, hemicellulose and lignins are broken down. The standing crop of the thermophilic micro-organisms is also exploited as a carbon and nitrogen source, and mycelial growth and fruit-body initiation are influenced by microbial activity in the compost and casing layers.

V. volvacea differs considerably from *A. bisporus* in a number of respects. Its hyphae extend at the rate of 1 mm h^{-1} at 35 °C, and it is capable of producing mature fruit bodies within 7–10 days of spawning, some three times faster than *A. bisporus*. Unlike *A. bisporus*, it is able to colonise uncomposted straw competitively, and this is the basis of traditional cultivation methods in South-east Asia. However, composting and pasteurisation seem to promote higher productivity, and evidence is presented in this chapter which indicates that the mycelium may be stimulated by the microbial environment of the compost, whilst fruit-body production may be increased by a casing layer. Unlike *A. bisporus, V. volvacea* seems to have little ability to degrade lignins, and may utilise hemicelluloses faster than cellulose.

Coprinus cinereus and the other *Coprinus* species which occur on rice-straw compost differ markedly from *A. bisporus* too. The growth rate of *C. cinereus* at 35 °C is similar to that of *V. volvacea*, and, like that fungus, it produces fruit bodies within a week of colonisation of the substrate. It has only a limited ability to degrade lignin, but utilises cellulose and hemicellulose at about the same rate. It is capable of very early colonisation of materials such as straw, especially at alkaline pH values and in the presence of free ammonia. In compost it appears to colonise later, owing to the high temperature phase, but rapidly replaces the thermophilic Hyphomycetes and ascomycetes as soon as a suitable temperature regime arises. Like *V. volvacea*, it can produce fruit bodies in axenic culture.

Clearly, *A. bisporus* corresponds to Swift's (1976) ideas of a *K*-selection life strategy, but what of *C. cinereus* and *V. volvacea*? These are both fast-growing fungi, even in comparison with such well-known fungal 'athletes' as *Mucor* and *Neurospora.* Both species also produce conidia on the vegetative mycelium. Neither fungus can be said to be typical of the *K*-selection strategy but nor do they really fit the *r*-selection, when a fungus such as *Rhizomucor pusillus,* which occurs in the same habitat, is compared with them.

Pugh (1980) has recently used the approach of Grime (1979) to analyse fungal colonisation strategies. He emphasised the importance of the nutritional status of the environment of the fungus (*stress*) and fluctuations in the environment, notably superimposition of a new environment (*disturbance*). Clearly, these relate to *K*-selection and *r*-selection strategies, and Pugh considers that *r*-selection fungi are adapted to a low-stress/high-disturbance environment, as exemplified by *Rhizomucor pusillus* in composts and by Garrett's category of primary saprophytic sugar fungi (Garrett, 1963). Fungi that represent the *K*-strategy may be growing in a high-stress/low-disturbance environment, and this is represented by colonisation of compost by *A. bisporus*, when freely available nutrient may have been removed in the previous colonisation phases, and when the fungus must be able to degrade a recalcitrant material, lignin.

Pugh also considered a third life strategy to be possible for fungi: low-stress/low-disturbance. These fungi he termed *competitors*, implying that they have the ability to replace other fungi in a uniform environment in which nutrient is freely available. *Coprinus cinereus, C. ephemeroides* and *C. patouillardii* illustrate that this strategy is possible for basidiomycetes in compost, since they replace the previous community of cellulolytic fungi during a period of decomposition of cellulose and hemicelluloses. This rapid replacement colonisation seems to occur by antagonism, rather than by nutrient competition, although the limited ability of these species to degrade lignin may also be of significance.

V. volvacea could also be considered to be a competitor as it is able to suppress the growth of some of the fungi which may confront it in rice-straw compost. However, although it can naturally colonise rice straw and other substrates in the tropics, it is unlikely to be as early a coloniser as *C. cinereus* and the other *Coprinus* species, and optimum growth seems to require some preconditioning of the material by composting. It therefore represents something of an intermediate between the *Coprinus* and *A. bisporus*.

The outcome of competition between the *Coprinus* species and *V. volvacea* is obviously of interest and seems to represent a deadlock of the type described by Rayner (1978) in interspecific basidiomycete competition during colonisation of hardwood. The outcome of competition between the basidiomycetes in the compost may depend on the amount and form of the nitrogen. In traditional cultivation methods, this competition presumably results in a balance between the two basidiomycete populations, but, in modern cultivation employing composting, mycelium of *V. volvacea* is able to dominate the straw completely following spawning. The *Coprinus* species then seem to invade, possibly assisted by changes in the C:N ratio of the bedding, which alter the balance between the two competitors.

Do other decomposer basidiomycetes show a similar range of life strategies? Clearly, many can be referred to the *K*-selection strategy as pointed out by Swift (1976), among them many of the litter-decomposing agarics, which occur at a late stage in the successional colonisation of leaf litter in a high-stress/low-disturbance habitat. These may have a similar physiological ecology to *A. bisporus*, and many are able to degrade lignins as well as cellulose and hemicelluloses. It would be instructive to know if they, like *A. bisporus*, require preconditioning of the litter by micro-organisms which have preceded them in the succession, and indeed whether they are also able to utilise these organisms. There seems to be little direct evidence that this is the case, although J. Holling (personal communication) found that eight litter-decomposing basidiomycetes which she studied were able to utilise hyphae and spores of other litter fungi as carbon and nitrogen sources on a low nitrogen and carbon medium; all produced active chitinase enzymes. Paul & Voroney (1980) drew attention to the amount of energy stored in the microbial standing crop of decomposer systems after some decomposition of the primary resource had occurred; basidiomycetes may be able to exploit this secondary resource. The stimulation of fruit-body production of *A. bisporus* by the casing microflora may also explain why it is so difficult to produce fruit bodies of many decomposer and mycorrhizal basidiomycetes in axenic culture.

Hering (Chapter 12) reviews the information available on the decomposer activities of basidiomycetes in the litter of mull and mor and concludes that examples of white rot ('bleaching') of litter are associated mostly with the genera *Mycena, Clitocybe, Marasmius* and *Collybia*. Possibly these and other genera, such as *Agaricus* itself, *Psathyrella, Tubaria* and *Inocybe*, contain species that show physiological parallels to *A. bisporus*. A particularly interesting group to examine would be the basidiomycetes that he considers to be typical of mull soils, such as *Pluteus*, which are growing on finely divided humic material, already considerably modified in the gut of earthworms and containing a large population of bacteria.

Competitive colonisation by basidiomycetes is exemplified by the coprophilous succession, in which basidiomycetes such as *Coprinus*, *Panaeolus* and *Stropharia* are able to suppress growth of other fungi

(Ikediugwu & Webster, 1970). Like compost, dung represents a low-stress/low-disturbance habitat. There are strong parallels between this succession and the compost succession, and indeed all three of the *Coprinus* species mentioned in this chapter can also be found on dung. The competitive colonisation strategy is also well developed in the wood-decomposing basidiomycetes, although, as with *C. cinereus* and *V. volvacea,* direct antagonism is better developed in some species than others in a colonisation sequence (Rayner, 1978). *Phlebia gigantea* affords an example of an antagonistic pioneer coloniser paralleling *C. cinereus* (Ikediugwu, Dennis & Webster, 1970), and Rayner (1978) has emphasised the importance of antagonism in colonisation of hardwood stumps. Like *C. cinereus* and *V. volvacea*, these pioneer basidiomycetes rapidly colonise the substrate which at this stage must be considered to represent a low-stress/low-disturbance habitat, but, even so, their invasion may be by displacement of the previous brief phase of non-basidiomycete fungi and prokaryotes (Käärik, 1975). However, basidiomycetes which occur very late in the decomposition sequence on wood (Mangenot, 1952) may require or tolerate a highly-conditioned substrate (high-stress/low-disturbance) and we are less clear about the physiological ecology of the fungi of this stage, often described as growing on 'well rotted' wood. These fungi may adopt an extreme form of the *A. bisporus* life strategy, and it could be that the standing microbial crop in this highly decomposed material provides an important carbon and nitrogen source for them. Käärik (1975) points out that many of the later-colonising basidiomycetes may be non-ligninolytic cellulose decomposers in contrast to pioneer colonisers which are white-rot fungi, so that such secondary resources could be of considerable importance to these fungi.

Table 3 illustrates a possible sequence of life strategies during colonisation and decomposition of a substrate such as compost or leaf litter by fungi, based on the concepts of Swift (1976) and Pugh (1980). Clearly, the *K*-selection life strategy for basidiomycetes includes a variety of different *colonisation* strategies, related to their position in the successional sequence. The pioneer *r*-selection fungi are *non-competitors* and may be quickly replaced by non-basidiomycete, *r*-selection, *nutrient competitors*, such as the thermophilic cellulose decomposers in compost. However, these in turn may be replaced, or even fail to establish, because of *antagonistic competitor basidiomycetes* such as *C. cinereus* in compost. Finally, however, the sequence will be dominated by the *K*-selection *nutrient competitor basidiomycetes* such as

Table 3. *A possible colonisation sequence during the decomposition of plant material*

Status of the material[a] (Pugh, 1980)	LOW STRESS HIGH DISTURBANCE	LOW STRESS LOW DISTURBANCE		HIGH STRESS LOW DISTURBANCE
Life strategy of colonisers (Swift, 1976)	*r*-selection → non-basidiomycete fungi (non-competitors)	*r*-selection → non-basidiomycete fungi (nutrient competitors) and protoco-operative fungi	*K*-selection → basidiomycetes (antagonistic competitors) and protoco-operative fungi	*K*-selection basidiomycetes (nutrient competitors) and protoco-operative fungi
Examples from compost successions (see text)	*Rhizomucor pusillus*, *Rhizopus* species	*Humicola grisea* var. *thermoidea*, *Chaetomium thermophilum*, *Thermomyces lanuginosus*	*Coprinus cinereus*, *Mortierella wolfii*	*Agaricus bisporus* + ?

[a] In terms of the nutrient and environmental conditions.

A. bisporus in compost. These decomposers may avoid competition by possessing the ability to degrade lignin, but may also utilise microbial tissue as well as responding to other living organisms, which could promote growth and fruit-body production. Obviously, these preliminary ideas are entirely hypothetical and may not stand the test of time. It could be argued that the compost succession and the related herbivore dung succession are so different from other decomposition systems, especially in their very large microbial populations, that parallels are not possible. The view taken here is that composts do represent a useful model system in which accelerated decomposition can be studied, and that the principles can be applied to other, slower, decomposition sequences.

Acknowledgements. The authors thank Dr D. A. Wood for useful discussion and permission to use unpublished data, also Drs R. F. O. Kemp, R. A. Samson and R. Watling for valuable advice.

References

Albersheim, P., Nevins, D. J., English, P. D. & Karr, A. (1967). A method for the analysis of sugars in plant cell polysaccharides by gas-liquid chromatography. *Carbohydrate Research*, **5**, 340–5.

Ander, P. & Eriksson, K. E. (1976). The importance of phenol oxidase activity in lignin degradation by the white rot fungus, *Sporotrichum pulverulentum*. *Archives of Microbiology*, **109**, 1–8.

Anonymous, (1978). Compost. An annotated bibliography on compost quality and composting, 1971–1977. WHO International Reference Centre for Waste Disposal, Dübendorf, Switzerland.

Apinis, A. E. (1965). A new thermophilous *Coprinus* species from coastal grasslands. *Transactions of the British Mycological Society*, **48**, 653–6.

de Bary, A. (1887). *Comparative Morphology and Biology of the Fungi, Mycetozoa and Bacteria*. London: Oxford University Press.

Basuki, T. (1981). Ecology and productivity of the padi straw mushroom (*Volvariella volvacea* (Bull. ex Fr.) Sing.) PhD thesis, University of Wales.

Bergman, O. & Nilsson, T. (1971). *Studies on outside storage of saw mill chips*. Research Notes of the Royal College of Forestry, Stockholm R71, 54 pp. and appendix.

Burdsall, H. H., Jr. & Eslyn, W. E. (1974). A new *Phanerochaete* with a *Chrysosporium* imperfect state. *Mycotaxon*, **1**, 123–33.

Burrows, I., Seal, K. J. & Eggins, H. O. W. (1979). The biodegradation of barley straw by *Coprinus cinereus* for the production of ruminant feed. In *Straw Decay and Its Effect on Disposal and Utilisation*, ed. E. Grossbard, pp. 147–54. Chichester: John Wiley.

Chang, S. T. (1972). *The Chinese Mushroom* (*Volvariella volvacea*): *Morphology, Cytology, Genetics, Nutrition and Cultivation*, pp. 1–99. Publication Office, the Chinese University of Hong Kong.

Chang, S. T. (1974). Production of the straw mushroom (*Volvariella volvacea*) from cotton wastes. *The Mushroom Journal*, **21**, 348–54.

Chang, S. T. (1978). *Volvariella volvacea.* In *The Biology and Cultivation of Edible Mushrooms*, ed. S. T. Chang & W. A. Hayes, pp. 573–603. London: Academic Press.

Chang, S. T. (1979). *The production of straw mushroom on agricultural wastes as a method of food protein recovery in south-east Asia.* ASAIHL lecture, 1978. Ramkhamhaeng University, Thailand.

Chang, Y. (1967). The fungi of wheat straw compost. II. Biochemical and physiological studies. *Transactions of the British Mycological Society*, **50,** 667–77.

Chang, Y. & Hudson, H. J. (1967). The fungi of wheat straw compost. I. Ecological studies. *Transactions of the British Mycological Society*, **50,** 649–66.

Chang-Ho, Y. & Ho, T. M. (1978). Effect of nitrogen amendment on the growth of *Volvariella volvacea. Mushroom Science*, **10,** 619–28.

Chang-Ho, Y. & Yee, N. T. (1977). Comparative study of the physiology of *Volvariella volvacea* and *Coprinus cinereus. Transactions of the British Mycological Society*, **68,** 167–72.

Cheng, S. & Mok, S. H. (1971). Preliminary experiment of water hyacinth used as a medium for the cultivation of the padi straw mushroom. *Journal of the Horticultural Society of China*, **17,** 194–7. (In Chinese.)

Chua, S. E. & Ho, S. Y. (1973). Fruiting on sterile agar and cultivation of straw mushrooms (*Volvariella* species) on padi straw, banana leaves and sawdust. *World Crops*, **25,** 90–1.

Cooney, D. G. & Emerson, R. (1964). *Thermophilic Fungi. An Account of Their Biology, Activities and Classification.* San Francisco: W. H. Freeman.

Crisan, E. V. (1973). Current concepts of thermophilism and thermophilic fungi. *Mycologia*, **65,** 1171–98.

Delcaire, J. R. (1978). Economics of cultivated mushrooms. In *The Biology and Cultivation of Edible Mushrooms*, ed. S. T. Chang & W. A. Hayes, pp. 727–93. London: Academic Press.

Dennis, C. & Webster, J. (1971). Antagonistic properties of species groups of *Trichoderma.* 3. Hyphal interaction. *Transactions of the British Mycological Society*, **57,** 363–9.

Eddy, B. P. & Jacobs, L. (1976). Mushroom compost as a source of food for *Agaricus bisporus. The Mushroom Journal,* **38,** 56–9.

Eicker, A. (1977). Thermophilic fungi associated with the cultivation of *Agaricus bisporus* (Lange) Singer. *Journal of S. African Botany*, **43,** 193–207.

Eicker, A. (1980). Mesophilic fungi associated with the cultivation of *Agaricus bisporus. Transactions of the British Mycological Society*, **74,** 465–70.

Evans, H. C. (1971). Thermophilous fungi of coal spoil tips. II. Occurrence, distribution and temperature relationships. *Transactions of the British Mycological Society*, **57,** 255–66.

van Faasen, H. G. & van Dijk, H. (1979). Nitrogen conversions during the composting of manure/straw mixtures. In *Straw Decay and its Effect on Disposal and Utilisation,* ed. E. Grossbard, pp. 113–20. Chichester: John Wiley.

Fergus, C. L. (1964). Thermophilic and thermotolerant molds and actinomycetes of mushroom compost during peak heating. *Mycologia,* **56,** 267–84.

Fergus, C. L. (1971). The temperature relationships and thermal resistance of a new thermophilic *Papulaspora* from mushroom compost. *Mycologia*, **63,** 426–31.

Fergus, C. L. (1978). The fungus flora of compost during mycelium colonisation by the cultivated mushroom, *Agaricus brunnescens. Mycologia*, **70,** 636–44.

Fermor, T. R., Smith, J. F. & Spencer, D. M. (1979). The microflora of experimental mushroom compost. *Journal of Horticultural Science,* **54,** 137–47.

Fermor, T. R. & Wood, D. A. (1979). The microbiology and enzymology of wheat straw mushroom compost production. In: *Straw Decay and Its Effect on Disposal and Utilisation*, ed. E. Grossbard, pp. 105–12. Chichester: John Wiley.

Fermor, T. R. & Wood, D. A. (1981). Degradation of bacteria by *Agaricus bisporus* and other fungi. *Journal of General Microbiology*, **126**, 377–87.

Fordyce, C. (1970). Relative numbers of certain microbial groups present in compost used for mushroom (*Agaricus bisporus*) cultivation. *Applied Microbiology*, **20**, 196–9.

Frankland, J. C. (1966). The succession of fungi on decaying petioles of *Pteridium aquilinum*. *Journal of Ecology*, **54**, 41–63.

Garrett, S. D. (1951). Ecological groups of soil fungi: a survey of substrate relationships. *New Phytologist*, **50**, 149–60.

Garrett, S. D. (1962). Decomposition of cellulose in soil by *Rhizoctonia solani* Kühn. *Transactions of the British Mycological Society*, **45**, 115–20.

Garrett, S. D. (1963). *Soil Fungi and Soil Fertility*. Oxford: Pergamon Press.

Gerrits, J. P. G. (1969). Organic compost constituents and water utilised by the cultivated mushroom during spawn run and cropping. *Mushroom Science*, **7**, 111–26.

Gerrits, J. P. G. (1977). The significance of gypsum applied to mushroom compost, in particular relation to the ammonia content. *Netherlands Journal of Agricultural Science*, **25**, 288–302.

Gerrits, J. P. G. (1978). Influence of pH and ammonia in mushroom compost. *Mushroom Science*, **10**, 15–29.

Gerrits, J. P. G., Bels-Koning, H. C. & Muller, F. M. (1967). Changes in compost constituents during composting, pasteurisation and cropping. *Mushroom Science*, **6**, 225–43.

Gibbs, J. N. (1967). A study of the epiphytic growth habit of *Fomes annosus*. *Annals of Botany*, **31**, 755–74.

Gray, K. R. & Biddlestone, A. J. (1974). Decomposition of urban waste. In *Biology of Plant Litter Decomposition*. ed. C. H. Dickinson & G. J. F. Pugh, vol. 2, pp. 743–75. London: Academic Press.

Gray, K. R., Sherman, K. & Biddlestone, A. J. (1971*a*). A review of composting, Part I. *Process Biochemistry*, **6**, 32–6.

Gray, K. R., Sherman, K. & Biddlestone, A. J. (1971*b*). A review of composting. Part II. *Process Biochemistry*, **6**, 22–8.

Grime, J. P. (1979). *Plant Strategies and Vegetation Processes*. Chichester: John Wiley.

Hayes, W. A. (1968). Microbiological changes in composting wheat straw/horse manure mixtures. *Mushroom Science*, **7**, 173–86.

Hayes, W. A. (1974). Microbiological activity in the casing layer and its relation to productivity and disease control. In *The Casing Layer*, ed. W. A. Hayes, pp. 27–48. Mushroom Growers Association, London.

Hayes, W. A. (1978*a*). *Agaricus bisporus*. Biological nature. In *The Biology and Cultivation of Edible Mushrooms*, ed. S. T. Chang & W. A. Hayes, pp. 191–217. London: Academic Press.

Hayes, W. A. (1978*b*). Nutrition, substrates, and principles of disease control. In *The Biology and Cultivation of Edible Mushrooms*, ed. S. T. Chang & W. A. Hayes, pp. 219–37. London: Academic Press.

Hayes, W. A. (1980). Solid state fermentation and the cultivation of edible fungi. In *Fungal Biotechnology*, ed. J. E. Smith, D. R. Berry & B. Kristiansen, pp. 175–202. London: Academic Press.

Hayes, W. A. & Lim, W. C. (1979). Wheat and rice straw composts and mushroom production. In *Straw Decay and Its Effect on Disposal and Utilisation*, ed. E. Grossbard, pp. 85–94. Chichester: John Wiley.

Hayes, W. A. & Nair, N. G. (1975). The cultivation of *Agaricus bisporus* and other edible mushrooms. In *The Filamentous Fungi*, ed. J. E. Smith & D. R. Berry, vol. 1, pp. 212–48.

Hayes, W. A., Randle, P. E. & Last, F. T. (1969). The nature of the microbial stimulus affecting sporophore formation on *Agaricus bisporus* (Lange) Sing. *Annals of Applied Biology*, **64**, 177–87.

Hedger, J. N. (1975). The ecology of thermophilic fungi in Indonesia. In *Biodégradation et Humification*, ed. G. Kilbertus, O. Reisinger, A. Mourey & J. Cancela da Fonseca, pp. 59–65. Sarreguemines: Editions Pierron.

Hedger, J. N. & Hudson, H. J. (1974). Nutritional studies of *Thermomyces lanuginosus* from wheat straw composts. *Transactions of the British Mycological Society*, **62**, 129–43.

von Hofsten, B. (1976). Cultivation of a thermotolerant basidiomycete on various carbohydrates. In *Food from Waste*, ed. G. G. Birch, K. J. Parker & J. T. Worgan, pp. 156–66. London: Applied Science Publishers.

Hu, K. J., Song, S. F., Liu, O. & Peng, J. T. (1974). Studies on sugar cane rubbish for Chinese mushroom culture and its growth factor. *Mushroom Science*, **9**, 691–700.

Hudson, H. J. (1968). The ecology of fungi in plant remains above the soil. *New Phytologist*, **67**, 837–74.

Hudson, H. J. (1975). Secondary saprophytic sugar fungi. In *Biodégradation et Humification*, ed. G. Kilbertus, O. Reisinger, A. Mourey & J. Cancela da Fonseca, pp. 15–18. Sarreguemines: Editions Pierron.

Hume, D. P. & Hayes, W. A. (1972). The production of fruit body primordia in *Agaricus bisporus* (Lange) Sing. on agar media. *Mushroom Science*, **8**, 527–32.

Ikediugwu, F. E. O., Dennis, C. & Webster, J. (1970). Hyphal interference by *Peniophora gigantea* against *Heterobasidion annosum. Transactions of the British Mycological Society*, **54**, 307–9.

Ikediugwu, F. E. O. & Webster, J. (1970). Hyphal interference in a range of coprophilous fungi. *Transactions of the British Mycological Society,* **54**, 205–10.

Käärik, A. (1975). Succession of microorganisms during wood decay. In *Biological Transformation of Wood by Microorganisms*, ed. W. Liese, pp. 39–51. Berlin: Springer-Verlag.

Kemp, R. F. O. (1974). Bifactorial incompatibility in the two-spored basidiomycetes *Coprinus sassii* and *C. bilanatus. Transactions of the British Mycological Society*, **62**, 547–55.

Kühner, R. & Romagnesi, H. (1953). *Flore analytique des champignons supérieures*. Paris: Editeurs Masson et Cie.

Kurtzman, R. H., Jr. (1978). *Coprinus fimetarius*. In *The Biology and Cultivation of Edible Mushrooms*, ed. S. T. Chang & W. A. Hayes, pp. 393–408. Chichester: John Wiley.

Kuthubutheen, A. J. (1980). Physiological studies on thermophilous fungi and their occurrence in the tropics. *Abstracts of the Second International Symposium on Microbial Ecology*, p. 67. University of Warwick, UK.

Kuthubutheen, A. J. & Pugh, G. J. F. (1977). The effects of fungicides on the occurrence of thermophilous leaf surface fungi. *Mycopathologia*, **62**, 131–41.

Lambert, E. B. (1941). Studies on the preparation of mushroom compost. *Journal of Agricultural Research*, **62**, 415–22.

Lindeberg, G. (1948). On the occurrence of polyphenol oxidases in soil inhabiting basidiomycetes. *Physiologia Plantarum*, **1**, 196–205.

Long, P. W. & Jacobs, L. (1974). Aseptic fruiting of the cultivated mushroom, *Agaricus bisporus. Transactions of the British Mycological Society*, **63**, 99–107.

MacArthur, R. H. & Wilson, E. O. (1967). *The Theory of Island Biogeography*. New Jersey: Princeton University Press.

Mangenot, M. F. (1952). Recherches méthodiques sur les champignons de certains bois en décomposition. *Revue générale de Botanique*, **59**, 381–99, 477–519.
Manning, K. & Wood, D. A. (1978). Cellulose production and regulation by *Agaricus bisporus. Proceedings of the Society for General Microbiology*, **5**, 99.
Miehe, H. (1907). *Die Selbsterhitzung des Heus, eine biologische Studie.* 127 pp. Jena: Gustav Fischer.
Ofosu-Asiedu, A. & Smith, R. S. (1973). Degradation of three softwoods by thermophilic and thermotolerant fungi. *Mycologia*, **65**, 240–4.
Olson, J. S. (1963). Energy storage and the balance of producers and decomposers in ecological systems. *Ecology*, **44**, 322–31.
Orton, P. D. (1957). Notes on British Agarics. 1–5 (Observations on the genus *Coprinus*). *Transactions of the British Mycological Society*, **40**, 263–276.
Orton, P. D. & Watling, R. (1979). *British Fungus Flora. Agarics and Boleti 2. Coprinaceae: Coprinus.* Edinburgh: Her Majesty's Stationery Office.
Park, J. Y. & Agnihotri, V. P. (1969). Bacterial metabolites trigger sporophore formation in *Agaricus bisporus. Nature, London*, **222**, 984.
Paul, E. A. & Voroney, R. P. (1980). Nutrient and energy flows through soil microbial biomass. In *Contemporary Microbial Ecology*, ed. D. C. Ellwood, J. N. Hedger, M. J. Latham, J. M. Lynch & J. H. Slater, pp. 215–37. London: Academic Press.
Penn, D. J. (1977). Some studies on the competitive colonisation of cellulosic substrates by microorganisms. M. Phil. thesis, University of Aston in Birmingham.
Perrier, A. (1929). Sur la présence de certains champignons thermophiles dans la fumier et les matières organiques en décomposition. *Comptes rendus Hebdomadaires des Séances del'Académie des Sciences*, **188**, 1426–9.
Pinto-Lopes, J. & Almeida, M. G. (1971). '*Coprinus lagopus*'. A confusing name as applied to several species. *Portugalia Acta Biologica*, **11** (B), 167–204.
Pugh, G. J. F. (1980). Presidential Address: Strategies in fungal ecology. *Transactions of the British Mycological Society*, **75**, 1–14.
Rao, S. N. & Block, S. S. (1962). Experiments on small scale composting. In *Developments in Industrial Microbiology*, vol. 3, pp. 326–40. Society for Industrial Microbiology. New York: Plenum Press.
Rayner, A. D. M. (1978). Interactions between fungi colonising hardwood stumps and their possible role in determining patterns of colonisation and succession. *Annals of Applied Biology*, **89**, 131–4.
Rege, R. D. (1927). Biochemical decomposition of cellulosic material with special reference to the action of fungi. *Annals of Applied Biology*, **14**, 1–14.
Saono, S. & Sastrapradja, D. (1979). Major crop residues in Indonesia and their potentialities as raw materials for bioconversion. *Workshop on bioconversion of lignocellulose and carbohydrate residues in rural communities*, Denpasar, Bali, pp. 1–13. Lembaga Biologi Nasional, Bogor, Java.
Sinden, J. W. (1971). Ecological control of pathogens and weed moulds in mushroom culture. *Annual Review of Phytopathology*, **9**, 411–32.
Sinden, J. W. & Hauser, E. (1950). The short method of composting. *Mushroom Science*, **1**, 52–9.
Skidmore, A. M. & Dickinson, C. H. (1976). Colony interactions and hyphal interference between *Septoria nodorum* and phylloplane fungi. *Transactions of the British Mycological Society*, **66**, 57–64.
Smith, J. F. & Spencer, D. M. (1976). Rapid preparation of composts suitable for the production of the cultivated mushroom. *Scientia Horticulturae*, **5**, 23–31.
Stanek, M. (1972). Microorganisms inhabiting mushroom compost during fermentation. *Mushroom Science*, **8**, 787–811.

Sundman, V. & Näse, L. (1971). A simple plate test for direct visualisation of biological lignin degradation. *Paper and Timber*, **53,** 67–71.

Swift, M. J. (1976). Species diversity and the structure of microbial communities in terrestrial habitats. In *The Role of Terrestrial and Aquatic Organisms in Decomposition Processes*, ed. J. M. Anderson & A. Macfadyen, pp. 185–222. Oxford: Blackwell Scientific Publications.

Swift, M. J., Heal, O. W. & Anderson, J. M. (1979). *Decomposition in Terrestrial Ecosystems.* Studies in Ecology, vol. 5. Oxford: Blackwell Scientific Publications.

Tansey, M. R. (1971). Isolation of thermophilic fungi from self-heating industrial wood chip piles. *Mycologia*, **58,** 537–47.

Tansey, M. R. (1973). Isolation of thermophilic fungi from alligator nesting material. *Mycologia*, **65,** 594–601.

Tansey, M. R. & Brock, T. D. (1978). Microbial life at high temperatures: ecological aspects. In *Microbial Life in Extreme Environments*, ed. D. Kushner, pp. 159–215. London: Academic Press.

Tansey, M. R. & Jack, M. S. (1975). *Thielavia australiensis* sp. nov. a new thermophilic fungus from incubator-bird (Mallee fowl) nesting material. *Canadian Journal of Botany*, **53,** 81–83.

Treschow, C. (1944). Nutrition of the cultivated mushroom. *Dansk Botanisk Arkiv*, **11,** 1–80.

Tribe, H. T. (1966). Interactions of soil fungi on cellulose film. *Transactions of the British Mycological Society*, **49,** 457–66.

Turner, E. M. (1974) Phenoloxidase activity in relation to substrate and development stage in the mushroom *Agaricus bisporus. Transactions of the British Mycological Society*, **63,** 541–7.

Turner, E. M., Wright, M., Ward, T., Osborne, D. J. & Self, R. (1975). Production of ethylene and other volatiles and changes in cellulase and laccase activities during the life cycle of the cultivated mushroom. *Agaricus bisporus. Journal of General Microbiology*, **91,** 167–76.

de Vries, C. D. (1973). Mushroom cultivation. *Tropical Abstracts*, **29,** 849–57.

Waksman, S. A. & Cordon, T. C. (1939). Thermophilic decomposition of plant residues in composts by pure and mixed cultures of organisms. *Soil Science*, **47,** 81–112.

Waksman, S. A. & McGrath, J. M. (1931). Preliminary study of chemical processes in the decomposition of manure by *Agaricus campestris. American Journal of Botany*, **18,** 573–81.

Waksman, S. A. & Nissen, W. (1932). On the nutrition of the cultivated mushroom and the chemical changes brought about by this organism in manure compost. *American Journal of Botany*, **19,** 514–37.

Watson, J. M. (1973). The beneficial effect of certain phosphate sources on commercial mushroom yields. *The Mushroom Journal*, **10,** 462–3, 466–8.

Wood, D. A. (1979). Degradation of composted straw by the edible mushroom, *Agaricus bisporus.* Enzyme activities associated with mycelial growth and fruit body function. In *Straw Decay and Its Effect on Disposal and Utilisation*, ed. E. Grossbard, pp. 95–104. Chichester: John Wiley.

Wood, D. A. & Goodenough, P. W. (1977). Fruiting of *Agaricus bisporus.* Changes in extracellular enzymes during growth and fruiting. *Archives of Microbiology*, **114,** 161–5.

Wood, D. A. & Hammond, J. B. W. (1977). Ethylene production by axenic cultures of *Agaricus bisporus. Applied and Environmental Microbiology*, **34,** 228–9.

Wood, D. A. & Matcham, S. E. (1980). Growth of *Agaricus bisporus* in composted straw.

Evidence for microbial interaction. *Abstracts of the 2nd International Symposium on Microbial Ecology*, p. 233. University of Warwick, UK.

Yee, N. T. & Chang-Ho, Y. (1980). Interaction between *Volvariella volvacea* and some weed fungi. *Transactions of the British Mycological Society*, **75,** 498–501.

Yong, Y. C. & Graham, K. M. (1973). Studies on the padi mushroom (*Volvariella volvacea*). 1. Use of oil palm pericarp waste as an alternative substrate. *Malaysian Agricultural Research*, **2,** 15–22.

Yong, Y. C. & Graham, K. M. (1974). Studies on the padi mushroom (*Volvariella volvacea*). 3. Effects of urea on mycelial development and upon yield in oil palm pericarp waste. *Malaysian Agricultural Research*, **3,** 7–11.

16 Basidiomycetes as components of forest ecosystems

M. J. SWIFT
Queen Mary College, University of London, Mile End Road, London E1 4NS, England

The decomposition subsystem

Decomposition processes are an essential functional component of all ecosystems with two main aspects of significance. Firstly, decomposition is the principal process whereby essential nutrient elements are made available to the primary producers; decomposer organisms are the main agents of transformation of nutrient elements from the organic (immobilised) state to the inorganic (mineralised) state. Secondly, decomposition processes play a major part in the formation of humus molecules.

The processes of decomposition may be regarded as an integrated subsystem intimately linked with other functional components of the ecosystem as illustrated in Fig. 1, where the relationships between the *decomposition subsystem*, the *plant subsystem* and the *herbivore subsystem* are shown. Decomposition and herbivore subsystems compete in the sense that they both depend on the input of carbon, energy and nutrient from the primary producers. Ultimately, however, products of the herbivore food chains, such as carcasses and faeces, enter the decomposition subsystem. In most terrestrial ecosystems, forest systems in particular, the herbivore subsystem is quantitatively relatively insignificant and the major proportion of energy released from the primary producers enters the decomposition subsystem directly as *plant litter*.

The decomposition of plant litter in terrestrial ecosystems is brought about by a diverse community of organisms which includes fungi, bacteria, protista and invertebrate animals. Detailed analysis of the nature of decomposition processes reveals a high degree of complexity in the structure of decomposer food webs, and subdivision into a series of well-defined trophic groups in the manner of, say, the herbivore –

carnivore food webs of marine ecosystems is rarely possible. Thus it is both unrewarding and misleading to attempt a general definition of the rôle of basidiomycetes in decomposition processes. Different basidiomycetes undoubtedly play different rôles, and under varying circumstances the same species may play different rôles. It is nevertheless possible to show that the basidiomycetes as a group exhibit some adaptive features which are of particular significance in decomposition processes. The specific features of these adaptations – physiological, biochemical, morphological and genetic – are reviewed elsewhere in this volume. My purpose is to relate these aspects to the functioning of the decomposition subsystem as a whole, and in particular to compare and contrast them to the rôles of the other members of the decomposer community.

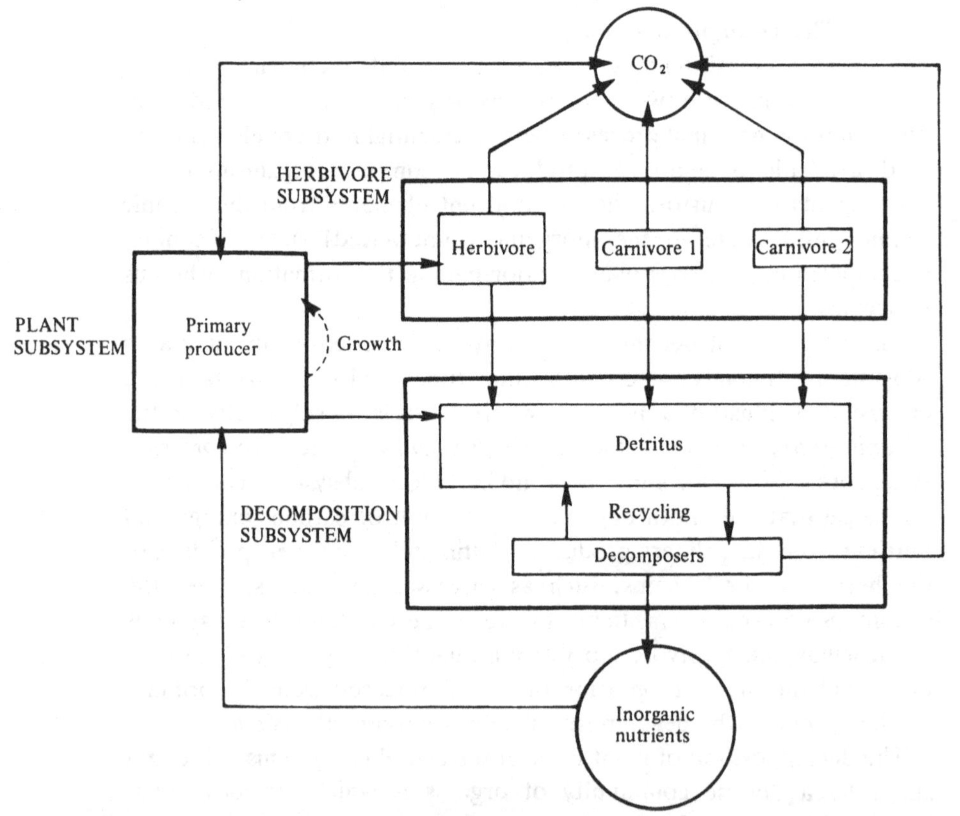

Fig. 1. The relationship of decomposition to other ecosystem components. The ecosystem is divided into three subsystems with the main flows of matter between and within them shown by the arrows. (After Swift, Heal & Anderson, 1979.)

Fungal rôles in ecosystems

An initial problem is deciding which of the wide diversity of basidiomycete fungi are decomposers *sensu stricto* (Watling: Chapter 1), and furthermore the extent to which the non-decomposers are nevertheless significant in decomposition processes. For these purposes, the ecological rôles of fungi, like those of other heterotrophic micro-organisms, may be considered under two broad headings: those that are *symbiotic* with plants or animals, and those that are *free-living*. Symbiotic fungi may be categorised further as *antagonistic* (e.g. parasites) or *mutualistic* according to the basis of their association with the host. Cooke (1977) listed a third category, *neutral*, where persistent associations exist which apparently do not affect the host in any beneficial or harmful way. The free-living fungi are saprotrophs, i.e. unequivocal decomposers obtaining their carbon, energy and nutrient from dead organic material whether it is of plant, animal or even microbial origin (see Swift, Heal & Anderson, 1979, p. 75).

Figure 2 depicts in a highly simplified form some of the general inter-relationships of the microbial, plant and animal components of ecosystems, as illustrated by linking them through transfers of matter and energy. There are two interesting features of this network that are relevant to a discussion of the rôle of basidiomycetes. The first is the clear status of micro-organisms as the primary processors of plant material. Even when plant material is taken directly by animals (e.g. herbivores), a major part of the chemical processing may be carried out by symbiotic micro-organisms. The second feature is the fact that

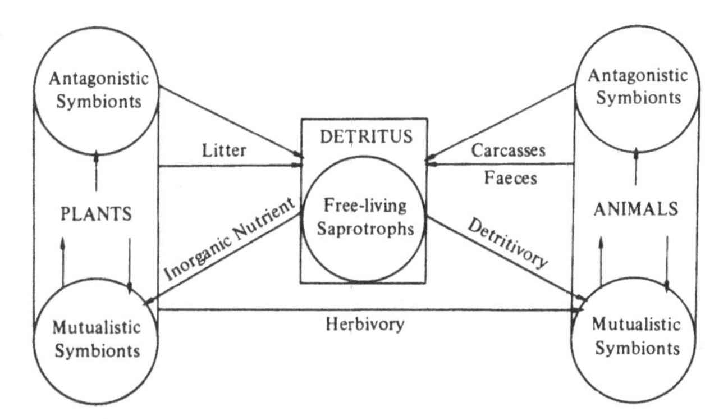

Fig. 2. Associations between plants, animals and micro-organisms. The arrows show some of the main pathways of matter exchange among the different groups of organisms. Note how the ecological rôles of the micro-organisms are such as to place them as intermediaries in the trophic links between plants and animals.

activity in decomposition processes is by no means confined to the saprotrophic free-living microbes. Many of the symbionts, for instance the necrotrophic antagonistic symbionts of plants and the mutualistic symbionts of animals, participate directly in decomposition, and even the mutualistic symbionts of plants such as mycorrhizal fungi influence and are influenced by decomposition processes because of their rôle in nutrient cycling. Representatives of the basidiomycete fungi can be demonstrated in all the ecological groups envisaged in Fig. 2 (Cooke, 1977). The distribution is uneven, however, and basidiomycetes are most commonly found as associates with forest trees – as saprotrophs, as mycorrhizal symbionts or as necrotrophic parasites. These may be contrasted with the basidiomycete associations with grasses or herbs which are more characteristic of the four orders of biotrophic parasites.

The characteristics of the necrotrophic parasites of trees have been reviewed by Mercer (Chapter 8). Reference to them raises the problem of defining a time-scale for decomposition. The processes of decomposition are initiated in many cases (perhaps indeed in all) by parasitic organisms. Those that are necrotrophic may persist significantly into the later decomposition phases (Hudson, 1968). On the other hand those that are biotrophic, such as the rusts, are rapidly succeeded by saprotrophs. For the necrotrophs in particular the point at which parasitism ceases and decomposition begins may be difficult to define. This is more than a semantic point because it is related to substantial practical problems in determining the initiation of the decomposition processes and thence the determination of the ecological significance of the organisms associated with it. For instance, litter-fall has often been taken as indicative of the start of decay of plant matter. With regard to wood decay this is clearly quite erroneous (Mercer: Chapter 8). Swift *et al.* (1976) provided quantitative evidence of this error when they showed, for a sample of 126 small branches, a mean weight loss of 43% before litter fall. The microflora of these branches was most commonly dominated by basidiomycetes and showed evidence of having passed through some of the earlier stages of succession reviewed by Mercer (see also Fig. 3). Leaves may also be subjected to decomposition before fall, but this is more difficult to determine quantitatively. A good example is the extensive colonisation and damage caused by *Lophodermium* species and other fungi on needles of *Pinus* species. Although there is little obvious cell-wall degradation, the attack affects the subsequent course of decomposition, including colonisation by basidiomycetes (Williamson, Mitchell & Millar, 1976; Mitchell & Millar, 1978).

Even more difficult to determine is the end-point of the decomposition process. One approach is to define it as the stage at which the original entity (e.g. the branch or the leaf) is no longer physically distinguishable. This concept correlates to some extent with that of the distinction between 'litter' and 'humus' horizons of soil (Hering: Chapter 12). In practice it is difficult to determine with any precision. 'When is a branch not a branch?' may well be asked when decomposition proceeds to a point where all that is left is a casing of bark filled with an amorphous particulate mass, visually indistinguishable from the humus layer of the soil on which it lies. Once again this poses problems of a practical as well as a conceptual nature in determining the limits of the rôle of any group of decomposers such as the basidiomycetes.

Basidiomycetes as decomposers

The decomposition of plant litter within a forest ecosystem may be due to the combined activities of several hundred species of organism. Within this diverse community, basidiomycetes are thought to play an important and central rôle. This conclusion derives from three main sources: the evidence of their physiological capacity for degradation of plant cell-wall polymers (Kirk & Fenn: Chapter 4; Montgomery: Chapter 3); the evidence of their status in the succession of organisms on decaying plant matter (Hering: Chapter 12; Mercer: Chapter 8), and the quantitative evidence from study of their production ecology (Frankland: Chapter 14).

The annual above-ground input of organic matter to the decomposer subsystem in forest ecosystems ranges from about 1 t ha^{-1} to about 20 t ha^{-1}. Of this input about 25% is woody and the remainder is leaves, fruit and other reproductive structures (Bray & Gorham, 1964). The input of wood to decomposers is actually underestimated by measuring litter-fall because, as mentioned above, it is about 40% decayed before reaching the forest floor. The correct proportion is thus probably nearer 35%. I estimate that over 70% of the total input of above-ground plant litter is in the form of cell-wall polymers, of which about 20–25% is lignin. The composition of the below-ground input is more difficult to calculate; there is a large input of soluble organic exudates, and there is also a substantial lignified input as dead roots. The main conclusion is inescapable, however: the bulk of the energy and carbon available to decomposers is in the form of cellulose, hemicelluloses and lignin.

Basidiomycetes are the best adapted of all decomposers to annex these materials. Whilst there are a wide range of organisms other than basidiomycetes with some degree of polysaccharase activity, and a small number with ligninolytic ability, there is no other group of organisms that possesses the full enzyme battery needed for complete degradation of secondary plant cell walls, nor is there any other group with these capacities so widely distributed. Perhaps more importantly, the detailed studies of enzyme mechanisms that are now being carried out (Kirk & Fenn: Chapter 4; Montgomery: Chapter 3) show that the basidiomycete systems may also be uniquely capable of initiating attack on *intact* cell walls. This appears to be particularly so in lignified walls where the attack of other organisms, such as soft-rot fungi or bacteria, is confined to the erosion of the wall surface or to the dissolution of limited zones within the cell walls.

On the basis of biochemical ability then, we can hypothesise that basidiomycetes should play a primary rôle in decomposition. This conclusion is not so readily substantiated by classical isolation studies. It has commonly been shown, in successional investigations for instance, that basidiomycetes are preceded by other fungi and even by some groups of bacteria. From this it is argued that they play only a secondary rôle (Hudson, 1968). It is well established, however, that studies based on isolation onto agar media will show a bias against the generally slow-growing basidiomycete mycelium, and in favour of rapidly growing and sporulating forms such as the Mucorales or Hyphomycetes (Warcup, 1965). It is thus relatively rare to isolate basidiomycetes from leaf litter, even when microscopic examination shows the mycelium to be abundant (Frankland: Chapter 14). The conclusion that basidiomycetes are only active in the late stages of leaf decomposition may therefore be misleading. The same problems occur with the alternative technical approach of describing succession in terms of fruiting, which may be promoted by incubating plant litter or other organic matter in a damp chamber. An elegant demonstration of how misleading these data can be was given by Harper & Webster (1964), who showed that although basidiomycetes fruit at the termination of the coprophilous succession their mycelium is active earlier.

Figure 3 shows isolation data for micro-organisms occupying small branches. Basidiomycetes were dominant at branch-fall, and for some time preceding and following it. Although it is not possible to determine with certainty the exact period of dominance, it probably encompasses a range of weight loss from about 20% to 60%, arguing again for a primary and important quantitative rôle. The terminal 40% of decomposition is largely influenced by the activities of wood-boring animals.

The quantitative rôle of basidiomycetes may also be assessed by looking at data for their biomass and production. Frankland (Chapter 14) has pointed out the difficulties associated with this and emphasised that current data must be viewed with caution. Her calculations show that basidiomycete mycelium is quantitatively the largest component of living microbial biomass in plant litters. Nevertheless, the estimated mycelial production of leaf-litter basidiomycetes (10 kg ha^{-1} yr^{-1}) would account for the decomposition of only about 0.3% of the annual input of cell-wall materials at the study site in question (Meathop Wood, UK). From this we may conclude either that our assumption that basidiomycetes are the primary decomposers of cell-wall materials is wrong or that the production value is an underestimate. The latter may be the case as the production estimate assumes that the living biomass only turns over twice in any one year. Evidence presented on p. 333 suggests that this may be a considerable underestimate.

The structure of basidiomycete communities

Before considering the ways in which basidiomycetes interact with other decomposer organisms, it is appropriate to review some of the main features of the basidiomycete communities of woodlands. In doing this we should take into account all phases of their life-cycle, including the vegetative mycelium, asexual multiplicative stages such as conidia, asexual survival stages such as chlamydospores, and sexual reproductive stages including the formation of fruit bodies and basidiospores. Swift (1976) suggested that the basidiomycete life-cycle represented a *K*-selection strategy in contrast to many other groups of fungi (notably the Mucorales) which are more characteristic of the *r*-selection type. Broadly speaking this distinguishes between organisms adapted to rapid response to rapidly varying circumstances (*r*-selection) and those adapted to exploitation of more stable habitats and resource opportunities over a long period of time (*K*-selection). The recognition of basidiomycetes as representing the latter was based on the supposed perennial nature of many basidiomycete mycelia, their slow growth rate, and the long period of time between basidiospore and basidiospore. Lussenhop (quoted by Wicklow, 1981) has questioned this distinction, and pointed out that many fungi may combine the traits of both *r* and *K*-strategists. For instance, the production of multiple generations of asexual spores provides a mechanism of rapid response

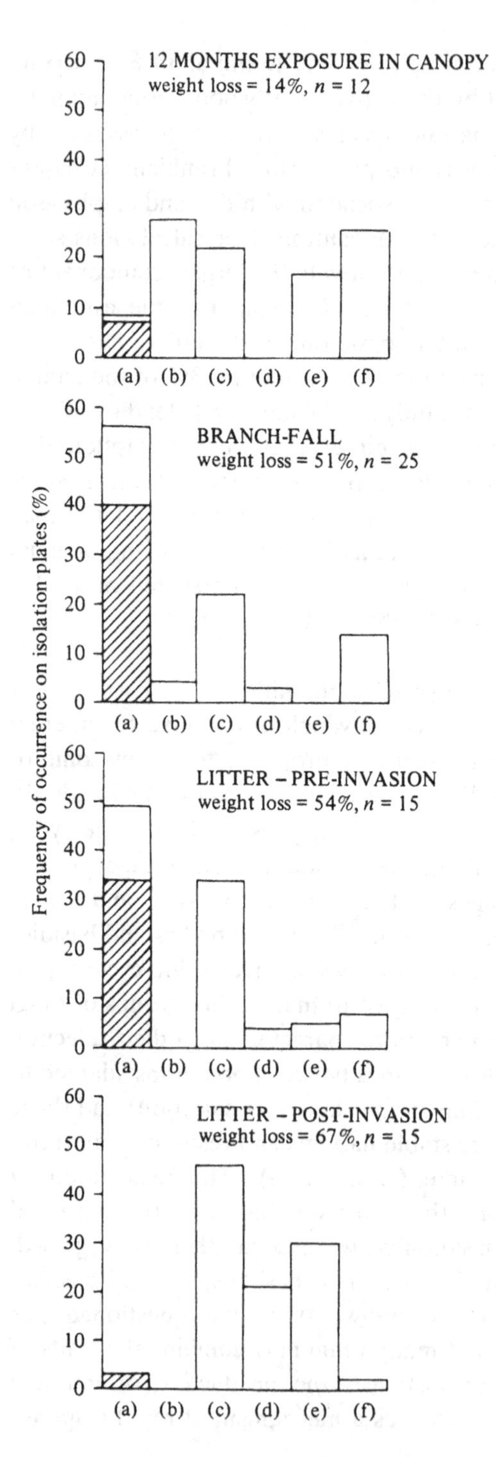
12 MONTHS EXPOSURE IN CANOPY
weight loss = 14%, n = 12
BRANCH-FALL
weight loss = 51%, n = 25
LITTER – PRE-INVASION
weight loss = 54%, n = 15
LITTER – POST-INVASION
weight loss = 67%, n = 15
Frequency of occurrence on isolation plates (%)
0
10
20
30
40
50
60
(a)
(b)
(c)
(d)
(e)
(f)

'nested' within the longer term sexual reproductive strategy. This is clearly a reasonable objection, because the 'dual strategy' is a great deal more common in the basidiomycetes than is generally realised, as Kendrick & Watling (1979) have recently pointed out. A good example of this, which is very familiar, is the well documented necrotroph of forest trees *Heterobasidion annosum*.

Patterns of mycelial distribution

The integrated functioning mycelium has always been the mycologist's concept of a fungal 'individual'. As Rayner & Todd (Chapter 6) pointed out, however, it has been difficult in the past to justify this idea in natural populations. Their elegant work has clarified the situation within one group of decomposer fungi where there are morphological, genetic, and presumably physiological correlations in the delimitations of mycelia. Populations of wood-decay fungi are indeed attractive sources of study in this respect, because the physical limits of the mycelial population are determined by those of the piece of wood it occupies. Other habitats such as leaf litter present more formidable sampling problems, for mycelia may stretch beyond the confines of a single leaf, occupying large and indeterminate areas of the litter layer and soil.

An alternative to the direct approach to the sampling of such mycelia by isolation or incubation techniques is to use fruit bodies as markers of

---

Fig. 3. Succession of micro-organisms in decomposing branch-wood. The histograms show the relative frequency of isolation of fungi from samples of *Quercus* branches at varying stages of decomposition. Also shown is the mean percentage weight loss measured in terms of the change in relative density of the branches and the number of branches in each sample. Isolations were made by placing small fragments of wood, dissected aseptically from the interior of the branch, onto plates of malt agar. Fifteen fragments were plated out for each branch (one per plate) so that, for instance, the total number of plates for the branch-fall sample was 375; an isolation frequency of 56% therefore implies that basidiomycetes occurred on 210 of the plates. The basidiomycete category includes those that were unequivocally identified as such by the presence of clamp connections on the mycelium or by comparison with cultures of known species (hatched) plus those that are putative basidiomycete mycelium only (unhatched). The histograms also show the frequency of bacterial colonies on the plates. This is a minimal figure as malt agar is not a satisfactory medium for bacteria. (a) Basidiomycetes; (b) Sphaeropsidales; (c) Moniliales; (d) Mucorales; (e) bacteria; (f) no microbial growth. (M. J. Swift & M. E. Nesbitt, unpublished.)

the mycelium. This has been done by using genetic markers, as in the work of Burnett & Evans (1966) where the genetic identity of widely spaced fruit bodies of *Marasmius oreades* in a fairy-ring confirmed the initial unity of the mycelium from which they were derived. Parker-Rhodes (1954) has similarly demonstrated the existence of polymorphism in populations of *Mycena galopus* within woodlands in Suffolk and Essex. By implication the distinct fruit body populations were derived from genetically (and thus ecologically) distinct mycelia.

A different approach, which gives some clues to the spatial distribution of mycelia in leaf litter, is shown in Fig. 4. Records were made at fortnightly intervals of fruiting for the two commonest leaf-litter basidiomycetes, *Collybia peronata* and *Mycena galopus*, at Blean Wood National Nature Reserve, Kent, during the autumns of 1967 and 1968. Fig. 4(*a*) shows the total record of fruiting occasions for *C. peronata* in 1968 as located within a 25×32 m grid of 1 m^2 quadrats. Fruiting occasions were used as the unit of counting in this species rather than total number of fruit bodies because of the tendency of the species to fruit in 'clusters'. As the intent was to show by the spatial patterns of fruiting the possible extent of mycelia, care was taken to distinguish between isolated fruiting and clustered fruiting within the same quadrat. Duplication of records was avoided by destroying fruit bodies after noting their appearance. The two-week interval between records probably gave an underestimate of the total fruiting occasions. Similarly the removal of fruit bodies was likely to affect the frequency of subsequent fruiting in the same mycelium, though with less predictable consequences. With these reservations in mind, the data can nevertheless be interpreted as showing evidence of a number of spatially separated mycelia all somewhat less than 3 m^2 in area. These fruiting patches may themselves have been heterogeneous (e.g. some of them may have been the opposite margins of larger mycelia analogous to 'fairy rings'). This could have been determined by genetical analysis in the manner of Rayner & Todd (Chapter 6), and the other workers cited by them. The linking of the techniques of quantitative ecology with an analysis of the morphological and genetic variation of the populations offers exciting possibilities for basidiomycete ecology. For instance, the recurrence of regular spatial features in fruiting may be detectable by pattern analysis (Kershaw, 1969; Greig-Smith, 1964) in appropriately designed sampling systems.

In Fig. 4(*b*) an example is shown of the mapping of both the above leaf-litter basidiomycetes, this time on a grid of 0.25 m^2 quadrats which

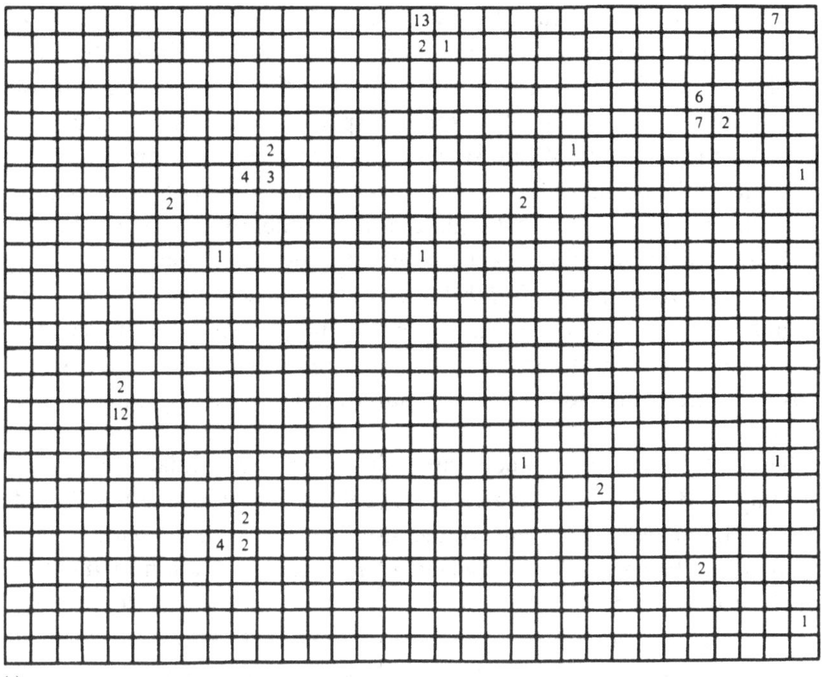

(*a*)

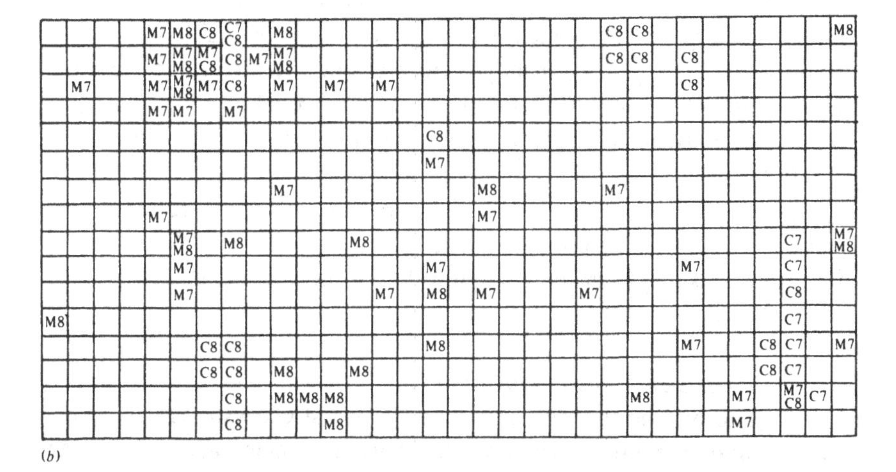

(*b*)

Fig. 4. Spatial distribution of fruit bodies. (*a*) Grid showing the numbers of occasions of fruiting of *Collybia peronata* within a series of 1 × 1 m quadrats during the autumn of 1968. (*b*) A grid of 0.5 × 0.5 m quadrats showing the positions of fruiting of *C. peronata* in 1967 (C7) and 1968 (C8) and of *Mycena galopus* in the same years (M7, M8). (M. J. Swift & A. Russell-Smith, unpublished.)

is 16 by 32 units in area. The numbered squares show the locations of fruiting of the two species (C and M) in 1967 and 1968. The contingency tables given in Table 1 provide the basis of analysis by the χ^2 test of two questions. The first is whether the two species showed any interaction, i.e. did the presence of one species in a particular location inhibit (or stimulate) the presence of the other? The χ^2 value is very low, indicating that the two species were distributed independently, i.e. at random in relation to one another. A word of caution is necessary here. The number of fruitings of *Collybia* was so low that the test is not very significant; this type of association analysis may however be of considerable value in other areas where fruiting frequency is higher. It would also be interesting to record the 'succession' of fruiting in the same location.

The left-hand contingency table refers to the spatial pattern of fruiting of *M. galopus* in the two years of study. The χ^2 is significant, and shows that in 1968 the fungus fruited more frequently in the same square as in 1967 than would be expected if the fruiting location in the two years was completely independent. This may be thought to provide evidence for the perennation of mycelia from year to year. Reference to the table, however, shows that the observed frequency of repeated fruiting was only five out of a possible total of twenty-one in 1968. Furthermore, fruiting in 1968 was only on a further four occasions adjacent to a quadrat in which fruiting occurred in 1967. This may argue for dispersal through time from year to year both by survival in the same location and following dispersal in space by mycelial growth or spores. Genetical analysis of the fruit bodies would again help to clarify the situation.

Population studies of this kind reinforce the evidence from production data that a large proportion of the total resource volume in the decomposer subsystem is occupied by a mosaic of basidiomycete mycelia.

Co-existence and competition

The basidiomycete community of forest ecosystems is species-diverse, and one might ask how the various species are enabled to co-exist when functionally their rôles may be very similar. In an earlier review I considered this problem for decomposer micro-organisms as a whole and suggested that there were four main categories of factor which defined decomposer niches: resource specificity; substrate specificity; environmental specificity, and colonisation strategy based on life-cycle patterns, particularly phenology (Swift, 1976).

Table 1. *Analysis of the spatial fruiting patterns of* Mycena galopus *and* Collybia peronata *by contingency tables and the χ^2-test.* See Fig. 4(*b*) and the text for details of the mapping procedure

M. GALOPUS: FRUITING SCORES BY YEAR

Number of quadrats with:	
fruit bodies in 1967 alone	30
fruit bodies in 1968 alone	16
fruit bodies in both years	5
fruit bodies in neither year	461
	$n = 512$

1968 \ 1967	+	−	
+	a 5 (E = 1.44)	b 16 (19.56)	21
−	c 30 (33.56)	d 461 (457.44)	491
	35	477	512

Using exact solution, $\chi^2 = 9.91$; $P = <0.01$

M. GALOPUS vs. *C. PERONATA*: YEARS COMBINED

Number of quadrats with:	
fruit bodies of *M. galopus* alone	49
fruit bodies of *C. peronata* alone	26
fruit bodies of both species	2
fruit bodies of neither species	435
	$n = 512$

C. peronata \ *M. galopus*	+	−	
+	a 2 (E = 2.79)	b 26 (25.21)	28
−	c 49 (48.21)	d 435 (434.79)	484
	51	461	512

$\chi^2 = 0.26$; $P = <0.5–0.3$

Resource-specificity. A *resource*, as used here, is simply any unit of organic matter occupied by a decomposer organism, and is contrasted with *substrate*, which is defined in this chapter as any specific chemical molecule utilised by the organism. Resource diversity may be expressed at a number of levels – the species of origin, and the variation in its physical and chemical composition being the most useful determinants. Thus we may speak of the resource-specific decomposer communities of twigs, leaves, branches, faeces, carcasses, etc., and moreover of fungi specific to particular taxa of such resource types, e.g. *Quercus* versus *Fraxinus* versus *Betula* leaves. Among the basidiomycetes we find both *resource-specific* types and others that are very widely distributed. For instance, the basidiomycete flora of branches is regarded in most cases as being distinct from that of leaves. Casual observation suggests, however, that in some cases extensive decomposition of leaves in the vicinity of a branch may occur by the agency of the fungus occupying the branch. This is often indicated by the presence of mycelial cords running between the two resource types (A. D. M. Rayner, personal communication). The phenomenon of resource-specificity needs more careful documentation, but clearly it may be one way whereby a large proportion of the basidiomycete community avoids interspecific competition.

Substrate specificity. Substrate specificity has been suggested as one of the ways in which microbes occupying the same resource (i.e. members of the same '*unit-community*', Swift, 1976) are enabled to co-exist by utilising different chemical components. It is doubtful whether this operates at all commonly in the saprotrophic basidiomycetes, most of which are probably 'after' the same substrates – the polysaccharides of the cell walls. However, there may be a degree of spatial separation on the basis of specific chemical relationships. For instance, fungi differ in their susceptibility to the various fungistatic and fungicidal compounds present in different plant tissues (e.g. polyphenols, tannins, terpenes, etc., see Scheffer & Cowling, 1966; Harrison, 1971; Hintikka: Chapter 13). It is not uncommon to note for instance that small agarics such as *Mycena* or *Marasmius* will fruit on the bark of branches, trunks, or stumps that are occupied internally by much more aggressive wood-rotters. It is possible that such spatial separation is determined by some chemical gradient within the resource.

Environmental specificity. Gradients of such factors as moisture content, pH or the internal atmosphere of resources are another source of possible niche separation within the unit-community occupying a single resource. The influence of the moisture content of wood on the activity of decay fungi is well documented and number of investigations have shown that wood of differing moisture content selects a different basidiomycete flora. Of particular interest is the comprehensive study by Käärik (1974). She showed that in poles placed vertically in the ground a marked gradient of moisture content developed with a closely correlated pattern of basidiomycete colonisation. She was able to distinguish species which occupied different parts of the same pole, the demarcation being determined apparently by the moisture content (see also Levy: Chapter 9).

Colonisation strategies. Clearly features such as those mentioned above may explain to some degree the niche structure of basidiomycete communities and account for some part of the observed diversity. It is possible to argue, however, that a good deal of the diversity can be accounted for simply on the basis of the sheer abundance of the resources available. The opportunities for colonisation of new resources – fresh leaves, recently exposed or killed branches or stumps – are constantly occurring. Even in highly seasonal environments such as that of the temperate deciduous forest, leaf fall may occupy a period of 12–16 weeks. The accessibility of twigs and branchwood to fungi is spread more widely through the year and is less seasonally predictable (Christensen, 1978). In some tropical forests, the input of plant litter to the decomposer subsystem may be truly continuous. Basidiomycetes and other fungi can exploit these patterns by 'adoption' of a suitable colonisation strategy.

Colonisation of leaf litter may occur by mycelial growth, particularly in the form of cords, after fall to the litter layer. This implies a 'waiting' strategy based on perennating mycelium or persistent propagules such as chlamydospores. Alternatively, colonisation may be by means of conidia or basidiospores. In the case of the latter, patterns of fruiting indicate that different species are adapted to exploit the same resource type at different times. This is of course particularly true of the smaller agarics like *Mycena* and *Marasmius*. For instance, Hintikka (1963) demonstrated marked differences in the fruiting period for twenty-nine species of *Mycena* many of which were potential competitors for the same resource. Dowding (1973) demonstrated a correlation between the periodicity of basidiospores in the air spora and success in colonisation of the freshly cut ends of pine logs exposed at different times of the year in Britain.

Many basidiomycetes, particularly among the Aphyllophorales, have extended fruiting periods during which there is a virtually continuous production of copious amounts of basidiospores (Gregory, 1973). This may be regarded as an 'opportunist' strategy (Kramer: Chapter 2). Potential colonisation sites on twigs, branches or stumps will become available in a relatively unpredictable (i.e. non-seasonal) manner. The strategy of 'keeping your troops in the air' will ensure maximum exploitation of that circumstance.

It should be noted that this discussion has been confined to the rôle of the basidiospore. Very little is known of the ecological significance of the asexual stages of most basidiomycetes, but conidia may serve as agents of dispersal and colonisation in circumstances unfavourable to the production of fruit bodies. Recently, for instance, Johnson, Thomas, Wood & Swift (1981) have shown that some species of *Termitomyces*, which is the obligate basidiomycete symbiont of the Macrotermitinae, are dispersed as conidia which are carried from nest to nest in the gut of the alate termites. In other species, basidiospores may be more significant agents of dispersal.

Patterns of change with time

Because of the relative abundance of the resources and the phenological differences in the patterns of colonisation there may be relatively little interspecific competition at the time of colonisation. As decomposition proceeds, however, the amount of available resource declines and species that have co-existed in the same unit-community may then begin to compete. Rayner & Todd (Chapter 6) have also clearly indicated that intra-specific competition between mycelia originating from different basidiospores may also occur at this time. Interspecific competition leads to replacement of one species by another and results in some of the patterns described as fungal *successions*. This has recently been extensively reviewed by Frankland (1981), including reference to examples of inter-basidiomycete competition.

One of the commonest causes of change in established basidiomycete communities is, however, the influence of decomposer animals. The data of Fig. 3 illustrate this quite clearly for the microflora of small branches. When branches occupied by basidiomycetes were invaded by wood-boring invertebrates (e.g. the dipteran larva, *Tipula flavolineata*, Swift *et al.*, 1976) a totally different microflora quickly became dominant. This community, dominated by moniliaceous and mucoraceous fungi and by a large bacterial population, more closely resembled the microflora of soil than the specialised wood-inhabiting community that preceded it.

What are the factors which lead to this dramatic change in the dominant decomposer microflora? Firstly there is the direct effect of the animal feeding on the basidiomycete mycelium. Microscopic examination of the faeces from tipulid larvae showed that among the particles of wood there were numerous short lengths of basidiomycete hyphae which were largely devoid of cytoplasmic contents. Even with the use of a medium selective for basidiomycetes (Russell, 1956), these fungi could be isolated only with a very low frequency from the faeces compared with the high frequency of isolation from unconsumed wood prior to invasion by the animal. Thus the effect of animal feeding may be to reduce the viable basidiomycete biomass to a very low or even non-sustainable level.

In other circumstances, however, the grazing activity of arthropods may stimulate fungal activity. Hanlon & Anderson (1979) demonstrated this for Collembola feeding on the hyphae of *Coriolus versicolor*, although they also showed that if the intensity of grazing was increased beyond a certain level the fungal activity (measured as respiration) was inhibited. In another example from leaf litter, Visser & Whittaker (1977) showed that basidiomycete hyphae were unpalatable to Collembola in comparison with dematiaceous Hyphomycetes. Selective grazing favoured the growth of the basidiomycete, which was thus enabled to dominate niches previously occupied by the Hyphomycetes (Parkinson, Visser & Whittaker, 1979).

A second possible factor in the diversion of the microfloral succession by animal activity is change in the physical character of the micro-environments within the resource. The particulate structure created by the comminutive action of the animal feeding sustains a moisture content higher even than that in decayed wood. This may be associated with a corresponding drop in the extent of oxygenation. It is also highly probable that this particulate structure is less attractive to the penetrative habit of the basidiomycete fungus than to the surface-dwelling unicells such as yeasts or bacteria, or to fungi such as the Mucorales which produce rapidly ramifying mycelium and a small sporangial apparatus.

The third, most straightforward, and perhaps most influential factor is the inoculation of the resource with the spores and cells of a diverse range of micro-organisms brought in by animals from the surrounding litter and soil. The agents here are probably not so much the pioneer specialised wood-boring insects as the secondary arthropods, such as Collembola and mites, which rapidly colonise the bore holes in large numbers (Fager, 1968). It should be noted that in other circumstances arthropods may be important agents of dispersal for basidiomycetes (Talbot, 1952).

The quantitative significance of basidiomycetes in decomposer communities

The hypothesis presented suggests that the basidiomycetes are responsible for the assimilation of a major proportion of the carbon, energy and nutrient entering the decomposition subsystem in forest ecosystems. The colonisation and annexation of these resources by basidiomycetes may be preceded by the activity of other organisms, but the quantitative significance of the latter in decomposition is regarded as lower. Indeed it has been suggested that their activity in some instances may serve to facilitate the colonisation process for basidiomycetes (Mercer: Chapter 8). If we therefore picture the basidiomycete mycelium as a major pool of carbon, energy and nutrient within the decomposition subsystem we can in turn ask what the fate of this pool may be. In Fig. 5, I have illustrated the probable major fluxes from basidiomycetes to other parts of the decomposition subsystem.

Production of soluble organics

Decomposition over a given period of time results in three main categories of product (other than the residue of undecayed resource): *fungal mycelium + inorganic molecules + extracellular low molecular weight organic molecules (LMWO)*. With the exception of the production of CO_2 and H_2O, mineralisation of fresh plant litter by basidiomycetes is probably relatively rare being confined to those circumstances where for some reason a nutrient element occurs in excess (Dowding, 1981). The production of LMWO may however be quantitatively more significant and also have considerable ecological importance. Such molecules arise from at least three different sources: extracellular enzymatic activity; secretion from living hyphae, and lytic release from dead hyphae.

Extracellular enzyme activity. Breakdown of the polymers of plant cell walls by the extracellular enzymes of a fungus may take place at a rate in

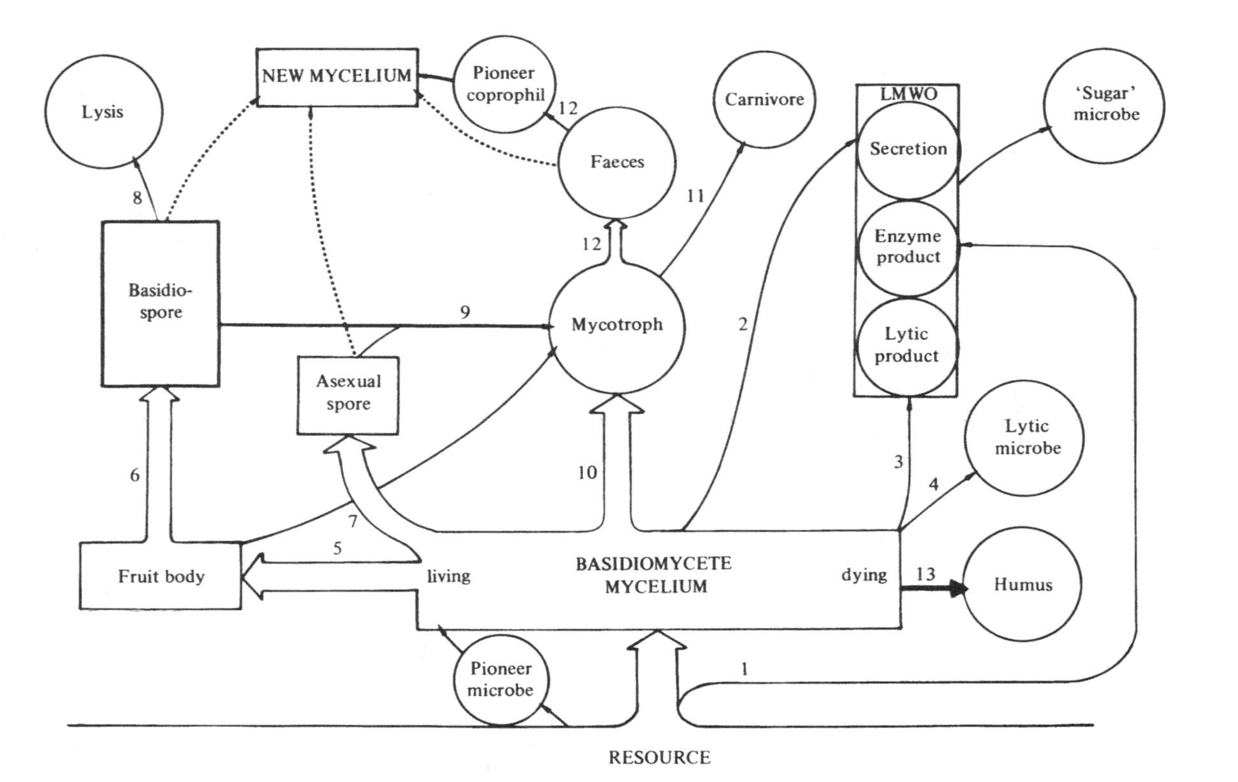

Fig. 5. The inter-relationships of basidiomycetes and other components of the decomposition subsystem. For details see text. LMWO, low molecular weight organic molecules. The dashed lines represent colonisation events.

excess of the uptake and utilisation of these products (Flux 1). This has been extensively documented for wood decomposition (e.g. Cowling, 1961). In wood decayed by brown-rot fungi the solubility in alkali may show a marked increase over the early period of decay. This is largely due to depolymerisation of polysaccharides although some increase in the solubility of lignin also occurs. In the later stages of decomposition, however, degradation products may be utilised as readily as they are produced. In contrast, the action of white-rot enzymes does not seem to result in the production of such an excess of low molecular weight compounds. This increased efficiency in the production:utilisation ratio may be attributable to a higher degree of localisation of polysaccharase enzymes on the cell walls of the fungal hyphae in the white-rot fungi (Cowling & Brown, 1969; Montgomery: Chapter 3). As the latter author remarks, the production of excess sugar at a distance from the site of uptake (by brown-rot fungi) invites competitive utilisation by other organisms.

The association of 'secondary sugar fungi' and bacteria with cellulolytic organisms has been documented on a number of occasions (Enebo, 1963; Tribe, 1966; Hedger & Hudson, 1974). More recently such an association between bacteria, yeasts and wood-decay basidiomycetes has been demonstrated by Blanchette & Shaw (1978). They showed that the rate of decomposition of wood by the basidiomycete was significantly stimulated by the presence of the secondary organisms. The stimulation was as great, however, for white-rot as for brown-rot basidiomycetes. The acceleration of depolymerising activity by the presence of secondary organisms may be explained by their co-operative rôle in removing catabolites and thus relieving enzyme repression (Hulme & Stranks, 1971). Alternatively, the secondary organisms may themselves have a limited capacity for depolymerisation which becomes possible once initial disruption of the cell-wall structure has been brought about by the basidiomycete.

Secretion. Further sources of LMWO compounds are the secondary metabolites released from the basidiomycete mycelium, either as secretory products of the actively growing hyphae (Flux 2) or as lytic products of senescent and dead hyphae (Flux 3 and 4). It is difficult in practice to distinguish between these pathways. A very diverse range of molecules are synthesised by fungi. Many, such as proteins, carbohydrates or organic acids, are components or products of normal metabolism and are most characteristic of active compartments in the growing tips of hyphae. In older non-proliferating parts of the mycelium, secondary metabolites may be formed. These include sterols, antibiotics, toxins, and pigment molecules such as carotenoids and melanins (Smith & Berry, 1975, 1976). Some of these products are known to be actively secreted by living hyphae (e.g. proteins, organic acids and some substances with antibiotic or toxic activity), whilst others are probably released only by lysis.

Carbon budgets for fungi grown in laboratory culture show that considerable amounts of some substances may be released from the mycelium. In general, however, basidiomycetes show lower tendencies in this direction than other fungi (Raistrick *et al.*, 1931; Whitaker, 1951). Laboratory cultures do not in any case give much indication of the pattern under natural circumstances for it is clear that the allocation of carbon between mycelium, carbon dioxide and secondary metabolites is highly dependent on the environmental circumstances. This factor has led some people to debate the reality and significance of antibiotic production in soil (e.g. see reviews by Gottlieb, 1976, and Wicklow, 1981). Nonetheless, substances with antibiotic or other toxic properties have a selective ecological impact which is out of proportion to their quantitative significance in energy flow.

Another excretory product of basidiomycete mycelium which is produced in large amounts under natural circumstances and which has far-reaching ecological effects is oxalic acid (Cromack *et al.*, 1977; Sollins *et al.*, 1981). These authors have shown that both mycorrhizal and saprotrophic basidiomycetes excrete large quantities of oxalic acid much of which then precipitates as the calcium salt on the surface of hyphae or rhizomorphs (see also Frankland: Chapter 14). They suggest that one possible function of this secretion is to maintain the ionic balance and electrical neutrality of the fungal cytoplasm. Thus, this particular organic secretion may play a rôle in both internal and external pH regulation and ionic equilibrium. It is interesting to note that oxalate is a highly oxidised molecule and has a very low energy content (less than one-third that of citric acid, Sollins *et al.*, 1981) which implies a low energy cost for this function.

Lysis. Release of organic materials by lysis may be the result of autolysis or be hastened by the activity of lytic micro-organisms. Chitinolytic activity is common amongst bacteria, in particular the Actinomycetales. To these organisms has been attributed, therefore, a major rôle in mycolysis (Bumbieris & Lloyd, 1967). Growth of bacteria at the expense of lysed mycelium results in the transfer of carbon, energy and nutrient from basidiomycete production to secondary decomposers (Flux 4). An interesting group of organisms that may intervene in this pathway are the mycoparasitic fungi. The quantitative significance of these organisms is uncharted as far as basidiomycetes are concerned, but the gradual realisation of their importance in regulating populations of soil fungi (Lumsden, 1981) suggests that a search might be made for them on the highly vulnerable basidiomycete mycelium in litter layers.

Frankland's tabulation (Chapter 14, Table 1) shows that there is an accumulation of dead basidiomycete mycelium many times in excess of the living biomass. This is mainly cell-wall material, representing the empty compartments of partially lysed cells. Much of the carbon, energy and nutrient lost from the cells may have been taken up by other decomposers by the pathways described in Fig. 5, but it is important to note that autolysis may play a key rôle in reallocating essential nutrients in short supply back to the growing mycelial tips (Dowding, 1976).

From the above discussion, it is clear that low molecular weight organic compounds released from the basidiomycete mycelium may provide an important link with other microbial decomposers of lesser enzymatic capacity. At present it is impossible to give a realistic quantitative estimate to this pathway. It is perhaps sufficient to note the enormous mass of dead bacterial cells in soil (Gray, Hissett & Duxbury, 1974; Frankland: Chapter 14). The majority of these are probably secondary decomposers dependent at some remove on the initial mobilisation of energy and nutrient by the basidiomycetes.

Reproduction

At various stages in the basidiomycete life cycle, reproduction of one kind or another takes place. The significance of this to the maintenance of community diversity and structure has already been discussed. The reproductive processes are also important steps in the flux of energy and nutrients both within basidiomycete populations and between the basidiomycetes and other organisms in the decomposer community (see Fig. 5).

The quantitative ecology of basidiomycete fruiting (Flux 5) is still in its infancy and few generalisations can be made with any degree of certainty. As Frankland (Chapter 14) has pointed out, whilst the fruit body represents only an insignificant proportion of the fungal biomass at any given time there may be considerable output of energy and nutrient from the basidiomycete pool by the agency of basidiospores (Flux 6). It is clear that for some species this may indeed be an almost continuous output over a large part of the year (Kramer: Chapter 2). To this must be added the output from asexual sporing structures, but at present there are no quantitative estimations of conidial production by basidiomycetes.

If reproductive processes do represent such an important link between basidiomycetes and the rest of the decomposition subsystem then it is clearly essential to understand more of the circumstances under which reproduction takes place. This is a complex subject on which much has been written (e.g. Taber, 1966; Smith & Berry, 1974; Manachère, 1980). A good deal of laboratory and other experimental evidence is emerging on the conditions that promote the initiation, the longevity and the termination of sexual reproduction in basidiomycetes. Similarly a good deal of understanding has been achieved of the morphogenetic processes involved and of the links between these processes and the external environmental triggers. However, laboratory studies are not enough for our purposes; we need to know not just what leads to and maintains fruiting in an individual but also what sustains the maximum spore production per unit area of woodland for a basidiomycete population. We need to know in particular how this relates to the availability of energy and nutrient and to the prior exploitation of the available resources by the mycelium. It appears that there are three possible types of regulation for the various phases of the reproductive process: a constitutive regulation based on some internally maintained circadian rhythms; regulation by factors of the external physical environment; regulation by the supply of necessary energy, carbon, nutrient elements and essential growth factors. All three of these categories of factor may operate on any single population, but it is to the last that I wish to turn attention because it represents the very mobilisation of organic resources that lies at the basis of our interest in these fungi as decomposers. Hora's work in 1959 and 1972, showing the way in which basidiomycete fruiting might be nutrient limited, has often been cited as pioneering in this respect but has not been followed up. The physiological ecology of basidiomycete populations should also be a high research priority for the future.

What then is the fate of this high reproductive production? A few spores, whether basidiospores or conidia, germinate and produce new mycelia (dotted lines in Fig. 5). The major proportion of spores do not survive to hand on their small reservoir of nutrient to a new generation of basidiomycetes but instead yield it up to other decomposers. A major link here must be with lytic micro-organisms (Flux 8) – bacteria, other fungi and even the intriguing spore-attacking protozoa described by Old (1977). A large fraction of the spore populations becomes the food of mycotrophic invertebrates (Flux 9). Indeed this may be a major pathway for, as Elton (1966) has pointed out, fungal fruit bodies are a favoured habitat and food source for many types of insect. For example, Paviour-Smith (1960) collected 346 species of beetle and 182 of fly from macrofungi in UK, and Russell-Smith (1979) recorded mycetophilid (Diptera) larvae in fifteen out of thirty-eight species of litter and mycorrhizal Agaricales during the autumn of 1968 for the study site described in Fig. 4(*a*). This is indeed an intriguing association. The suitability of fruit bodies as habitats is obvious because of their high nutritional value (see Frankland: Chapter 14, Table 3), but specialisation for this habitat must have its hazards. In addition to the opportunistic colonisation of resources by the fungus there are also the uncertainties inherent in fruiting before the potential habitat is likely to appear in any given place. Again, unfortunately, quantitative evidence is lacking, but this may be a major path of energy and nutrient flow in woodlands at certain times of the year, for, in addition to the decomposer fungi, the fruit bodies of many mycorrhizal fungi harbour insects. This then represents an input from the primary producers via mycorrhizal symbionts to the decomposer subsystem.

Mycotrophy

The association of invertebrates with basidiomycetes is even more widespread at the mycelial level than after reproduction (Flux 10). The effects of arthropod grazing on mycelium have already been discussed. Quantitative estimates of the extent of this activity are again lacking. R. Moore & M. J. Swift (unpublished) have shown that hyphomycete mycelium growing on the surface of cellophane films may be largely stripped off within 24 h of exposure to the grazing activity of micro-arthropods in tropical agricultural soils. There is no reason to suppose that basidiomycete mycelium does not suffer the same fate. It is now generally accepted that many so-called detritivore animals are in fact *mycotrophs*. In consuming decomposed plant litter, they ingest both plant matter and the fungal mycelium growing in it. It is probably from the latter that they obtain their main assimilate both of energy and also, perhaps more importantly, of nutrient elements such as nitrogen and phosphorus. This association with fungi, in which again basidiomycetes are probably central, is based both on the ability of the fungi to digest the cell-wall polymers which otherwise would be unavailable nutriment for the animals, and on the tendency of fungi to accumulate concentrations of nutrient elements far in excess of those in the original resources (Frankland: Chapter 14, Table 4). The large body of evidence that suggests that, with a few exceptions, wood-boring insects are similarly dependent on the prior activity of basidiomycete fungi has been summarised by Campbell (1952) and Swift (1977). Perhaps the most dramatic examples are the mutualistic associations between social insects and basidiomycete fungi found in the ants of the family Attinae and in termites of the family Macrotermitinae (Wood, 1976; Stradling, 1977; Swift, Heal & Anderson, 1979). In each case the decomposer community is reduced to virtually two species. The animal collects the plant food and may subject it to some pre-treatment, both comminution and some digestive and secretory action; the fungus then carries out the main processes of cellulolytic and ligninolytic digestion and the insect reconsumes the decomposed litter, obtaining its main nutrition from the fungal hyphae. In these specialised mycotrophs there is tight re-cycling of faeces from the animal to the basidiomycete fungus and back to the animal. This 'model' may be of more general application for it is becoming clear that coprotrophy is quite common among decomposer animals and represents a means of utilising microbial food to the greatest efficiency (Burnett, Mason & Rhodes, 1969).

Decomposer systems are not closed and loss of nutrient and energy to other parts of the ecosystem occurs mainly through predation of the animal populations (Flux 11), for instance, in the case of the termites cited above by ponerine ants (Wood & Sands, 1977). Faeces also represent one of the pathways by which, in free-living decomposer animals, energy and nutrient fluxes are channelled away from the basidiomycetes to other fungi and bacteria (Flux 12).

Humus formation

The chemistry of humus synthesis is complex and still poorly understood but there now seems to be general agreement that the decomposer fungi play a crucial rôle in the formation of the aromatic core of molecules such as humic and fulvic acids (Flux 13). A number of sources have been proposed for these moieties, but two modes of formation are currently given particular significance – aromatic molecules derived from the breakdown products of lignin, and aromatic pigment molecules synthesised or modified from plant products by microbial action. To the microfungi has been attributed the major function in the latter case, but basidiomycetes are clearly of importance in that they are responsible for much of the initial release of the lignin units. It is unlikely, however, that they are deeply involved in the later modifications that result in the final humus products (Haider, Martin & Filip, 1975).

Some authors, however (e.g. Handley, 1954, and see Hering: Chapter 12; Hintikka: Chapter 13), have given special status to the basidiomycetes' rôle in humus formation. Hintikka (1970) pointed out that the humus layer below the fruit bodies of many ligninolytic litter fungi is different in character to that surrounding it. This pale-coloured humus he has termed 'white-rot' humus. It is not clear, however, whether this is distinct from humus formed by other agencies or whether it is merely at an earlier stage in the formation processes. Humus, as used here, refers to the amorphous chemical compounds such as humic and fulvic acids, which are the product of chemical reactions, both enzymatic (described above) and non-enzymatic (Felbeck, 1971). White-rot humus may be a preliminary stage in the formation of these molecules; this question cannot be resolved except by the chemical analysis of the aromatic components of the different types of humus. Nonetheless, it is intriguing to speculate whether the occurrence of highly active lignin-degrading activity in a particular locality may affect the nature of the humus formed there.

Conclusion

The brief analysis given above, and the more detailed one contained in this book as a whole, is consistent with the hypothesis that saprotrophic basidiomycetes play a primary and central rôle in the decomposition subsystem of forest ecosystems. A picture is slowly emerging of the complex patterns in time and space formed by the basidiomycete community. This is built at present from only fragments of the jigsaw; a comprehensive study in both dimensions which could confirm and expand our concept of the whole has not yet been attempted. This will surely come, as will the genetic and physiological evidence that will enable us to understand better the nature, origin and sources of change in the community structure.

Any understanding of the rôle of basidiomycetes as decomposers starts and finishes with the question of their productivity. Frankland (Chapter 14) has pointed out how inadequate current methods are in providing satisfactory estimates of this essential parameter. An alternative spur to further investigation might be to look at the problem the other way round. If we accept the hypothesis of the basidiomycetes as primary decomposers, then what level of production and nutrient capture would we expect? Frankland has provided data for microbial biomass at Meathop Wood (Chapter 14, Table 1). This woodland has an annual input by litter fall of leaves, twigs and branches of approximately 5000 kg ha^{-1}; of this about 75% by weight can be attributed to cell-wall polymers. On the modest assumption that the basidiomycetes annex 50% of this polymer input and that their yield efficiency (including the production of secretory organics) is about 0.5 g g^{-1}, then the production of basidiomycete mycelium and its organic secretions would approximate to 938 kg ha^{-1} yr^{-1}. Frankland's value for basidiomycete biomass in the litter layer is about 43 kg ha^{-1} (we should ignore the mycelium of the lower horizons for much of this will be derived from roots). This would imply a turnover rate for the mycelium of about twenty-two times per year. Whether this is reasonable and can be explained in terms of reproductive production, lysis and animal grazing, I leave the reader to decide. The annual input given above suggests that about 48 kg ha^{-1} of nitrogen and about 2.3 kg ha^{-1} of phosphorus become available to the decomposers each year. Taking the above estimate of basidiomycete production and the estimates of the tendency of the mycelium to concentrate nutrients given in Table 3 of Frankland's chapter, we can calculate what fraction of the nutrient input is required to satisfy this estimated production. This calculation yields values of 45 and 2.5 kg ha^{-1} yr^{-1} for nitrogen and phosphorus respectively; thus the estimated level of activity for the basidiomycetes would result in the mobilisation of 93 and 110% respectively of the available pools of the two nutrients during the year.

In the analysis of the decomposition subsystem given in earlier sections, the basidiomycetes were pictured as forming a complex interface with the other decomposer organisms. If the speculative calculation of the previous paragraph is anywhere near correct then the most important feature of the 'basidiomycete connection' is that it provides a major route for the channelling of nutrient elements from the plant litter to the decomposition subsystem. Ultimately this nutrient path leads back to the plant. Here again the basidiomycetes have an important rôle to play; in most forests of the temperate and boreal zones and in some tropical forests they are the mycorrhizal symbionts that act as nutrient-capture agencies for the tree roots. The interface between basidiomycete decomposer and basidiomycete symbiont in litter and soil may yet prove to be the most interesting of all decomposition problems to solve.

References

Blanchette, R. A. & Shaw, C. G. (1978). Associations among bacteria, yeasts, and basidiomycetes during wood decay. *Phytopathology*, **68,** 631–7.

Bray, J. R. & Gorham, G. (1964). Litter production in forests of the world. *Advances in Ecological Research*, **2,** 101–57.

Bumbieris, M. & Lloyd, A. B. (1967). Influence of soil fertility and moisture on lysis of fungal hyphae. *Australian Journal of Biological Sciences*, **20,** 103–12.

Burnett, J. H. & Evans, J. (1966). Genetical homogeneity and the stability of the mating-type factors of 'fairy rings' of *Marasmius oreades*. *Nature, London*, **210,** 1368–9.

Burnett, A. M., Mason, W. H. & Rhodes, S. T. (1969). Re-ingestion of faeces and excretion rates of Zn^{65} in *Popilius disjunctus* versus *Cryptocercus punctulatus*. *Ecology*, **50,** 109–96.

Campbell, W. G. (1952). The biological decomposition of wood. In *Wood Chemistry*, ed. L. E. Wise & E. C. Jahn, 2nd edn, pp. 1061–118. New York: Reinhold Publishing Corporation.

Christensen, O. (1978). The dynamics of wood litter fall in a Danish oak forest. *Natura Jutlandica*, **20,** 155–62.

Cooke, R. (1977). *The Biology of Symbiotic Fungi*. London: John Wiley.

Cowling, E. B. (1961). Comparative biochemistry of the decay of Sweetgum sapwood by white-rot and brown-rot fungi. *United States Department of Agriculture Technical Bulletin*, No. 1258. Washington, DC: Government Printing Office.

Cowling, E. B. & Brown, W. (1969). Structural features of cellulosic materials in relation to enzymatic hydrolysis. *Advances in Chemistry*, **95,** 152–87.

Cromack, K., Sollins, P., Todd, R. L., Fogel, R., Todd, A. W., Fender, W. M. & Crossley, D. A. (1977). The role of oxalic acid and bicarbonate in calcium cycling by fungi and bacteria: some possible implications for soil animals. In *Soil Organisms as Components of Ecosystems*, ed. U. Lohm & T. Persson, pp. 246–52. Ecological Bulletins (Stockholm) 25. Swedish Natural Science Research Council.

Dowding, P. (1973). Effects of felling time and insecticide treatment on the inter-relationships of fungi and arthropods in pine logs. *Oikos*, **24,** 422–9.

Dowding, P. (1976). Allocation of resources; nutrient uptake and release by decomposer organisms. In *The Role of Terrestrial and Aquatic Organisms in Decomposition Processes*, ed. J. M. Anderson & A. Macfadyen, pp. 169–83. Oxford: Blackwell Scientific Publications.

Dowding, P. (1981). Nutrient uptake and allocation, during substrate exploitation by fungi. In *The Fungal Community: its Organisation and Role in the Ecosystem*, ed. D. T. Wicklow & G. C. Carroll, pp. 621–36. New York: Marcel Dekker.

Elton, C. S. (1966). *The Pattern of Animal Communities*. London: Methuen.

Enebo, L. (1963). Symbiosis in thermophilic cellulose fermentation. *Nature, London*, **163,** 805.

Fager, E. W. (1968). The community of invertebrates in decaying oak wood. *Journal of Animal Ecology*, **37,** 121–42.

Felbeck, G. T. (1971). Structural hypotheses of soil humic acids. *Soil Science*, **111,** 42–8.

Frankland, J. C. (1981). Mechanisms in fungal succession. In *The Fungal Community: its Organisation and Role in the Ecosystem*, ed. D. T. Wicklow & G. C. Carroll, pp. 403–26. New York: Marcel Dekker.

Gottlieb, D. (1976). The production and role of antibiotics in the soil. *Journal of Antibiotics*, **29,** 987–1000.

Gray, T. R. G., Hissett, R. & Duxbury, T. (1974). Bacterial populations of litter and soil

in a deciduous woodland. II. Numbers, biomass and growth rates. *Revue d'Ecologie et de Biologie du Sol*, **11**, 15–26.
Gregory, P. H. (1973). *The Microbiology of the Atmosphere*, 2nd edn. London: Leonard Hill.
Greig-Smith, R. (1964). *Quantitative Plant Ecology*, 2nd edn. London: Butterworths.
Haider, K., Martin, J. P. & Filip, Z. (1975). Humus biochemistry. In *Soil Biochemistry*, ed. E. A. Paul & A. D. McLaren, vol. 4, pp. 195–244. New York: Marcel Dekker.
Handley, W. R. C. (1954). Mull and mor formation in relation to forest soils. *Bulletin of the Forestry Commission*, No. 23, 1–115.
Hanlon, R. D. G. & Anderson, J. M. (1979). The effects of Collembola grazing on microbial activity in decomposing leaf litter. *Oecologia*, **38**, 93–9.
Harper, J. E. & Webster, J. (1964). An experimental analysis of the coprophilous fungal succession. *Transactions of the British Mycological Society*, **47**, 511–30.
Harrison, A. F. (1971). The inhibitory effect of oak leaf litter tannins on the growth of fungi, in relation to litter decomposition. *Soil Biology and Biochemistry*, **3**, 167–72.
Hedger, J. N. & Hudson, H. J. (1974). Nutritional studies of *Thermomyces lanuginosus* from wheat straw compost. *Transactions of the British Mycological Society*, **62**, 129–43.
Hintikka, V. (1963). Studies in the genus *Mycena* in Finland. *Karstenia*, **6/7**, 77–87.
Hintikka, V. (1970). Studies on white-rot humus formed by higher fungi in forest soils. *Communicationes Instituti Forestalis Fenniae*, **69**, 1–68.
Hora, F. B. (1959). Quantitative experiments on toadstool production in woods. *Transactions of the British Mycological Society*, **42**, 1–14.
Hora, F. B. (1972). Productivity of toadstools in coniferous plantations – natural and experimental. *Mycopathologia et Mycologia Applicata*, **48**, 35–42.
Hudson, H. J. (1968). The ecology of fungi on plant remains above the soil. *New Phytologist*, **67**, 837–74.
Hulme, M. A. & Stranks, D. W. (1971). Regulation of cellulase production by *Myrothecium verrucaria* grown on non-cellulosic substrates. *Journal of General Microbiology*, **69**, 145–55.
Johnson, R. A., Thomas, R. J., Wood, T. G. & Swift, M. J. (1981). The inoculation of the fungus comb in newly founded colonies of some species of the Macrotermitinae (Isoptera) from Nigeria. *Journal of Natural History*, **15**, 751–6.
Käärik, A. (1974). Decomposition of wood. In *Biology of Plant Litter Decomposition*, ed. C. H. Dickinson & G. J. F. Pugh, vol. 1, pp. 129–74. London and New York: Academic Press.
Kendrick, B. & Watling, R. (1979). Mitospores in Basidiomycetes. In *The Whole Fungus*, ed. B. Kendrick, vol. 2, pp. 473–545. National Museum of Natural Science, National Museum of Canada, and the Kananaskis Foundation.
Kershaw, K. A. (1969). The detection of pattern and association. *Journal of Ecology*, **48**, 232–42.
Lumsden, R. D. (1981). The ecology of mycoparasitism. In *The Fungal Community: its Organisation and Role in the Ecosystem*, ed. D. T. Wicklow & G. C. Carroll, pp. 295–318. New York: Marcel Dekker.
Manachère, G. (1980). Conditions essential for controlled fruiting of macromycetes – a review. *Transactions of the British Mycological Society*, **75**, 255–70.
Mitchell, C. P. & Millar, C. S. (1978). Mycofloral successions on Corsican Pine needles colonised on the tree by three different fungi. *Transactions of the British Mycological Society*, **71**, 303–17.

Old, K. M. (1977). Giant soil amoebae cause perforation of conidia of *Cochliobolus sativus*. *Transactions of the British Mycological Society*, **68**, 277–81.

Parker-Rhodes, A. F. (1954). Deme structure in higher fungi: *Mycena galopus*. *Transactions of the British Mycological Society*, **37**, 314–20.

Parkinson, D., Visser, S. & Whittaker, J. B. (1979). Effects of collembolan grazing on fungal colonization of leaf litter. *Soil Biology and Biochemistry*, **11**, 529–35.

Paviour-Smith, K. (1960). The fruiting bodies of macrofungi as habitats for the beetles of the family Ciidae (Coleoptera). *Oikos*, **11**, 43–71.

Raistrick, H., Birkinshaw, J. H., Charles, J. H. V., Clutterbuck, P. W., Coyne, P. O., Hetherington, A. C., Lilly, C. H., Rintoul, M. L., Rintoul, W., Robinson, R., Stoyle, J. A. R., Thom, C. & Young, W. (1931). Studies in the biochemistry of micro-organisms. *Philosophical Transactions of the Royal Society*, **B220**, 1–367.

Russell, P. (1956). A selective medium for the isolation of basidiomycetes. *Nature, London*, **178**, 1038–9

Russell-Smith, A. (1979). A study of fungus flies (Diptera: Mycetophilidae) in beech woodland. *Ecological Entomology*, **4**, 355–64.

Scheffer, T. C. & Cowling, E. B. (1966). Natural resistance of wood to microbial deterioration. *Annual Review of Phytopathology*, **4**, 147–70.

Smith, J. E. & Berry, D. R. (1974). *An Introduction to the Biochemistry of Fungal Development*. London and New York: Academic Press.

Smith, J. E. & Berry, D. R. (1975). *The Filamentous Fungi*, vol. 1. *Industrial Mycology*. London: Edward Arnold.

Smith, J. E. & Berry, D. R. (1976). *The Filamentous Fungi*, vol. 2, *Biosynthesis and Metabolism*. London: Edward Arnold.

Sollins, P., Cromack, K., Fogel, R. & Chin Yan Li (1981). Role of low molecular weight organic acids in the inorganic nutrition of fungi and higher plants. In *The Fungal Community: its Organisation and Role in the Ecosystem*, ed. D. T. Wicklow & G. C. Carroll, pp. 607–20. New York: Marcel Dekker.

Stradling, D. J. (1977). Food and feeding habits of ants. In *Production Ecology of Ants and Termites*, ed. M. V. Brian, pp. 81–106. Cambridge University Press.

Swift, M. J. (1976). Species diversity and the structure of microbial communities. In *The Role of Aquatic and Terrestrial Organisms in Decomposition Processes*, ed. J. M. Anderson & A. Macfadyen, pp. 185–222. Oxford: Blackwell Scientific Publications.

Swift, M. J. (1977). The ecology of wood decomposition. *Science Progress, Oxford*, **64**, 175–99.

Swift, M. J., Heal, O. W. & Anderson, J. M. (1979). *Decomposition in Terrestrial Ecosystems*. Oxford: Blackwell Scientific Publications.

Swift, M. J., Healey, I. N., Hibberd, J. K., Sykes, J. M., Bampoe, V. & Nesbitt, M. E. (1976). The decomposition of branch-wood in the canopy and floor of a mixed deciduous woodland. *Oecologia*, **26**, 139–49.

Taber, W. A. (1966). Morphogenesis in basidiomycetes. In *The Fungi*, ed. G. C. Ainsworth & A. S. Sussman, vol. 2, pp. 387–412. London and New York: Academic Press.

Talbot, P. H. B. (1952). Dispersal of fungus spores by small animals inhabiting wood and bark. *Transactions of the British Mycological Society*, **35**, 123–8.

Tribe, H. (1966). Interactions of soil fungi on cellulose film. *Transactions of the British Mycological Society*, **49**, 457–66.

Visser, S. & Whittaker, J. B. (1977). Feeding preferences for certain litter fungi by *Onychiurus subtenuis* (Collembola). *Oikos*, **29**, 320–5.

Warcup, J. H. (1965). Growth and reproduction of soil micro-organisms in relation to substrate. In *Ecology of Soil-borne Plant Pathogens*, ed. K. F. Baker & W. C. Snyder, pp. 52–60. London: John Murray.

Whitaker, D. R. (1951). Studies in the biochemistry of cellulolytic fungi. 1. Carbon balances of wood-rotting fungi in surface culture. *Canadian Journal of Botany*, **29**, 159–75.

Wicklow, D. T. (1981). Interference competition and the organisation of fungal communities. In *The Fungal Community: its Organisation and Role in the Ecosystem*, ed. D. T. Wicklow & G. C. Carroll, pp. 351–75. New York: Marcel Dekker.

Williamson, B., Mitchell, C. P. & Millar, C. S. (1976). Histochemistry of Corsican Pine needles infected by *Lophodermella sulcigena* (Rostr.) v. Höhn. *Annals of Botany, London*, N.S., **40**, 281–8.

Wood, T. G. (1976). The role of termites (Isoptera) in decomposition processes. In *The Role of Terrestrial and Aquatic Organisms in Decomposition Processes*, ed. J. M. Anderson & A. Macfadyen, pp. 45–168. Oxford: Blackwell Scientific Publications.

Wood, T. G. & Sands, W. A. (1977). The role of termites in ecosystems. In *Production Ecology of Ants and Termites*, ed. M. V. Brian, pp. 245–92. Cambridge University Press.

Systematic index

A formal classification of organisms referred to in this volume is given here with some of the more important synonyms. 'Trivial', superseded terms such as 'basidiomycetes', which are still in common parlance even among taxonomists, occur frequently in the chapters, but they have been printed in lower case (compare Basidiomycotina). The term 'homobasidiomycete' refers to fungi with a single-celled basidium, and 'heterobasidiomycete' to fungi with a basidium divided either by a longitudinal septum or one to three transverse septa (see Gareth Jones: Chapter 11). The Hymenomycetes is a grouping covering all fungi in which the basidia are borne in a hymenium exposed before maturity; it includes a few heterobasidiomycetous fungi (hymenomycetous Heterobasidiae, Donk (1966)). The basidiospores of fungi belonging to the Gasteromycetes are usually enclosed within the fruit body and are borne on apobasidia from which they are not forcibly discharged. The Teliomycetes is retained here for the rusts and smuts; the former with a terminal teliospore and the latter with an intercalary teliospore producing sessile basidiospores. Authorities for most of the fungal genera are cited by Ainsworth, James & Hawksworth (1971).

BACTERIA & ACTINOMYCETES

FUNGI

ZYGOMYCOTINA

Mucorales

ASCOMYCOTINA

Microascales

Hypocreales

Clavicipitales

Sphaeriales

References

Ainsworth, G. C., James, P. W. & Hawksworth, D. L. (1971). *Ainsworth & Bisby's Dictionary of the Fungi*. Kew: Commonwealth Mycological Institute.
Donk, M. A. (1966). Check list of European hymenomycetous Heterobasidiae. *Persoonia*, **4**, 145–335.

Subject index

passim, denotes scattered references

References added in proof:

Since completion of the original manuscripts for Chapters 6 and 7 on basidiomycete populations, there have been a number of further developments in the authors' and other laboratories which they were unable to summarise here, but there follows a list of pertinent references.

Adams, T. J. H., Todd, N. K. & Rayner, A. D. M. (1981). Antagonism between dikaryons of *Piptoporus betulinus*. *Transactions of the British Mycological Society*, **76**, 510–13.

Anderson, J. B., Ullrich, R. C., Roth, L. F. & Filip, G. M. (1979). Genetic identification of clones of *Armillaria mellea* in coniferous forests in Washington. *Phytopathology*, **69**, 1109–11.

Boddy, L. & Rayner, A. D. M. (1981). Fungal communities and formation of heartwood wings in attached oak branches undergoing decay. *Annals of Botany*, **47**, 271–4.

Boddy, L. & Rayner, A. D. M. (1982). Population structure, inter-mycelial interactions and infection biology of *Stereum gausapatum*. *Transactions of the British Mycological Society*, **78**, 337–51.

Goldstein, D. & Gilbertson, R. L. (1981). Cultural morphology and sexuality of *Inonotus arizonicus*. *Mycologia*, **73**, 167–80.

Hansen, E. M. (1979). Sexual and vegetative incompatibility reactions in *Phellinus weirii*. *Canadian Journal of Botany*, **57**, 1573–8.

Rayner, A. D. M. & Turton, M. N. (1982). Mycelial interactions and population structure in the genus *Stereum: S. rugosum, S. sanguinolentum* and *S. rameale*. *Transactions of the British Mycological Society* (in press).

Williams, E. N. D., Todd, N. K. & Rayner, A. D. M. (1981). Spatial development of populations of *Coriolus versicolor*. *New Phytologist*, **89**, 307–20.